Research and Development in Intelligent Systems XXXI

Incorporating Applications and Innovations
in Intelligent Systems XXII

T0137828

Max Bramer · Miltos Petridis
Editors

Research and Development in Intelligent Systems XXXI

Incorporating Applications and
Innovations in Intelligent Systems XXII

Proceedings of AI-2014, The Thirty-fourth SGAI
International Conference on Innovative Techniques
and Applications of Artificial Intelligence

 Springer

Editors

Max Bramer
School of Computing
University of Portsmouth
Portsmouth
UK

Miltos Petridis
School of Computing, Engineering
 and Mathematics
University of Brighton
Brighton
UK

ISBN 978-3-319-12068-3   ISBN 978-3-319-12069-0   (eBook)
DOI 10.1007/978-3-319-12069-0

Library of Congress Control Number: 2014951661

Springer Cham Heidelberg New York Dordrecht London

Printed on acid-free paper

Springer is part of Springer Science+Business Media (www.springer.com)

Programme Chairs' Introduction

This volume comprises the refereed papers presented at AI-2014, the Thirty-fourth SGAI International Conference on Innovative Techniques and Applications of Artificial Intelligence, held in Cambridge in December 2014 in both the technical and the application streams. The conference was organised by SGAI, the British Computer Society Specialist Group on Artificial Intelligence.

The technical papers included new and innovative developments in the field, divided into sections on Knowledge Discovery and Data Mining, Machine Learning, and Agents, Ontologies and Genetic Programming. This year's Donald Michie Memorial Award for the best refereed technical paper was won by a paper entitled "On Ontological Expressivity and Modelling Argumentation Schemes using COGUI" by Wael Hamdan, Rady Khazem and Ghaida Rebdawi (Higher Institute of Applied Science and Technology, Syria), Madalina Croitoru and Alain Gutierrez (University Montpellier 2, France) and Patrice Buche (IATE, INRA, France).

The application papers included present innovative applications of AI techniques in a number of subject domains. This year, the papers are divided into sections on Evolutionary Algorithms/Dynamic Modelling, Planning and Optimisation, and Machine Learning and Data Mining. This year's Rob Milne Memorial Award for the best refereed application paper was won by a paper entitled "Combining Semantic Web Technologies with Evolving Fuzzy Classifier eClass for EHR-based Phenotyping: a feasibility study" by M. Arguello and S. Lekkas (University of Manchester, UK), J. Des (SERGAS, Spain), M.J. Fernandez-Prieto (University of Salford, UK) and L. Mikhailov (University of Manchester, UK).

The volume also includes the text of short papers presented as posters at the conference.

On behalf of the conference organising committee we would like to thank all those who contributed to the organisation of this year's programme, in particular the programme committee members, the executive programme committees and our administrators Mandy Bauer and Bryony Bramer.

Max Bramer, Technical Programme Chair, AI-2014.

Miltos Petridis, Application Programme Chair, AI-2014.

Acknowledgments

AI-2014 Conference Committee

Conference Chair
Prof. Max Bramer   University of Portsmouth

Technical Programme Chair
Prof. Max Bramer   University of Portsmouth

Application Programme Chair
Prof. Miltos Petridis   University of Brighton

Deputy Application Programme Chair
Dr. Jixin Ma   University of Greenwich

Workshop Organiser
Prof. Adrian Hopgood   Sheffield Hallam University

Treasurer
Rosemary Gilligan   University of Hertfordshire

Poster Session Organiser
Dr. Nirmalie Wiratunga   The Robert Gordon University

FAIRS 2014

Giovanna Martinez   Nottingham Trent University

UK CBR Organisers

Prof. Miltos Petridis   University of Brighton
Prof. Thomas Roth-Berghofer   University of West London

Conference Administrator

Mandy Bauer   BCS

Paper Administrator

Bryony Bramer

Technical Executive Programme Committee

Prof. Max Bramer   University of Portsmouth (Chair)
Dr. John Kingston   Tribal Group
Prof. Thomas Roth-Berghofer   University of West London
Dr. Nirmalie Wiratunga   Robert Gordon University, Aberdeen

Applications Executive Programme Committee

Prof. Miltos Petridis   University of Brighton (Chair)
Mr. Richard Ellis   Helyx SIS Ltd
Ms. Rosemary Gilligan   University of Hertfordshire
Dr. Jixin Ma   University of Greenwich (Vice-Chair)
Dr. Richard Wheeler   University of Edinburgh

Technical Programme Committee

Andreas Albrecht	Middlesex University
Ali Orhan Aydin	Istanbul Gelisim University
Yaxin Bi	University of Ulster
Mirko Boettcher	University of Magdeburg, Germany
Max Bramer	University of Portsmouth
Krysia Broda	Imperial College, University of London
Ken Brown	University College Cork
Frans Coenen	University of Liverpool
Madalina Croitoru	University of Montpellier, France
Bertrand Cuissart	Universite de Caen
Ireneusz Czarnowski	Gdynia Maritime University, Poland
Nicolas Durand	University of Aix-Marseille
Frank Eichinger	Karlsruhe Institute of Technology (KIT), Germany
Adriana Giret	Universidad Politécnica de Valencia
Nadim Haque	Thunderhead.com
Arjen Hommersom	University of Nijmegen, The Netherlands
Adrian Hopgood	Sheffield Hallam University, UK
John Kingston	Tribal Group
Konstantinos Kotis	University of Piraeus
Ivan Koychev	Bulgarian Academy of Science
Fernando Lopes	LNEG-National Research Institute, Portugal
Jixin Ma	University of Greenwich
Stephen G. Matthews	De Montfort University, UK
Roberto Micalizio	Universita' di Torino
Lars Nolle	Jade Hochschule, Germany
Dan O'Leary	University of Southern California
María Dolores Rodríguez-Moreno	Universidad de Alcalá
Thomas Roth-Berghofer	University of West London
Fernando Sáenz-Pérez	Universidad Complutense de Madrid
Miguel A. Salido	Universidad Politécnica de Valencia
Rainer Schmidt	University of Rostock, Germany
Sid Shakya	BT Innovate and Design
Frederic Stahl	University of Reading
Simon Thompson	BT Innovate
Jon Timmis	University of York
Andrew Tuson	City University London
M.R.C. van Dongen	University College Cork
Graham Winstanley	University of Brighton
Nirmalie Wiratunga	Robert Gordon University

Application Programme Committee

Hatem Ahriz	Robert Gordon University
Tony Allen	Nottingham Trent University
Ines Arana	Robert Gordon University
Mercedes Argüello Casteleiro	The University of Manchester
Ken Brown	University College Cork
Richard Ellis	Helyx SIS Ltd
Roger Evans	University of Brighton
Rosemary Gilligan	University of Hertfordshire
John Gordon	AKRI Ltd
Chris Hinde	Loughborough University
Adrian Hopgood	De Montfort University
Stelios Kapetanakis	University of Brighton
Jixin Ma	University of Greenwich
Miltos Petridis	University of Brighton
Miguel A. Salido	Universidad Politécnica de Valencia
Roger Tait	University of Cambridge
Wamberto Vasconcelos	University of Aberdeen
Richard Wheeler	Edinburgh Scientific

Contents

Machine Learning

Agents, Ontologies and Genetic Programming

Short Papers

Applications and Innovations in Intelligent Systems XXII

Best Application Paper

Evolutionary Algorithms/Dynamic Modelling

Planning and Optimisation

Machine Learning and Data Mining

Research and Development in Intelligent Systems XXXI

Best Technical Paper

On Ontological Expressivity and Modelling Argumentation Schemes Using COGUI

Wael Hamdan, Rady Khazem, Ghaida Rebdawi, Madalina Croitoru, Alain Gutierrez and Patrice Buche

Abstract Knowledge elicitation, representation and reasoning explanation by/to non computing experts has always been considered as a crafty task due to difficulty of expressing logical statements by non logicians. In this paper, we use the COGUI editor in order to elicit and represent Argumentation Schemes within an inconsistent knowledge base. COGUI is a visual, graph based knowledge representation editor compatible with main Semantic Web languages. COGUI allows for default reasoning on top of ontologies. We investigate its use for modelling and reasoning using Argumentation Schemes and discuss the advantages of such representation. We show how this approach can be useful in the practical setting of EcoBioCap where the different Argumentation Schemes can be used to lead reasoning.

1 Introduction

COGUI[1] (COnceptual Graphs User Interface) is a knowledge base editor in which knowledge is encoded as graphs and that supports sound and complete graph based reasoning operations. The COGUI editor will allow to encode knowledge bases expressed in a logical formalism encompassing Semantic Web main languages: RDF/S, OWL and Datalog+ [12]. COGUI graphs have a semantics in first-order logic (FOL) and reasoning tasks operate directly on the knowledge defined by the user (the graphs) and not on their translation into logical formulas [5]. COGUI can

---

[1] http://www.lirmm.fr/cogui/.

W. Hamdan · R. Khazem · G. Rebdawi
Higher Institute of Applied Science and Technology (HIAST), Damascus, Syria
e-mail: waelHamdan1977@gmail.com

M. Croitoru (✉) · A. Gutierrez
University Montpellier 2, Montpellier, France
e-mail: croitoru@lirmm.fr

P. Buche
IATE, INRA, Montpellier, France
e-mail: patrice.buche@supagro.inra.fr

© Springer International Publishing Switzerland 2014
M. Bramer and M. Petridis (eds.), *Research and Development in Intelligent Systems XXXI*, DOI 10.1007/978-3-319-12069-0_1

import and export all major Semantic Web main languages (RDF/S, OWL, Datalog, CGIF, CogXML) and has been recently extended to support non-monotonic reasoning using default rules. This extension was developed given the need induced by practical applications to support inconsistent ontology based reasoning [2, 4].

Argumentation Schemes (AS) are used to classify forms of arguments people exchange in their daily life discourses. They are used to identify and evaluate common and stereotypical forms of arguments [13]. The Argument Interchange Format (AIF; [6]), largely based on AS, proposes an "unified Ontology" for argumentation. The first AIF Ontology was proposed by [10] and was based on Resource Description Framework Schema RDFS, this AIF-RDF Ontology was implemented in a Semantic Web-based system named ArgDF. This work was extended in [9] by introducing OWL-based AIF Ontology in Description Logic DL [1]. This ontology enabled automatic scheme classifications, instance classification, inference of indirect support in chained argument structures, and inference of critical questions. The model focused on typology and overall structure of the arguments and did not enable argument acceptability. Furthermore, the type of reasoning in AS is non-monotonic. The reasoning in this OWL Ontology is based on a subset of first order predicate logic thus non-monotonic reasoning is not supported (for more details please see the expressive overlaps among knowledge representation languages illustrated in [7]).

In this article we present an AIF compatible Ontology for modelling AS that extends the expressivity of the existing work to default rule base reasoning. The ontology is available in RDFS, OWL, Datalog+ or CogXML and has been built using COGUI. The model extends the various types of inference supported in [9] by supporting argument acceptability and enabling non-monotonic reasoning. We model the following AS: argument from expert opinion, argument from analogy, and argument from popular opinion. Our work distinguishes itself from the AIF Ontology not only by (1) the expressivity brought by default rules but also by (2) its practical application. Indeed, we model the various domain application statements (we will later explain this feature in more details) using logical facts. This distinguishes us from the model introduced by [9] which deals with statements as black-boxes and not logical facts made of grounded atoms one can reason upon. We will showcase next an example based on the COGUI to illustrate types of inference supported by our AIF Ontology and which are not supported in [9].

2 Motivating Example

The "argument from position to know" [14] has the following elements:

- **Position to know premise**: E is in a position to know whether A is true (false).
- **Assertion premise**: E asserts that A is true (false).
- **Conclusion**: A may plausibly be taken to be true (false).

The scheme has a set of critical questions, we mention for example the trustworthiness question: "Is E reliable?".

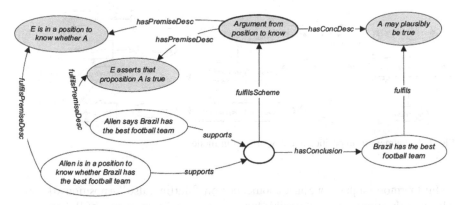

Fig. 1 "Argument from position to know" using ArgDF ontology

Figure 1 shows an argument network for this scheme using the underlying ontology of ArgDF [9]. This follows a graph based depiction of RDF and namely the nodes of the network representing subjects or objects of the RDF triple while the edges are labelled with the predicate. According to RDF/S semantics two nodes s and o linked by an edge labelled with p have the logical semantics of $p(s, o)$. In the case of Fig. 1 the nodes can represent either domain statements such as: "Brazil is the best football team in the world" or generic arguments such as: "E is in position to know whether A is true or false". This means that we cannot reason further about Brazil being the best football team in the world (for instance inferring that Brazil won the World Cup). The statements are seen as black-boxes and we cannot reason about the knowledge contained in the black-boxes.

Rahwan et al. [9] proposed ontology does reasoning in order to retrieve argument instances semantically. More precisely, in this particular example, the necessary and sufficient conditions for an instance to be classified as an argument from position to know are as follows:

ArgFromPositionToKnow(PresumptiveArgument ∩
∃hasConclusion.KnowledgePositionStmnt ∩
∃hasPremise.PositionToHaveKnowledgeStmnt ∩
∃hasPremise.KnowledgeAssertionStmnt)

Our proposed model (illustrated in Fig. 2) models all the grounded atoms as different pieces of knowledge, practically splitting the black-boxes of the previous model. This means that when we apply the rule of inference associated with this scheme we do not loose any expressivity as per existing work. However we additionally model the exceptions, and link them directly to the premises of the argument. In this example we link "LackOfReliabilityStmnt" to "Allen". Exceptions in work of Rahwan are not treated in a non monotonic manner but they are pure syntactic flags not handled by any reasoning engine. This type of reasoning is supported in our model using the default rule associated with the scheme.

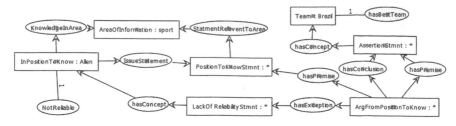

Fig. 2 "Argument from position to know" using our model

Furthermore, suppose we have another person "Martin", also "in position to know" who says the opposite: i.e. "Brazil is not the best team". The model of Rahwan will not capture the conflict unless we explicitly state (and thus not infer) that there is a conflict between theses two arguments. In our model, and thanks to the rule defined in Fig. 6, we conclude that the conclusion statements of the two arguments (issued by "Allen" and "Martin") are contradictory. Thereafter, the rule (depicted in Fig. 5) will infer the conflict scheme and the attack relations.

3 Background Notions

3.1 AIF Ontology

The AIF model was introduced by the research community to represent a consensus "abstract model" for argumentation in order to facilitate the exchange of semi-structured arguments among different argumentation systems. References [9, 10] illustrated the use of the proposed abstract model in argumentations systems by introducing concrete realizations. The AIF model is based on AS, and each AS in AIF has a name, set of premises, conclusion and a set of predefined critical questions. Critical questions are used to identify potential weaknesses of the argument and thus possibilities for the proponent to "attack" this argument.

3.2 Conceptual Graphs Knowledge Bases and Cogui

The Conceptual Graphs (CGs) formalism introduced by [8, 12] is a knowledge representation and reasoning formalism representing a subset of first order logic compatible with the major Semantic Web languages. A CGs knowledge base comprises the vocabulary part (also called support) that represents the domain ontology (equivalently the TBox in Description Logics, Datalog rules or RDF/S schema), and the assertions part called basic CGs (BCGs) that represents facts or assertions

(ABox, DB instances, RDF files etc.). The vocabulary is composed of two partially ordered sets by a specialization relation: a set of concepts and a set of relations of any arity (the arity is the number of arguments of the relation). The specialization relation is defined as: x is a specialization of X, if x and X are concepts, specialisation means that every instance (individual) of the concept x is also an instance of the concept X. A basic conceptual graph (BG) is a bipartite graph composed of: (i) a set of concept nodes, represents entities; (ii) a set of relation nodes, represents relationships between these entities or properties of them; (iii) a set of edges linking relation nodes to concept nodes. A concept node is labeled by a couple t :m where t is a concept (and more generally, a list of concepts) called the type of the node, and m is called the marker of this node: this marker is either the generic marker, denoted by *, if the node refers to an unspecified entity, otherwise this marker is a specific individual name.

The CGs model comprises also more complex constructs such as complex first order rules (equivalent to tuple generating dependencies in databases or Datalog+ rules) and default rules (which allow for non-monotonic reasoning).

Rules: a rule expresses implicit knowledge of the form "if hypothesis then conclusion", where hypothesis and conclusion are both basic graphs. This knowledge can be made explicit by applying the rule to a specific fact: intuitively, when the hypothesis graph is found in a fact, then the conclusion graph is added to this fact.

Default rules: CGs default rules are based on Reiters default logics in [3, 11]. They are defined by a tuple $DR = (H, C, J_1 \ldots J_k)$, where H is called the hypothesis, C the conclusion and $J_1, \ldots, J_k$ are called justifications of the default. All components of DR are themselves basic CGs. The intuitive meaning of a CG default is: if H holds for all individuals, then C can be inferred, provided that no justification J_i (for all i from 1 to k) holds.

Negation in Conceptual Graphs is represented by the means of the negative constrains which are basic graphs with the semantic that if they occur in the knowledge base the knowledge base is inconsistent. Please note that this specific kind of negation is equivalent to the negation used by the OWL and Description Logics as well as the integrity constrains used by databases. We can also impose positive constraints, that is, pieces of knowledge that need to appear in the graph (and the fact that it does not appear renders the graph inconsistent). Both constraints will be used later on for modelling purposes.

In the following section let us introduce the basic ontology which will include the support of the CGs, rules of inference and constraints.

4 AIF Ontology for Argumentation Schemes

The backbone of the AIF Ontology is shown in Fig. 3 and follows the model of the ontology defined in [9]. The hierarchy of concepts includes on the top level: *Statements* that describe statements which could be issued, *Schemes* which describe arguments made up of statements. Three type of schemes are defined. The first

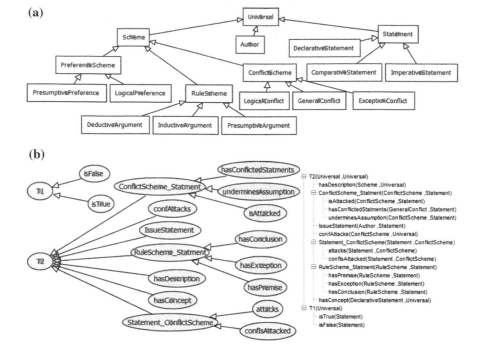

Fig. 3 AIF ontology. **a** Concept types. **b** Relation types

scheme is the rule scheme that defines types of arguments that could by defined. The second is a conflict scheme that represents the attack relations between arguments. The third is the preference scheme and it includes logical and presumptive preference. The relation types defined between arguments in *ruleScheme* and *Statements* are: *hasPremise*, *hasConclusion* and *hasException*. These relations denote that an argument may have premises, conclusions and exceptions respectively. A statement could attack a *conflictScheme* through the relation *attack* and could be attacked by this scheme with the relation *conflictAttacked*. Other relation types denote the facts that a statement may be true (*isTrue*) or false (*isFalse*).

After defining the backbone of concepts and relations, we need to impose the constraints to ensure model consistency. We use a positive constraint to say that every argument has at least one premise and one conclusion as illustrated in Fig. 4a. In order to ensure that every argument has at most one conclusion we use a negative constraint as shown in Fig. 4b.

Attacks among Arguments: Specializations of the concept *conflictScheme* are used to represent attacks among different arguments, *GeneralConflict* instances capture simple symmetric and asymmetric attacks among arguments, while *ExceptionConflict* instances represent exceptions to rules of inference.

(a)

(b)

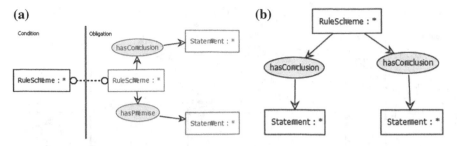

Fig. 4 Conclusion constraints. **a** Positive constraint. **b** Negative constraints

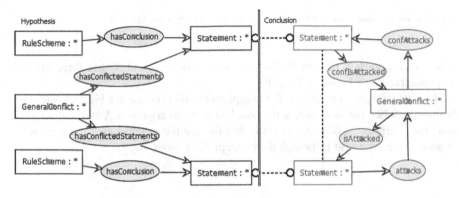

Fig. 5 Symmetric conflict rule

We first define the rule for the symmetric attack between two arguments as follows: if two statements S_1, S_2 belong to the concept *GeneralConflict* through the relation *hasConflictedStatements(GeneralConflict, Statement)*, i.e. there is a conflict between these two statements, and if one statement appears in the conclusion of argument A and the other appears in the conclusion of argument B, we say that there is a symmetric attack between the arguments A and B. Figure 5 illustrates the symmetric conflict rule. The rule semantics is as follows: *if $\forall S_1, S_2 \in$ Statement, $\exists GF \in$ GeneralConflict, such that hasConflictedStatements(GF, S_1), hasConflictedStatements(GF, S_2) and if $\exists A_1, A_2 \in$ RuleScheme, such that has Conclusion(A_1, S_1) and hasConclusion(A_2, S_2)* then there is a symmetric conflict defined by the relations: *confAttacks, confAttacked, attacks and isAttacked*.

In addition to the general conflict rule defined above, we define in Fig. 6a the rule that models the relation *hasConflictedStatements(GeneralConflict, Statement)* as follows: when a statement S is plausible to be evaluated "true" and "false" at the same time, then the two instances of S: S_1 (evaluated "true") and S_2 (evaluated "false") belong to the relation *hasConflictedStatements*. Thus, having S_1, S_2 belong to the relation *hasConflictedStatments* and using the previous rule we conclude that

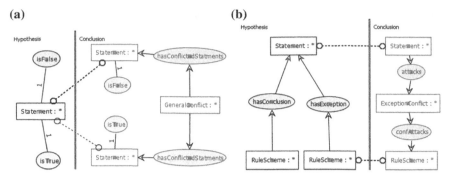

Fig. 6 Conflict constraints. **a** Statement contradiction. **b** Except conflict rule

if S_1 and S_2 appear in the conclusions of two arguments A and B, then there will be a symmetric attack between A and B.

The second type of conflict is the exception conflict (please see Fig. 6b), and this is the case when the statement is the conclusion of an argument A and at the same time the exception of other argument B. In this case the conclusion of the argument A attacks the argument B through the concept *ExceptionConflict*.

5 Modelling Argumentation Schemes

Our methodology for modelling new AS includes the following three steps:

- STEP 1—Ontology enriching. We enrich the vocabulary of our AIF Ontology with new concepts and relation types, add a concept that represent the new scheme as descendent of rule scheme and add concepts according to critical questions.
- STEP 2—Rules definition. We define the rules of inference that will enable AS inference (i.e. identifying the schemes) and critical questions inference.
- STEP 3—Default rules definition. We introduce the default rules that will define the non-monotonic reasoning. The generic inference rule used is defined as follows: "If an argument conforms to an AS, we conclude that its conclusion is true, unless one of its exceptions holds".

Let us now explain how we apply this methodology in order to define three Argumentations Schemes.

6 COGUI Implementation of Argumentation Schemes

6.1 Argument from Expert Opinion

In many forms of arguments people cite an authority to back up their point of view. In other words they indicate that someone (the authority cited) could give reasons that back up the statements they defend. This form of arguments is called "appeal to authority" or "argument from position to know". Depending on the type of the cited authority, the AS would be "Argument from Expert Opinion" if the authority is an expert, or "Argument from Witness Testimony", if the person is a witness in the situation at hand. This type of arguments comes with common critical questions such as questioning the reliability of the authority, more precisely: is the cited person an authority? or is the authority an authority in the domain under discussion?

We will model the scheme "argument from expert opinion" as a sub-scheme of the scheme "argument from position to know" (for the lack of space, we will not present the full modeling of "argument from position to know", and only a part of it is depicted in the motivating example). The Scheme [14] has the following elements:

- **Expertise premise**: The source E is an expert in the domain D that is containing proposition A.
- **Assertion premise**: E asserts that A is true (false).
- **Conclusion**: A may plausibly be taken to be true (false).

Critical questions are:

1. Expertise: How credible is expert E?
2. Trustworthiness: Is E reliable?
3. Consistency: Is A consistent with the testimony of other experts?
4. Backup evidence: Is A supported by evidence?

We model this scheme as follows:

- STEP 1—Ontology enriching. We add a concept named *ArgFromExpert Opinion* to represent the scheme as descendent of rule scheme. In order to model the expertise premise, we add the concept *ExpertStmnt* as descendent of *DeclarativeStatement*. This statement is translated as an *Expert* E issues a *Statement* through the relation *IssueStatement*, and the issued statement belongs to a *domainOfExperience* in which E has enough expertise. The statement *Assertion Stmnt* denotes that E asserts that A is true (false) which is the assertion premise, and also denotes the conclusion of the argument which is A may plausibly be taken to be true (false).
- STEP 2—Rules definition. The rule could be written as: "if a rule scheme *R* has a premise of type *ExpertStmnt* and has a premise of type *AssertionStmnt* which is also its conclusion then *R* belongs to *ArgFromExpertOpinion*". This is formally depicted in Fig. 7.

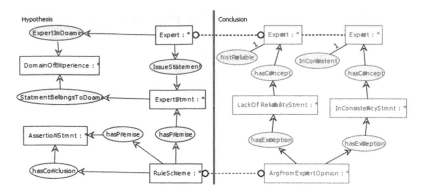

Fig. 7 Expert rule

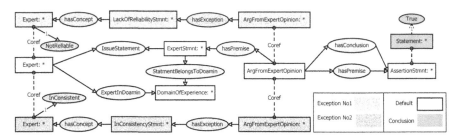

Fig. 8 Expert default rule

- STEP 3—Default rules definition. The default rule is formulated as: "If a statement *S* is issued by an expert *E* in the domain of the statement *S*, we conclude that *S* is true, unless for example the expert *E* was not reliable or inconsistent with other experts". Figure 8 depicts this rule.

6.2 Argument from Analogy

The AS for argument from analogy can be represented as follows (please see [15] for more details):

- **Similarity Premise**: Generally, case C1 is similar to case C2
- **Base Premise**: A is true (false) in case C1.
- **Conclusion**: A is true (false) in case C2.

  The critical questions which are:

1. Are C1 and C2 similar in the respect cited?
2. Are there differences between C1 and C2 that would tend to undermine the force of the similarity cited?

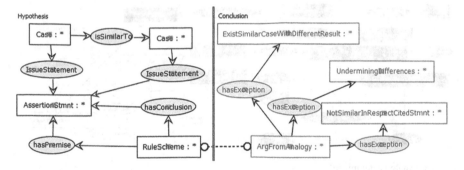

Fig. 9 Analogy rule

3. Is there some other case C3 that is also similar to C1, where A is false (true)?

   The COGUI methodology to implement this scheme is as follows:

- STEP 1—Ontology enriching. Since we consider two similar cases C_1, C_2, we obviously need to add the concept *Case* and the relation *isSimilarTo*. We also refer to the fact that a statement S is true in case C by stating that: S is issued by the *Case C*. Thus, the argument could be written as follows: if a statement S is issued by a case C_1, and if C_2 is similar to C_1, then S will be also considered as issued by the case C_2. We finally need to add the concepts *NotSimilarInRespectCited*, *UnderminingDifferences*, and *ExistSimilarCaseWithDifferentResult* for representing the three critical questions respectively.

- STEP 2—Rules definition. The rule could be written as: "if a rule scheme R has first premise: statement S is issued by case C_1, and if R has a second premise: there exists a case C_2 which is similar to C_1, and has a conclusion: the statement S is is also issued by case C_2, then the scheme R belongs to *ArgFromAnalogy*". This is formally depicted in Fig. 9.

- STEP 3—Default rules definition. The default rule for this scheme could be formulated as: "if a statement S is issued by a *Case* C_1 and if exists a case C_2 which is similar to C_1, then we conclude that S is true in case C_2, unless the similarity was not in the respect cited, or there were undermining differences, or if there exists a case C_3 which is similar to C_1 and in which S is not true (false)". Figure 10 depicts a COGUI modelling of this rule (in order to have a readable and less complicated diagram we consider only two of the exceptions: *NotSimilarInRespectCited* and *ExistSimilarCaseWithDifferentResult*.

6.3 Argument from Popular Opinion

The argument from popular opinion as described by [14] is: If a large majority (everyone, nearly everyone, etc.) accepts A as true, this would be evidence that A is generally accepted. The structure of the scheme include the following elements:

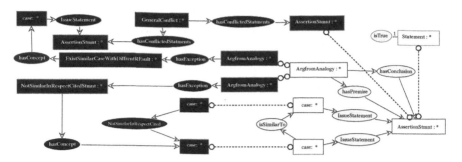

Fig. 10 Analogy default rule

- **General acceptance premise**: A is generally accepted as true.
- **Presumption premise**: If A is generally accepted as true, it gives reason for A.
- **Conclusion**: There is a reason in favor of A.

The following two critical questions match the scheme:

1. What evidence, such an appeal to common knowledge, supports the claim that A is generally accepted as true?
2. Even if A is generally accepted as true, are there any good reasons for doubting that it is true?

The COGUI methodology to represent this scheme is as follows:

- STEP 1—Ontology enriching. We refer to the fact that the statement S is generally accepted by public opinion by stating that: S is issued by *PublicOpinion*. Therefore, the statement S is the premise and the conclusion of the rule scheme. Consequently we enrich our ontology with the concepts *ArgFromPublicOpinion* and *PublicOpinion*. We need also to add the concept *LackOfEvidence* and the relation *issuedWithLackOfEvidence(statement)* for the first critical question, and the concept *ExistGoodReasonsForDoubting* and relation *hasGoodReasonsFor Doubting(statement)* for the second critical question.
- STEP 2—Rules definition. The rule could be written as: "if a statement S is issued by a *PublicOpinion* and if S is the premise and the conclusion of a rule scheme R, then R belongs to *ArgFromPublicOpinion*". This is formally depicted in Fig. 11.
- STEP 3—Default rules definition. The default rule for this scheme could be formulated as: "if a statement S is issued by a *PublicOpinion*, we conclude that S is true, unless there is a lack of evidences or there is good reasons for doubting". Figure 12 includes a modeling of this rule.

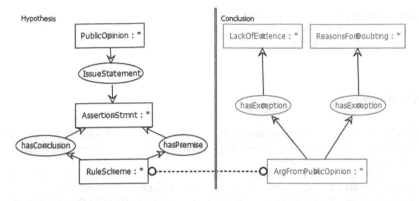

Fig. 11 Public opinion rule

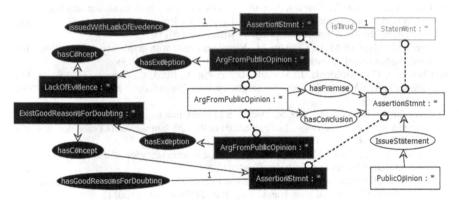

Fig. 12 Public opinion default rule

7 Conclusions

This paper presented a methodology and the modelling of AS using COGUI (graphical knowledge modelling and querying environment) to represent and use knowledge about arguments. This approach showed how a generally useful and accessible tool can be used to do non-monotonic reasoning which could be characterized as represent common knowledge. Our work distinguishes itself from the state of the art due to the expressiveness of the rules (that allow for non monotonic reasoning) as well as the fine grain representation of arguments (as opposed to black-box approaches taken by previous work). The ontology is publicly available and we are currently investigating its use for reasoning about decision making in agri food scenarios. This is an ongoing ontological modelling effort and we are currently extending our work with other schemes in the literature.

Acknowledgments The research leading to these results has received funding from the European Community's Seventh Framework Programme (FP7/2007–2013) under the grant agreement no FP7-265669-EcoBioCAP project.

References

1. Baader, F., Calvanese, D., McGuinness, D.L., Nardi, D., Patel-Schneider, P.F. (eds.): The Description Logic Handbook: Theory, Implementation, and Applications. Cambridge University Press, Cambridge (2003)
2. Baget, J.-F., Croitoru, M., Fortin, J., Thomopoulos, R.: Default conceptual graph rules: preliminary results for an agronomy application. In: ICCS. Springer, Berlin (2009)
3. Boley, H., Kifer, M., Patranjan, P.-L., Polleres, A.: Rule interchange on the web. Reasoning Web, pp. 269–309. Springer, Berlin (2007)
4. Buche, P., Cucheval, V., Diattara, A., Fortin, J., Gutierrez, A.: Implementation of a knowledge representation and reasoning tool using default rules for a decision support system in agronomy applications. In: GKR, pp. 1–12. Springer, Switzerland (2013)
5. Chein, M., Mugnier, M.-L., Croitoru, M.: Visual reasoning with graph-based mechanisms: the good, the better and the best. Knowl. Eng. Rev. **28**(3), 249–271 (2013)
6. Chesñevar, C.I., McGinnis, J., Modgil, S., Rahwan, I., Reed, C., Simari, G.R., South, M., Vreeswijk, G., Willmott, S.: Towards an argument interchange format. Knowl. Eng. Rev. **21**(4), 293–316 (2006)
7. Grosof, B.N., Horrocks, I., Volz, R., Decker, S.: Description logic programs: combining logic programs with description logic. In: Proceedings of the 12th International Conference on World Wide Web, pp. 48–57 (2003)
8. Mugnier, M.-L., Chein, M.: Graph-based knowledge representation. Advanced Information and Knowledge Processing. Springer, London (2009)
9. Rahwan, I., Banihashemi, B., Reed, C., Walton, D., Abdallah, S.: Representing and classifying arguments on the semantic web. Knowl. Eng. Rev. **26**(4), 487–511 (2011)
10. Rahwan, I., Zablith, F., Reed, C.: Laying the foundations for a world wide argument web. Artif. Intell. **171**(10–15), 897–921 (2007)
11. Reiter, R.: A logic for default reasoning. Artif. intell. **13**(1), 81–132 (1980)
12. Sowa, J.F.: Conceptual Structures: Information Processing in Mind and Machine. Addison-Wesley, Boston (1984)
13. Walton, D.: Argumentation Schemes for Presumptive Reasoning. Psychology Press, London (1996)
14. Walton, D.: Fundamentals of Critical Argumentation. Cambridge University Press, Cambridge (2006)
15. Walton, D., Reed, C., Macagno, F.: Argumentation Schemes. Cambridge University Press, Cambridge (2008)

Knowledge Discovery
and Data Mining

Computationally Efficient Rule-Based Classification for Continuous Streaming Data

Thien Le, Frederic Stahl, João Bártolo Gomes,
Mohamed Medhat Gaber and Giuseppe Di Fatta

Abstract Advances in hardware and software technologies allow to capture streaming data. The area of Data Stream Mining (DSM) is concerned with the analysis of these vast amounts of data as it is generated in real-time. Data stream classification is one of the most important DSM techniques allowing to classify previously unseen data instances. Different to traditional classifiers for static data, data stream classifiers need to adapt to concept changes (concept drift) in the stream in real-time in order to reflect the most recent concept in the data as accurately as possible. A recent addition to the data stream classifier toolbox is *eRules* which induces and updates a set of expressive rules that can easily be interpreted by humans. However, like most rule-based data stream classifiers, eRules exhibits a poor computational performance when confronted with continuous attributes. In this work, we propose an approach to deal with continuous data effectively and accurately in rule-based classifiers by using the Gaussian distribution as heuristic for building rule terms on continuous attributes. We show on the example of eRules that incorporating our method for continuous attributes indeed speeds up the real-time rule induction process while maintaining a similar level of accuracy compared with the original eRules classifier. We termed this new version of eRules with our approach G-eRules.

T. Le (✉) · F. Stahl · G.D. Fatta
School of Systems Engineering, University of Reading, Whiteknights, PO Box 225,
Reading RG6 6AY, UK
e-mail: t.d.le@pgr.reading.ac.uk

F. Stahl
e-mail: f.t.stahl@reading.ac.uk

G.D. Fatta
e-mail: g.difatta@reading.ac.uk

J.B. Gomes
Institute for Infocomm Research (I2R), A*STAR 1 Fusionopolis Way Connexis,
Singapore 138632, Singapore
e-mail: bartologjp@i2r.a-star.edu.sg

M.M. Gaber
School of Computing Science and Digital Media, Robert Gordon University, Riverside East,
Garthdee Road, Aberdeen AB10 7GJ, UK
e-mail: m.gaber1@rgu.ac.uk

© Springer International Publishing Switzerland 2014   21
M. Bramer and M. Petridis (eds.), *Research and Development
in Intelligent Systems XXXI*, DOI 10.1007/978-3-319-12069-0_2

1 Introduction

One of the problems in the area of Big Data Analytics is concerned with is the analysis of high velocity data streams. A data stream is defined as data instances that are generated at a high speed and thus challenges our computational capabilities for the processing of this data [1]. There are many applications that generate such fast and possibly infinite data streams, such as sensor networks, communication networks, Internet traffic, stock markets [2, 3].

Unlike traditional data mining systems that process static (batch) data, data stream mining algorithms analyse the streaming data on the fly in order to avoid storing these possibly infinite amounts of data. Data stream classification is only one but very important data mining task on streaming data. Data stream classifiers learn and adapt to changes in the data (concept drifts) in real-time and thus require only a single pass through the training data. A concept drift occurs if the current data mining model (i.e. the classifier) is no longer valid through a change of the distribution in the data stream.

A recent data stream classifier development is eRules [4]. eRules is based on a sliding window approach [5] and uses the rule-based Prism classifier [6] to work on streaming data. eRules has shown good classification accuracy and adaptability to concept drifts [4]. eRules induces rules of the form *IF condition THEN classification* that are expressive and compact and effectively represent information for classification. The condition is a conjunction of rule terms of the form (*Attribute* = *Value*) for categorical attributes and (*Attribute* < *Value*) or (*Attribute* $\geq$ *Value*) for continuous data. Compared with other stream classifiers (such as Hoeffding Trees) eRules tends to leave a data instance unclassified rather than forcing a possibly wrong classification. This feature is highly desirable in many applications where the results from miss-classification are costly and un-reversible, e.g.: medical and financial applications. Further weaknesses of the decision tree structure in comparison with rulesets have been analysed, and the reader is referred to [4] for further reading.

However, eRules method of processing continuous attributes is computationally very expensive and hence results in long processing times, which is a disadvantage on the application on data streams. In this paper, we propose a new heuristic method based on the Gaussian distribution in order to make eRules computationally more efficient. We show empirically that our method improves eRules' processing times and competes well with other data stream classifiers. We termed this new version of the classifier G-eRules, where G stands for the use of the **G**aussian distribution. Our method is not only applicable to eRules but could potentially be adopted in other rule and decision tree based stream classifiers as well.

The paper is organised as follows. Section 2 highlights related work in the field of rule-based data stream classification. Section 3 discusses the principles of the eRules classifier and our new approach for dealing with continuous attributes based on the Gaussian distribution. An empirical analysis of eRules classifier with our approach, termed G-eRules is presented in Sect. 4. Concluding remarks and a brief description of ongoing work are given in Sect. 5.

2 Related Work

Many algorithms for data stream classification have been proposed. Gaber et al. have identified such techniques and their features in [7]. Among these techniques, possibly the best established methods, are the Hoeffding tree based approaches [8]. Nevertheless, tree based classification approaches have been criticised in the past for both static data classification [6, 9] as well as for data stream classification [4, 10]. In particular, the drawbacks of decision tree based techniques in dealing with concept drift are mentioned in [10] and the fact that decision trees cannot abstain from a potentially wrong classification are discussed in [4]. Hence, alternative rule-based approaches have been developed for data stream classification, such as the aforementioned eRules algorithm [4] or Very Fast Decision Rules (VFDR) [11], both based on the separate and conquer rule induction approach.

eRules induces expressive *IF-THEN* rules and performs well in terms of classification accuracy and adaptation to concept drifts, also it is able to abstain from classification. However, eRules drawback is its computational efficiency when dealing with continuous attributes. To the best of our knowledge, there is no single optimised method to deal with continuous attributes for rule induction based algorithms. Methods for working with continuous attributes have been studied extensively in the past 2 decades for batch data. An overview of well-known methods for data discretisation in batch data is described in [12] and most of these methods cannot be adopted for data streams without significant modifications, and consequently they are unlikely to perform as good as they do on batch data. The work presented in this paper addresses the problem of low computational efficiency in generating rules from continuous attributes, especially for eRules as a representative technique, by developing a heuristic method based on the Gaussian distribution.

3 Computationally Efficient Rule Term Induction for Continuous Attributes in Data Streams

This section presents a new method for inducing rule terms from continuous attributes that is computationally more efficient, assuming a Gaussian distribution of the values of continuous attributes. We first highlight eRules conceptually and discuss its underlying rule induction mechanism. Then we present an alternative approach of dealing with continuous attributes with a different rule term structure. We argue and show empirically that this approach is computationally more efficient.

Fig. 1 The three main processes in eRules

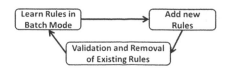

3.1 eRules and Its Rule Induction Approach

The eRules classifier has three main processes as illustrated in Fig. 1.

Essentially eRules makes use of Cendrowska's Prism algorithm [6] to learn classification rules from data streams by using a sliding window approach [5]. The three main processes of eRules illustrated in Fig. 1 are outlined as follow:

Learn Rules in Batch Mode—An initial ruleset is learnt using Prism on the first batch of incoming data instances. Later instances are added to a buffer if they are not covered by the current ruleset. If the number of instances in the buffer reaches a user defined threshold then its instances are used to induce new rules (using Prism) and the buffer is emptied.

Add New Rules—Whenever new rules are generated from the aforementioned buffer then they are added to the current ruleset and thus adapt to new emerging concepts in the data stream.

Validate and Remove Existing Rules—eRules also removes learnt rules from the model if the concerning rules are not relevant to the current concept anymore, i.e. if a concept drift occurred. This is quantified by the deterioration of the individual rule's classification accuracy over time.

Algorithm 1 shows the basic Prism rule induction method employed by eRules. ω_i denotes a target class for a rule and C is the number of classes; α_j is a rule term of the form $(\alpha < v)$ or $(\alpha \geq v)$ for continuous attributes or $(\alpha = v)$ for categorical attributes, where α denotes an attribute name and v a valid value for this attribute.

Algorithm 1: Prism classification rule induction algorithm

1 **for** $i = 1 \to C$ **do**
2   D ← original Dataset
3   **while** *D contains classes other than* ω_i **do**
4   **forall the** *attributes* α *in D* **do**
5   Calculate probability of occurrence, $p(\omega_i|\alpha_j)$ for each possible rule term α_j;
         end
7   Select the α_j with the maximum probability of occurrence, $p(\omega_i|\alpha_j)$ as rule term;
8   Create subset S of D containing all the instances covered by α_j;
9   D ← S;
     end
11   The induced rule, R is a conjunction of all α_j at line 7;
12   Remove all instances covered by rule R from original Dataset;
13   **repeat**
14   lines 2 to 12
     until *all instances of* ω_i *have been removed from original Dataset*;
16   Reset original Dataset to its initial state;
end

3.2 Current Rule Term Induction Method Based on Conditional Probability

The original Prism approach [6] does only work on categorical data for inducing rule terms of the form ($\alpha = v$). On the other hand, eRules is also able to work on continuous attributes in a very similar way compared with the VFDR [11] algorithm.

For continuous attributes eRules produces rule terms of the form ($\alpha < v$), or ($\alpha \geq v$) and VFDR produces rule terms of the form ($\alpha \leq v$), or ($\alpha > v$). The process of eRules dealing with a continuous attribute is outlined in the following steps:

1. Sort the dataset according to the attribute value.
2. For each possible value v of the continuous attribute α, calculate the probability for a target class for both terms ($\alpha < v$) and ($\alpha \geq v$).
3. Return the term, which has overall highest probability for the target class.

It is evident that this method of dealing with continuous attributes in eRules requires many cutpoint calculations for the conditional probabilities $p(\omega_i | \alpha < v)$ and $p(\omega_i | \alpha \geq v)$ for each attribute value v, where ω_i is the target class. Thus, this method decreases the computational efficiency of eRules and VFDR considerably. We propose to use the Gaussian distribution of the attribute values associated with the target class in order to create rule terms of the form ($x < \alpha \leq y$), and thus avoid frequent cutpoint calculations, as described in Sect. 3.3.

3.3 Using Gaussian Distribution to Discriminant Classification

For each continuous attribute in a dataset, we can generate a Gaussian distribution as shown in Fig. 2 to represent all possible values of that continuous attribute for a target class.

Assume a dataset with classifications, $\omega_1, \ldots, \omega_i$. If we have a measurement vector (of attribute values) x then we can compute the attribute value that is the most relevant one to a particular classification based on the Gaussian distribution of the values associated with this particular classification.

Fig. 2 Gaussian distribution of a classification from a continuous attribute

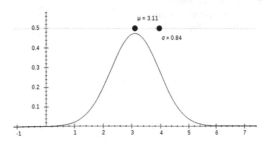

The Gaussian distribution is calculated for a continuous attribute α with mean μ and variance σ^2 from all attribute values with classification, ω_i. The class conditional density probability is then given by:

$$p(\alpha_j|\omega_i) = p(\alpha_j|\mu, \sigma^2) = \frac{1}{\sqrt{2r\sigma^2}}\exp(-\frac{(\alpha_j - \mu)^2}{2\sigma^2}) \tag{1}$$

Then a heuristic measurement of posterior class probability, $p(\omega_i|\alpha_j)$, or equivalently $\log(p(\omega_i|\alpha_j))$ can be calculated and used to determine the probability of a target class for a valid value of a continuous attribute.

$$\log(p(\omega_i|\alpha_j)) = \log(p(\alpha_j|\omega_i)) + \log(p(\omega_i)) - \log(p(\alpha_j)) \tag{2}$$

We calculate the probability of regions Ω_i for these attribute values such that if $x \in \Omega_i$ then x belongs to class ω_i. This approach may not necessarily capture the full details of the intricate continuous distribution, but it is highly efficient in computation and memory perspectives. This is because the Gauss distribution only needs to be calculated once and can then be updated when new data stream instances are received, by simply recalculating mean μ and variance σ^2.

The range of values, which extends to both sides from the mean μ of the distribution should represent the most common values of attribute α for a target class ω_i. A candidate rule term can be generated by selecting an area under the curve for a range of values for which the density class probability $p(x < \alpha \leq y|\omega_i)$ is the highest, where x and y are valid values of attribute α from the Gauss distribution for the target class ω_i. As shown in Fig. 3, the shaded area represents the highest density class probability $p(x < \alpha \leq y|\omega_i)$ of a subset from the training dataset.

Globally, the area right in the middle under the curve represents the most common values of the attribute for a target class. For example, the shaded area of one standard deviation of the mean ($\mu \pm 1\sigma$), as illustrated in Fig. 4a, covers 68 % of all possible values of the attribute for the target class; or as illustrated in Fig. 4b 95 % of all possible values of the attribute for the area of ($\mu \pm 1.96\sigma$) [13].

However, distributions for different classifications can overlap each other and an area of a distribution for a target class sometimes cannot be used to precisely distinguish a classification as shown in Fig. 5.

Fig. 3 *Shaded area represents a range of values of attribute α for class ω_i*

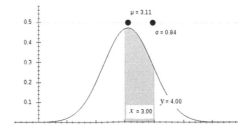

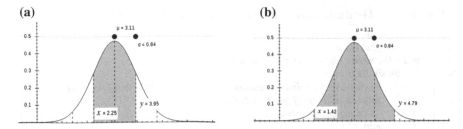

Fig. 4 *Shaded area* represents a range of values of attribute α for class ω_i. **a** 68 % of all possible values. **b** 95 % of all possible values

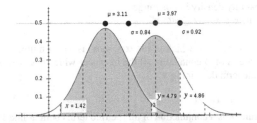

Fig. 5 Distributions of different classification overlap each other

However, we are interested to find a rule term, which can maximise the coverage of the rule for a target class. Therefore, our approach uses density estimation to discover a rule term in the form of $(x < \alpha \le y)$ by selecting only a highly relevant range of values from a continuous attribute, which can then be used to represent a subset of instances for the target class along with other rule terms. Our approach for rule induction from continuous attributes can be described with the following steps:

1. For each continuous attribute, calculate a Gaussian distribution with mean μ and variance σ^2 for each classification.
2. Calculate class conditional density and posterior class probability for each continuous attribute value for the target class, using Eqs. (1) and (2).
3. Select the value of the attribute with greater posterior class probability.
4. Select the next smaller and larger values from the value chosen in **step 3** which have the greatest posterior class probability.
5. Calculate density probability with two values from **step 4** from the normal distribution for the target class.
6. Select the range of the attribute $(x < \alpha \le y)$ as the rule term for which density class probability is the maximum.

Algorithm 2 outlines eRules using our approach for extracting rule terms from continuous attributes. The resulting algorithm is termed G-eRules, where G stands for Gauss distribution.

Algorithm 2: G-eRules induction approach for continuous attributes

---

1 **for** $i = 1 \rightarrow C$ **do**
2   D ← original Dataset
3   **while** D *contains classes other than* ω_i **do**
4   **forall the** α *in D* **do**
5   calculate mean μ and variance σ^2 of continuous attribute α for class ω_i;
6   **foreach** *value* α_j *of attribute* α **do**
7 | Calculate $p(\alpha_j | \omega_i)$;
                    end
9   Select α_j of attribute α, which has highest value of $p(\alpha_j | \omega_i)$;
10   For values $< \alpha_j$ select the value x that has the highest probability density in this range, and for values $\geq \alpha_j$ select the value y that has the highest probability density in this range;
11   Calculate $p(x < \alpha \leq y | \omega_i)$;
              end
13   Select $(x < \alpha \leq y)$ for which $p(x < \alpha \leq y | \omega_i)$ is a maximum;
14   Create subset S of D containing all the instances which has $(x < \alpha \leq y)$;
15   Build a rule term describing S;
16   D ← S;
       end
18   The induced rule, R is a conjunction of all the rule terms built at line 15;
19   Remove all instances covered by rule R from original Dataset.;
20   **repeat**
21 | lines 2 to 19;
       until *all instances of* ω_i *have been removed*;
23   Reset original Dataset to its initial state;
   end

---

4 Evaluation

The main aim of our experimental evaluation is to study the effectiveness of our proposed approach in terms of its computational performance but also its competitiveness in terms of its accuracy compared with other established data stream classifiers. The implementation of our experiments were realised in the Java based Massive Online Analysis (MOA) [14] framework, a workbench for evaluating data stream mining algorithms. MOA was chosen as it already implements a wide range of data stream classifiers.

4.1 Experimental Setup

In our experiments, we used four different classifiers for a comparative analysis:

1. *Very Fast Decision Rules (VFDR)*—[11] a rule-based data stream classifier available in MOA.

2. *Hoeffding Tree*—[8] is the state-of-art decision tree classifier implemented in MOA for data streams.
3. *eRules*—is outlined in Sect. 3.1, it is the only data stream classifier that is able to abstain from classifying when uncertain.
4. *G-eRules*—inherited from original eRules but this version of the classifier uses our continuous rule term format and induction approach outlined in Sect. 3.3.

The reason for these choices is that Hoeffding tree has been studied by the community [15] for the last decade as the state-of-the-art in data stream classification and VFDR is one of few rule-based classifiers, which can generate rules directly from a data stream, and thus shares similarities with eRules. We used the default settings for eRules and G-eRules in the experiments, which are: a sliding window of size 500; a particular rule is removed if it falls below an individual accuracy 0.8 and a has had a minimum number of 5 classification attempts. We use both, artificial (stream generators in MOA) and real datasets in our experiments:

SEA Generator—This artificial dataset is introduced in [16], it generates an artificial data stream with 2 classes and 3 continuous attributes, whereas one attribute is irrelevant for distinguishing between the 2 classes. The interested reader is referred to [16] for more information. This data generator has been used in empirical studies in [4, 11, 17] amongst others. For our evaluation we used the default generator settings which are concept function 1, a random instance seed of 1, allowing unbalanced classes and a noise level of 10 %. We generated 500,000 instances.

Random Tree Generator—This generator is introduced in [8] and is based on a randomly generated decision tree and the user can specify the number of attributes and classes. New instances are generated by assigning uniformly distributed random values to attributes and the class label is determined using the tree. Because of the underlying tree structure, decision tree based classifiers should perform better on this stream. We have generated 2 versions of this data stream, one with 5 categorical and 5 continuous attributes, and 3 classifications called RT Generator 5-5-3; and the other one also with 3 classifications but no categorical and only 4 continuous attributes called RT Generator 0-4-3. Both versions comprised 500,000 instances. The remaining settings were left at their default values, which are a tree random seed of 1, instance random seed option of 1, 5 possible values for each categorical attribute, a maximum tree depth of 5; minimum tree level for leaf nodes of 3, and the fraction of leaves per level from first leaf Level onwards is 0.15.

Covertype—is a dataset from US Forest Service (USFS) Region 2 Resource Information System (RIS) which contains the forest cover type for 30×30 m cells. There are 581,012 instances with 54 attributes and 10 of them are continuous attributes. This dataset has been used in several papers on data stream classification, i.e. in [18]. This real dataset was downloaded from the MOA website [19] and used without any modifications.

Airlines—this data source was created based on the regression dataset from Elena Ikonomovska, which was extracted from Data Expo 2009, which consists of about 500,000 flight records and the task is to use the information of scheduled departure to predict whether a given flight will be delayed. The datset comprises 3 continuous and 4 categorical attributes. The dataset was also used in one of Elena's studies about data streams [20]. This real dataset was downloaded from the MOA website [19] and used without any modifications.

Instead of initialising each artificial generator for each experiment anew, we have created static datasets and streamed these datasets to the classifiers. We did not initialise each artificial data stream anew as the generators are non-deterministic. This way we ensure that each classifier is presented with exactly the same instances and the same order of instances in the stream.

4.2 Results

The evaluation is focussed on the scalability of G-eRules to fast data streams but also shows its competitiveness in terms of classification accuracy. We used 'Inter-leaved Test-Then-Train' strategy [14] to calculate the mean-accuracy of each of the algorithms. All parameters for artificial stream generators were left at default values unless stated otherwise.

4.2.1 Overall Accuracy, Tentative Accuracy and Learning Time

One feature of eRules is the ability to abstain from classification and G-eRules inherited this feature of eRules. Therefore, we also record tentative accuracy for eRules and G-eRules, which is the accuracy for instances where the classifier is confident.

We can see that G-eRules achieves a very similar classification accuracy compared with its original eRules classifier, however, it is significantly faster. Both, eRules and G-eRules suffer from higher abstain rates when confronted with data streams with many continuous attributes (i.e. Random Tree Generator datasets and SEA), however, G-eRules has a lower abstain rate on continuous data. Nevertheless, G-eRules not only has a lower abstain rate when confronted with continuous attributes, but also a higher accuracy.

G-eRules generally outperforms, in terms of accuracy and running time, existing rule-based classifier VFDR. In terms of accuracy G-eRules outperforms VFDR in 3 out of 4 cases and in general processes the data stream much faster. For the RT-Generator 0-4-3 stream VFDR is unexpectedly slow. This could be explained by VFDR not reaching a stable ruleset that keeps changing over time. Regarding the Airlines stream eRules and G-eRules need much longer to process the stream, however, again, this could be due to the fact that both algorithms are unstable on this particular dataset due to a fixed sliding window size. However, this problem can be addressed by using a sliding window that adapts its size dynamically to concept changes (see Sect. 5).

In comparison to Hoeffding trees, we achieve a comparable accuracy on the two real datasets Covertype and Airlines, on the MOA data generators Hoeffding tree

Table 1 G-eRules compared with eRules, Hoeffding trees and VFDR

Dataset	Classifier									
	eRules			G-eRules			Hoeffding tree		VFDR	
	Ab (%)	T.Ac (%)	T (ms)	Ab (%)	T.Ac (%)	T (ms)	Ac (%)	T (ms)	Ac (%)	T (ms)
CoverType	27.993	81.12	90,098	35.77	80.68	12,484	80.61	10,631	61.29	20,722
Airlines	9.63	61.76	366,859	16.68	62.55	116,975	65.20	4,289	61.02	4,818
RT generator 5-5-3	10.52	69.28	333,163	26.47	64.77	36,486	89.78	2,422	55.51	79,153
RT generator 0-4-3	96.35	34.94	47,790	65.65	81.21	4,283	96.39	2,391	90.10	3,883,291
SEA generator	94.06	68.58	22,043	40.84	95.87	3,227	99.9.3	1,265	62.70	176,705

The abbreviation *Ab* stands for abstain rate, *T.Ac* stands for tentative accuracy, *T* stands for execution time and *Ac* stands for accuracy

performs better. However, one needs to note that Hoeffding trees are less expressive classifiers than G-eRules and eRules. The rulesets generated by our techniques can easily be interpreted and examined by the users. On the other had, Hoeffding trees need a further processing step to explain the single path that led to a particular decision that may well be quite long for interpretation by decision takers. As such we argue that G-eRules is the most expressive rule-based technique for data stream classification.

In order to examine G-eRules' computational advantage on continuous attributes we have compared eRules and G-eRules on the same base datasets as in Table 1, but with increasing numbers of continuous attributes. SEA and Random Tree Generator (with continuous attributes only) were used, both with a gradual concept drift lasting 1,000 instances starting at the 250,000th instance in order to trigger adaptation and thus make the problem computationally more challenging. The two real datasets Covertype and Airlines have been chosen. For all datasets the continuous attributes were gradually increased by duplicating the existing attributes. The reason for duplication is that the concept encoded in the stream stays the same, but the computational effort needed to find the rules is increased.

As expected, Figs. 6a, b and 7a, b show that the execution times of G-eRules do increase at a much smaller rate than the execution times of eRules do. In fact, G-eRules' increase in execution time when confronted with an increasing number of continuous attributes seems to be very close to that of Hoeffding trees. The reason for this is that eRules has to work out $p(\omega_i|\alpha < v)$ and $p(\omega_i|\alpha \geq v)$ for each value of a continuous attribute and then this process is repeated during the learning.

On the other hand, for each continuous attribute, G-eRules only has to calculate Gaussian distributions for each classification once and then the classifier can update and look up $p(\omega_i|x < \alpha \leq y)$ value from created distributions. Please note we have omitted the execution times for VFDR in Figs. 6a, b and 7a, b, as they are generally

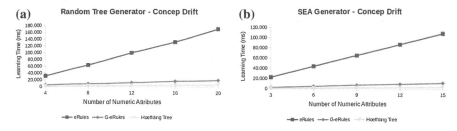

Fig. 6 Learning time of random tree and SEA generators with concept drift

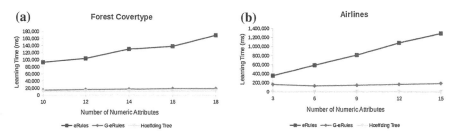

Fig. 7 Learning time of real datasets, CoverType and airlines

much greater compared with eRules and G-eRules on these particular cases, as shown in Table 1.

Moreover, we can observe that eRules tends to produce rules with irrelevant terms for continuous attributes, resulting in some of the rules produced by eRules being too specific. In other words, eRules encounters the problem of overfitting in dealing datasets with lots of continuous attributes. But rule terms in form of $(x < \alpha \leq y)$ produced from G-eRules for continuous attributes tend to cover more instances matching the target class and thus need to produce less rule terms. Although, G-eRules may not dramatically improve the levels of accuracy from original eRules, the processing time is clearly improved.

5 Conclusion

In this paper we presented a new computationally efficient method of extracting rule terms in the form of $(x < \alpha \leq y)$ from continuous attributes in a data stream for classification algorithms. The new method is based on the density class probability from Gaussian distribution. This generic way of extracting continuous rule terms can potentially be used in any rule-based data stream classifier in order to increase the data throughput of the algorithm. We have implemented this approach in the eRules algorithm, a simple yet competitive rule-based data stream classifier, that allows abstaining from a classification, but is computationally inefficient when dealing with continuous attributes. We termed this new version of the classifier G-eRules. Our evaluation shows that G-eRules with the new method for generating continuous rule terms achieves comparable levels of accuracy compared with original eRules, but is much faster when dealing with continuous attributes, which is desired in mining high velocity data streams.

We are currently developing a new rule-based classifier motivated by the scalability of our density class probability based approach for inducing continuous rule terms. Currently eRules and G-eRules are using a fixed size sliding window, however, this does not take the length of a concept drift into account. We are currently exploring metrics that could be used to dynamically re-size the sliding window to the optimal number of data instances for generating rules. For this the Hoeffding bound is considered similarly to the Hoeffding Tree algorithm [8]. However, we are also planning to change the nature of the rules from conjunctive combinations of rule terms to also allow disjunctive combinations. Thus the classifier could naturally capture concepts for classification that encode not only 'AND' but also 'OR' relationships between different attributes.

References

1. Gaber, M.M., Zaslavsky, A., Krishnaswamy, S.: Mining data streams: a review. SIGMOD Rec. **34**(2), 18–26 (2005)
2. de Aquino, A.L.L., Figueiredo, C.M.S., Nakamura, E.F., Buriol, L.S., Loureiro, A.A.F., Fernandes, A.O., Coelho, Jr. C.J.N.: Data stream based algorithms for wireless sensor network applications. In: AINA, pp. 869–876. IEEE Computer Society, New York (2007)
3. Thuraisingham, B.M.: Data mining for security applications: Mining concept-drifting data streams to detect peer to peer botnet traffic. In: ISI, IEEE (2008)
4. Stahl, F., Gaber, M.M., Salvador, M.M.: eRules: amodular adaptive classification rule learning algorithm for data streams. In: Bramer, M., Petridis, M., (eds.) SGAI Conference, pp. 65–78. Springer, Berlin (2012)
5. Babcock, B., Babu, S., Datar, M., Motwani, R., Widom, J.: Models and issues in data stream systems. In: Proceedings of the Twenty-first ACM SIGMOD-SIGACT-SIGART Symposium on Principles of Database Systems, PODS'02, pp. 1–16. ACM, New York (2002)
6. Cendrowska, J.: PRISM: an algorithm for inducing modular rules. Int. J. Man-Mach. Stud. **27**(4), 349–370 (1987)
7. Gaber, M.M., Zaslavsky, A., Krishnaswamy, S.: A survey of classification methods in data streams. In: Data Streams, pp. 39–59. Springer, Berlin (2007)
8. Domingos, P., Hulten, G., Mining high-speed data streams. In: Ramakrishnan, R., Stolfo, S.J., Bayardo, R.J., Parsa, I. (eds.) KDD, pp. 71–80. ACM (2000)
9. Witten, I.H., Eibe, F.: Data Mining: Practical Machine Learning Tools and Techniques with Java Implementations, 2nd edn. In: Kaufmann, M. (2005)
10. Wang, H., Fan, W., Yu, P.S., Han, J.: Mining concept-drifting data streams using ensemble classifiers. In: Getoor, L., Senator, T.E., Domingos, P., Faloutsos, C. (eds.) KDD, pp. 226–235. ACM (2003)
11. Gama, J., Kosina, P.: Learning decision rules from data streams. In: Walsh, T. (ed.) IJCAI 23, pp. 1255–1260. IJCAI/AAAI (2011)
12. Han, J., Kamber, M.: Data Mining. Concepts and Techniques, 2nd edn. In: Kaufmann, M. (2006)
13. Bruce, N., Pope, D., Stanistreet, D.: Quantitative methods for health research: a practical interactive guide to epidemiology and statistics. Wiley, Chichester (2008), 2013

14. Bifet, A., Holmes, G., Kirkby, R., Pfahringer, B.: MOA: massive online analysis. J. Mach. Learning Res. **11**, 1601–1604 (2010)
15. Hoeglinger, S., Pears, R.: Use of hoeffding trees in concept based data stream mining. In: Third International Conference on Information and automation for sustainability, ICIAFS 2007, pp. 57–62. IEEE, New York (2007)
16. Street, W.N., Kim, Y.: A streaming ensemble algorithm (SEA) for large-scale classification. In: Lee, D., Schkolnick, M., Provost, F.J., Srikant, R. (eds.) KDD, pp. 377–382. ACM (2001)
17. Bifet, A., Holmes, G., Pfahringer, B., Kirkby, R., Gavaldà, R.: New ensemble methods for evolving data streams. In: Elder IV, J.F., Fogelman-Soulié, F., Flach, P.A., Zaki, M., (eds.) KDD, pp. 139–148. ACM, New York (2009)
18. Abdulsalam, H., Skillicorn, D.B., Martin, P.: Streaming random forests. In: IDEAS, pp. 225–232. IEEE Computer Society, Washington, DC (2007)
19. Datasets from moa (massive online analysis) website (online). Accessed Apr 2014
20. Ikonomovska, E., Gama, J., Dzeroski, S.: Learning model trees from evolving data streams. Data Min. Knowl. Discov. **23**(1), 128168 (2011)

Improved Stability of Feature Selection by Combining Instance and Feature Weighting

Gabriel Prat and Lluís A. Belanche

Abstract The current study presents a technique that aims at improving *stability* of feature subset selection by means of a combined instance and feature weighting process. Both types of weights are based on margin concepts and can therefore be naturally interlaced. We report experiments performed on both synthetic and real data (including microarray data) showing improvements in selection stability at a similar level of prediction performance.

1 Introduction

The *feature subset selection* (FSS) problem has been studied for many years by the statistical as well as the machine learning communities [3]. In the last years FSS has become a hot topic, boosted by the introduction of new application domains and the growth of the number of features involved [6]. Problems with many features and a limited number of observations are nowadays very common, e.g. in molecule classification, genomics, DNA-microarrays, proteomics, etc.

The confidence in a model highly depends on its *stability* with respect to changes in the data used to obtain it, either in the instances themselves or in the features used to describe them. For example, in a typical FSS process using cross-validation, different features are typically selected in every validation fold. This is particularly important in greedy wrapper sequential FSS, where classification accuracy is a random variable that depends on the training sub-sample. A different feature choice at an early step may completely change the search path and lead to a substantially different set of features. Stability is particularly crucial when the aim is not (or not only) good accuracy but also *inference*: domain experts tend to have less confidence in features that change radically with slight variations in the training data.

G. Prat · L.A. Belanche (✉)
School of Computer Science, Universitat Politècnica de Catalunya,
Jordi Girona, 1-3, 08034 Barcelona, Catalonia, Spain
e-mail: belanche@lsi.upc.edu

G. Prat
e-mail: gprat@lsi.upc.edu

© Springer International Publishing Switzerland 2014
M. Bramer and M. Petridis (eds.), *Research and Development in Intelligent Systems XXXI*, DOI 10.1007/978-3-319-12069-0_3

The stability of FSS has been relatively neglected in the literature until very recently—see e.g. [8, 10, 11, 14]. Previous research aimed to measure (i.e., to *quantify*) stability, rather than to enhance it, leading to the development of stability measures [16]; few works address the explicit *improvement* of such stability, notably [7, 14].

In the current study we present a novel method that aims at providing a more stable selection of feature subsets when variations in the training process occur. This is accomplished by weighting the *instances* according to their outlying behavior; this weighting is a preprocessing step that is independent of the learner or the specific FSS algorithm. We report performance in two series of experiments: first using well-known benchmarking datasets and then some challenging microarray gene expression problems. Our results show increases in FSS stability for most subset sizes and most problems, without compromising prediction accuracy.

2 Preliminaries

Let $D = \{(\mathbf{x}_1, t_1), \ldots, (\mathbf{x}_N, t_N)\}$ be a training data set of length N, each instance $\mathbf{x}_n \in \mathbb{R}^d$ with its corresponding class label t_n. The *margin* of an instance with respect to a hypothesis (a classification rule, in this case) measures the confidence of the classifier when making its prediction [1]. There are two possible approaches to define the margin of a particular instance $\mathbf{x}$ to a set of points D:

Definition 1 The *sample margin* of $\mathbf{x}$ with respect to a hypothesis is the distance between $\mathbf{x}$ and the decision boundary induced by the hypothesis.

Definition 2 The *hypothesis margin* of $\mathbf{x}$ is the distance between the hypothesis and the closest hypothesis that assigns an alternative label to $\mathbf{x}$.

For 1-Nearest Neighbours (1NN), the classifier is defined by a set of training points and the decision boundary is the Voronoi tessellation. The sample margin in this case is the distance between the instance and the Voronoi tessellation, and hence it measures the sensitivity to small changes in the values of the instance. For the case of 1NN one has the following results [2]:

1. The hypothesis margin lower bounds the sample margin
2. The hypothesis margin of an instance $\mathbf{x}$ to a set of data points D is given by:

$$\theta_D(\mathbf{x}) = \frac{1}{2}\left(\|\mathbf{x} - m(\mathbf{x})\| - \|\mathbf{x} - h(\mathbf{x})\|\right) \tag{1}$$

were $m(\mathbf{x})$ and $h(\mathbf{x})$ are the **near hit** and **near miss**: the instances in $D\backslash\{(\mathbf{x}, \cdot)\}$ nearest to $\mathbf{x}$ with the same and with a different class label, respectively.

Relief is a filter algorithm that uses the hypothesis-margin concept in Eq. 1 to assess the importance of each feature in a dataset D as the accumulated influence that each feature has in computing the margin of every instance in D [9]. The deterministic variant of the algorithm picks one instance at a time and computes the hypothesis margin of each feature independently, accumulating the feature-wise distances to its nearest hit and nearest miss. As a result, the weight w_j given to feature X_j is its average distance to the selected neighbors:

$$w_j = \sum_{n=1}^{N} \left(|(\mathbf{x}_n)_j - m(\mathbf{x}_n)_j| - |(\mathbf{x}_n)_j - h(\mathbf{x}_n)_j| \right), \qquad j \in \{1, \dots, d\}.$$

Using this weighting strategy, features are ranked from largest to smallest weight, and those with lowest ranks can be good candidates for elimination.

Simba is a more recent feature weighing algorithm that assigns weights to features based on their contributions to the hypothesis margins of the instances [1]. Since better generalization is expected if instances have larger margins, one should favour features that *contribute* more to these margins.

Definition 3 Consider $\mathbf{x} = \{x_1, \dots, x_d\}$ be a particular instance in D. Given $\mathbf{w} = \{w_1, \dots, w_d\}$ be a weight vector over the feature set, the *weighted margin* of $\mathbf{x}$ is

$$\theta_D^{\mathbf{w}}(\mathbf{x}) = \frac{1}{2} \left(\|\mathbf{x} - m(\mathbf{x})\|_{\mathbf{w}} - \|\mathbf{x} - h(\mathbf{x})\|_{\mathbf{w}} \right) \qquad (2)$$

where the **weighted norm** of a vector $\mathbf{z}$ is $\|\mathbf{z}\|_{\mathbf{w}} = \sqrt{\sum_{i=1}^{d} w_i^2 z_i^2}$. Given a data sample, these margins are integrated into an overall score:

Definition 4 Given sample D and the weight vector $\mathbf{w}$, the *evaluation function* is:

$$e(\mathbf{w}) = \sum_{\mathbf{x} \in D} u \left(\theta_{D \setminus \{\mathbf{x}\}}^{\mathbf{w}}(\mathbf{x}) \right) \qquad (3)$$

where $u(\cdot)$ is a differentiable non-decreasing *utility function* (since larger margins entail larger utility). **Simba** uses by default a linear utility function $u(\theta) = \theta$ (hence the evaluation function is simply the sum of the margins) and normalizes the weight vector (using the supremum norm) at the end of the process. In this case, the gradient of $e(\mathbf{w})$ when evaluated on a data sample D is:

$$(\nabla e(\mathbf{w}))_i = \sum_{\mathbf{x} \in D} \frac{\partial u(\theta(\mathbf{x}))}{\partial \theta(\mathbf{x})} \frac{\partial \theta(\mathbf{x})}{\partial w_i} = \frac{1}{2} \sum_{\mathbf{x} \in D} \left(\frac{(x_i - (m(\mathbf{x}))_i)^2}{\|\mathbf{x} - m(\mathbf{x})\|_{\mathbf{w}}} - \frac{(x_i - (h(\mathbf{x}))_i)^2}{\|\mathbf{x} - h(\mathbf{x})\|_{\mathbf{w}}} \right) w_i.$$

3 Combining Instance and Feature Weighting

One problem with the definition of the hypothesis-margin concept in Eq. 1 is the presence of outliers, or redundant or noisy features, which might mislead the margin calculus of an instance. With the purpose of obtaining a more robust evaluation, the average margin between every instance in D and all the rest can be calculated:

$$\theta(\mathbf{x}) = \frac{1}{M} \sum_{i=1}^{M} \|\mathbf{x} - m_i(\mathbf{x})\| - \frac{1}{H} \sum_{i=1}^{H} \|\mathbf{x} - h_i(\mathbf{x})\|, \tag{4}$$

being $m_i(\mathbf{x})$ and $h_i(\mathbf{x})$ the i-th nearest *miss* (instance of different class) and i-th nearest hit (instance of same class) in D, respectively and M, H are the total number of misses and hits for $\mathbf{x}$ (such that $M + H + 1 = N$).

Instances $\mathbf{x}$ achieving highly positive $\theta(\mathbf{x})$ present good modeling behavior (being far from misses and close to hits), while those with highly negative $\theta(\mathbf{x})$ become outlying ones (surrounded by misses and far from hits). The presence or absence of these latter instances in a training sub-sample is therefore a source of instability. In order to obtain a bounded positive weight in $(0, 1)$, we use a logistic function:

$$\omega(\mathbf{x}) = \frac{1}{1 + \exp\{-\alpha z(\theta(\mathbf{x}))\}}, \tag{5}$$

where α is a parameter controlling the slope, and $z(\cdot)$ is the *standard score* $z(x) = (x - \hat{\mu}_D)/\hat{\sigma}_D$, being $\hat{\mu}_D$ and $\hat{\sigma}_D$ the sample mean and standard deviation of $\theta(\mathbf{x})$, for all $\mathbf{x} \in D$, respectively. A suitable value for α will depend on the problem and the user's needs. As a default value, under the assumption that hypothesis margins loosely follow a Gaussian distribution, we propose to set $\alpha = 3.03$, which corresponds to assign a weight of 0.954 to an instance whose average margin is two standard deviations from the mean, that is $\theta(\mathbf{x}) = 2\hat{\sigma}_D$.

In order to illustrate the procedure, a simple example is provided. Consider a 2D synthetic dataset containing $N = 30$ instances, obtained by equally sampling from one of two distributions: either $\mathbf{x} \sim \mathcal{N}(\mu_1, \Sigma)$ or $\mathbf{x} \sim \mathcal{N}(\mu_2, \Sigma)$, where $\mu_1 = [0, 0]$, $\mu_2 = [0, 0.25]$ and $\Sigma = diag(0.01, 0.01)$. Figure 1 shows the weighted dataset, which clearly assigns low values to instances close to the boundary between classes and those inside opposite-class region; and assigns higher values the farther from the boundary inside the proper-class region. This is consistent with the intuition that outlying instances are a source of instability and therefore, must be lowly rated.

The proposed method extends **Simba** to incorporate the instance weights, obtained in Eq. 5, into the feature weights, to influence the way **Simba** picks instances. Two different approaches can be devised towards this integration:

Sample: Based selection on a probability distribution, according to the obtained weights (some instances may be selected more than once and some none).

Order: Iterate over every instance, and base the iteration order directly on the instance weights, from the instance with the largest weight downwards (all instances are selected exactly once).

In any case, we call the method **SimbaLIW**: Simba with Logistic Instance Weighting (pseudo-code is shown in Algorithm 1).

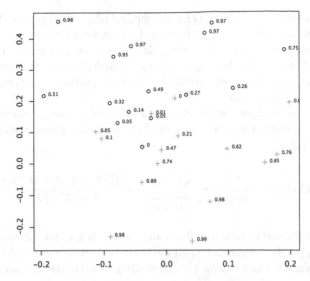

Fig. 1 Importances assigned to the synthetic example data. These importances are computed using the formula in Eq. 5. The '+' and 'o' symbols denote the two classes

---

Algorithm 1: SimbaLIW (D, ω) (strategy can be either **sample** or **order**)

---

1   $\mathbf{w} \leftarrow (1, 1, \dots, 1)$ // Feature weights
2   **for** $n \leftarrow 1$ **to** N **do**
3   **if** *strategy is sample* **then**
4   draw an instance $\mathbf{x}$ from D, according to the distribution $\omega / \|\omega\|_1$
5   **end**
6   **else**
7   let $\mathbf{x}$ be the instance ranked in position n according to ω
8   **end**
9   calculate $m(\mathbf{x})$ and $h(\mathbf{x})$ with respect to $D \setminus \{\mathbf{x}\}$ and the weight vector $\mathbf{w}$;
10   **for** $i \leftarrow 1$ **to** d **do**
11       $\Delta_i \leftarrow \frac{1}{2}\left(\frac{(x_i - (m(\mathbf{x}))_i)^2}{\|\mathbf{x} - m(\mathbf{x})\|_\mathbf{w}} - \frac{(x_i - (h(\mathbf{x}))_i)^2}{\|\mathbf{x} - h(\mathbf{x})\|_\mathbf{w}} \right) w_i$
12   **end**
13    $\mathbf{w} \leftarrow \mathbf{w} + \Delta$;
14 **end**
15 $\mathbf{x} \leftarrow \mathbf{x}^2 / \|\mathbf{x}^2\|_\infty$ where $(\mathbf{x}^2)_i := (x_i)^2$

---

4 Experimental Work

This section provides empirical evaluation of the proposed method. First, we illustrate **SimbaLIW** using a synthetically generated dataset; given that the truth behind the features is perfectly known, one has the possibility of assessing true performance.

Then, the algorithm is tested to verify its real applicability, in two groups of problems: first using some well-known datasets from the UCI machine learning repository [12], and then in widely-used cancer microarray data. These are different problems: first, *the number of features* is in the range of tens to a hundred for the former, and in the range of thousands for the latter; second, *the number of instances* is generally much lower for the microarray data.

The stability of an algorithm in selecting a subset of k features out of the initial full feature size d over a batch of M runs can be evaluated using the *Kuncheva index*, defined as in [11]:

$$KI(\mathscr{E}(k)) = \frac{2}{M(M-1)} \sum_{i=1}^{M-1} \sum_{j=i+1}^{M} \frac{|S_i(k) \cap S_j(k)| - (k^2/d)}{k - (k^2/d)}$$

where $S_i(k)$ is the subset of selected features of length k in the ith run, and $\mathscr{E} = \{S_1, S_2, ..., S_M\}$ is the set containing all the retrieved feature subsets. KI values are contained between -1 and 1, with 1 corresponding to the maximum stability.

4.1 Synthetic Data

We first use a synthetic dataset designed to verify the performance of stable feature subset strategies [7]. It consists of $M = 500$ training sets, each of the form $\mathbf{x}^m \in \mathbb{R}^{N \times d}$, with $N = 100$ instances and $d = 1,000$ features, for $m = 1, \ldots, M$. Every instance is equiprobably drawn from one of two distributions: $\mathbf{x} \sim \mathscr{N}(\mu_1, \Sigma)$ or $\mathbf{x} \sim \mathscr{N}(\mu_2, \Sigma)$, where

$$\mu_1 = (\underbrace{0.5, \ldots, 0.5,}_{50} \underbrace{0, \ldots, 0}_{950}), \quad \mu_2 = -\mu_1,$$

and

$$\Sigma = \begin{bmatrix} \Sigma_1 & 0 & \cdots & 0 \\ 0 & \Sigma_2 & \cdots & 0 \\ \vdots & \vdots & \ddots & \vdots \\ 0 & 0 & \cdots & \Sigma_{100} \end{bmatrix},$$

being $\Sigma_i \in \mathbb{R}^{10 \times 10}$, with 1 in its diagonal elements and 0.8 elsewhere. Class labels are assigned according to the expression:

$$y_i = \text{sgn}\left(\sum_{j=1}^{d} x_{i,j} r_j\right), \quad \mathbf{r} = (\underbrace{0.02, ..., 0.02,}_{50} \underbrace{0, ..., 0}_{950}).$$

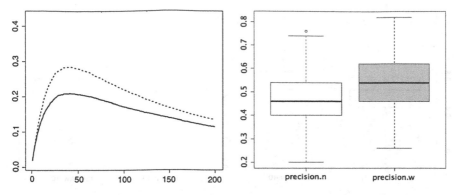

Fig. 2 Feature stability on Han and Yu synthetic data. *Left plot* shows the average KI over 500 repetitions of the process, as a function of increasing subset size (the *bold line* is the normal **Simba**, the *dashed line* is the weighted **SimbaLIW** version). *Right plot* shows the corresponding average precisions (n for normal, w for weighted)

The plots in Fig. 2 shows the stability and accuracy results obtained by averaging 500 runs of independent artificial data generation, comparing **Simba** against **SimbaLIW**. The left plot is the average KI as a function of subset size (size has been cut at 200 for clarity, the rest of the plot being similar to the shown slice). It can be seen the stability is increased at all subset sizes, topping in the 40–50 range (recall that this problem has exactly 50 relevant features). The plot on the right shows boxplots of the distribution of *precision* over the 500 runs when the first 50 features are kept (thus we use the knowledge that only 50 features are relevant). Welch's t-test for a difference on the means (0.4722 vs. 0.5414) over the 500 independent results is highly significant (p-value $< 2.2e-16$). Notice the *recall* is the same for both methods because there are exactly 50 relevant variables (out of the 1,000).

The plots in Fig. 3 do not use this knowledge. Rather, for both methods, we select the feature subset showing *maximum stability* and compute the corresponding average precision (left) and recall (right).[1]

4.2 Real Data

A collection of four UCI and four microarray datasets presenting a variety of diseases is used. These problems are difficult for several reasons, in particular the sparsity of the data, the high dimensionality of the feature (gene) space, and the fact that

---

[1] Recall = **True positives/(True Positives + False Negatives)**; Precision = **True positives/(True Positives + False Positives)**. A **True Positive** is a selected and relevant feature, a **False Negative** is a discarded and relevant feature, etc.

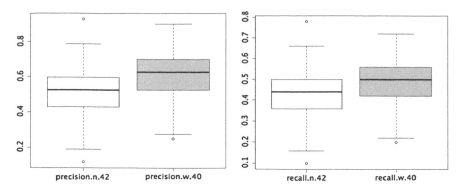

Fig. 3 Boxplots of precision and recall at the point of maximum stability (42 features for **Simba** and 40 for **SimbaLIW**) on Han and Yu synthetic data

Table 1 Detailed dataset description

Problem	Features	Classes	Instances
Ionosphere	34	2	351
Mammogram	65	2	86
Musk	166	2	476
SPECTF	44	2	267
Breast cancer	24,481	2	97
GCM	16,063	14	190
Leukemia	7,129	2	72
Lung cancer	12,533	2	181

The top four problems come from UCI; the bottom four are microarray problems

many features are irrelevant or redundant. A summary of the main characteristics can be found in Table 1. For microarray data, a preliminary selection of genes was performed on the basis of the ratio of their between-groups to within-groups sum of square, keeping the top 200 genes for each dataset [5].

4.3 Experimental Setting

We use a carefully designed resampling methodology in order to avoid feature selection bias [13, 15], specially serious in high-dimensional biomedical data, such as gene expression microarrays. Each experiment consists of an *outer* loop of 5 × 2-fold feature selection [4]. This algorithm makes 5 repetitions of a 2-fold cross-validation. It keeps half of the instances out of the feature selection process and uses them as a test set to evaluate the quality of the selected features. For every fold and repetition of the outer cross-validation loop, two feature-weighting processes are conducted with the same instances: one with the original **Simba** algorithm and one with our modified version taking instance weights into account (the *order* variant for microarray problems and the *sample* variant for all others). The KI is computed for every subset length at every partition loop and then averaged over the 10 times. Once the features have been obtained we test various possible numbers of feature subsets using an SVM classifier, recording prediction accuracy on the leftout test parts. Specifically, we select the first $n \in \{2, 10, 20, 30, 40, 50\}$ features (if present, and ranked by the feature weights) and perform an **inner** 5×2-fold cross-validation with the purpose of estimating the prediction error of each classifier. This error is then computed for each fold to compare the feature sets selected by both **Simba** and **SimbaLIW**. A detailed pseudo-code for the full experimental setup is shown in Algorithm 2.

Algorithm 2: Experimental setup

---

1 **for** *OUTER* ← 1 **to** *5* **do**   // Outer cross-validation
2   split the data randomly into two equal-sized **fs-train** and **fs-test** sets;
3   compute the ω weights for **fs-train** using Eq. 5 ;
4   $\mathbf{w}_S$ ← Simba (**fs-train**) ;
5   $\mathbf{w}_{SLIW}$ ← SimbaLIW (**fs-train**, ω) ;
6   **for** *INNER* ← 1 **to** *5* **do**   // Inner cross-validation
7   split **fs-test** randomly into two equal-sized **class-train, class-test** sets;
8   **foreach** n **in** $\{2, 10, 20, 30, 40, 50\}$ **do**
9   normal-model ← train SVM (**class-train**) using $\mathbf{w}_S^n[1 \ldots n]$;
10   weighted-model ← train SVM (**class-train**) using $\mathbf{w}_{SLIW}^n[1 \ldots n]$;
11   normal-prediction ← predict (normal-model, **class-test**);
12   weighted-prediction ← predict (weighted-model, **class-test**);
13   normal-error ← error (normal-prediction, **class-test**) ;
14   weighted-error ← error (weighted-prediction, **class-test**)
15   **end**
16 **end**
17 **end**

Figures 4, 5, 6 and 7 show feature subset selection stability on UCI and microarray data, respectively, against number of selected features. It also displays average test errors for some fixed numbers of features, according to the ranking given by both methods (normal vs. weighted). Excluding small variations at the very first and last iterations, the same general trend can be observed in both sets of plots, showing that stability is enhanced for most subset sizes and virtually all problems. It is interesting to note that the small or null gains correspond to very small (sizes 1–5) or large (near full size) subsets, where the set of possible combinations is much smaller.

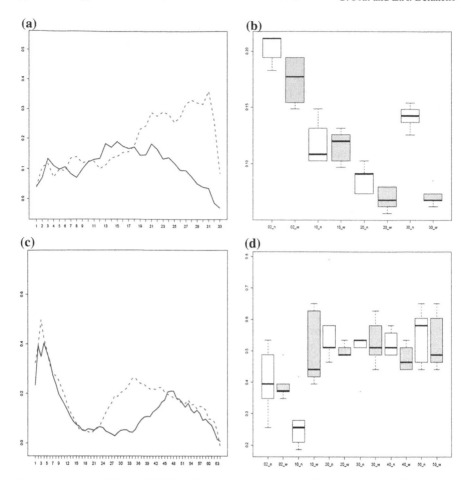

Fig. 4 Feature stability on UCI data. *Left plots* show average KI against number of subset size (the *bold line* is the normal **Simba**, the *dashed line* is the weighted **SimbaLIW** version). *Right plot* shows the average test errors (*n* for normal, *w* for weighted—*shaded*) for different numbers of selected features. **a** Ionosphere (stability). **b** Ionosphere (test error). **c** Mammogram (stability). **d** Mammogram (test error)

Remarkably, on all other sizes, the method chooses feature subsets in a more stable way when the training set changes.

It is beyond the intention of this study to search for the best absolute subset (or best subset size) for each problem. Rather, the motivation is to study stability for all sizes and a glimpse at possible differences in test accuracy. In this sense, as it could reasonably be expected, performance varies for the different chosen sizes. Some problems seem to benefit from an aggresive FSS process (like *Mammogram*

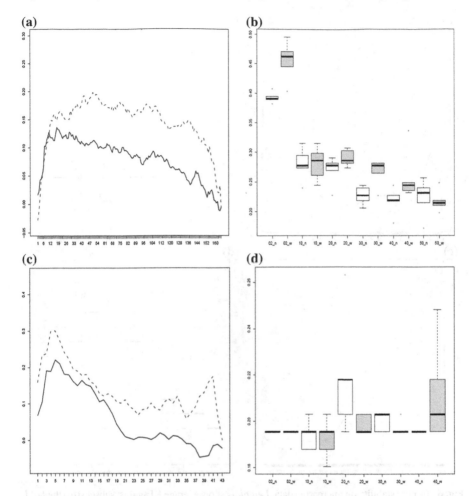

Fig. 5 Feature stability on UCI data (continued). *Left plots* show average KI against subset size (the *bold line* is the normal **Simba**, the *dashed line* is the weighted **SimbaLIW** version). *Right plot* shows the average test errors (*n* for normal, *w* for weighted—*shaded*) for different numbers of selected features. **a** Musk (stability). **b** Musk (test error). **c** SPECTF (stability). **d** SPECTF (test error)

and *Leukemia*), and others the other way around (this being the case of *Ionosphere*, *Musk* and *Breast cancer*). For the rest, drawing a conclusion on approximate best size is difficult. Comparing both approaches, the weighted **SimbaLIW** version seems better for *Ionosphere*, *Mammogram*, *Musk* and *Breast cancer*, and similar for the rest. Specially good are results for the GCM microarray dataset, one showing 14 classes

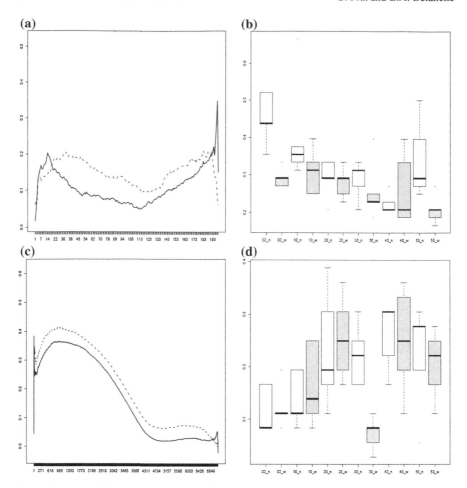

Fig. 6 Feature stability on microarray data. *Left plots* show average KI against subset size (the *bold line* is the normal **Simba**, the *dashed line* is the weighted **SimbaLIW** version). *Right plot* shows the average test errors (*n* for normal, *w* for weighted—*shaded*) for different numbers of selected features. **a** Breast cancer (stability). **b** Breast cancer (test error). **c** Leukemia (stability). **d** Leukemia (test error)

(a large source of unstability, given the double partitionings in a small sample size). For this problem, the gains in selection stability are larger, and the variances of the prediction errors (as given by the boxplots) are smaller.

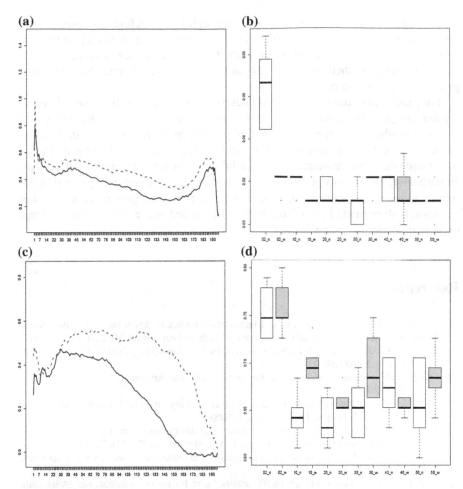

Fig. 7 Feature stability on microarray data (continued). *Left plots* show average KI against subset size (the *bold line* is the normal **Simba**, the *dashed line* is the weighted **SimbaLIW** version). *Right plot* shows the average test errors (*n* for normal, *w* for weighted—*shaded*) for different numbers of selected features. **a** Lung cancer (stability). **b** Lung cancer (test error). **c** GCM (stability). **d** GCM (test error)

5 Conclusions

The present work has introduced **SimbaLIW**, a new method for improving the stability of feature subset selection algorithms, which draws upon previous algorithmic work on feature weighting and hypothesis margins for instances. Our strategy uses a double set of weights, one for the features and another one for the instances. The suitability for standard feature selection practice has been assessed using data from two different environments: microarray gene expression and real-world datasets from the UCI repository. The reported results, although being preliminary, suggest that there is room for stability enhancement, very specially in small-sized problems. It could be argued that models showing high stability are inherently more robust to input variance, and therefore more advisable to rely upon, despite showing lower prediction accuracy in certain cases.

The present work offers a number of interesting avenues for further research. First, we are interested in quantifying and improving *prediction* stability: the ability of a classifier in labelling each instance coherently (independently of its correctness), over variations in the sample used to train it. Second, there are several alternative ways to combine the instance weighting idea and the **Simba** feature weighting algorithm. In particular, the instance weights can be updated at each iteration, given that the feature weights are re-computed, which would lead to a synergetic process. Last, but not least, both computed sets of weights can be handed over a learner able to accept them to influence the model it learns.

References

1. Bachrach, R.G., Navot, A., Tishby, N.: Margin based feature selection-theory and algorithms. In: Proceedings of International Conference on Machine Learning (ICML), pp. 43–50 (2004)
2. Crammer, K., Gilad-Bachrach, R., Navot, A., Tishby, N.: Margin analysis of the LVQ algorithm. Adv. NIPS, 462–469 (2002)
3. Devijver, P.A., Kittler, J.: Pattern Recognition: A Statistical Approach. Prentice Hall, Englewood Cliffs (2002)
4. Dietterich, T.G.: Approximate statistical tests for comparing supervised classification learning algorithms. Neural Comput. **10**, 1895–1923 (1998)
5. Dudoit, S., Fridlyand, J., Speed, T.P.: Comparison of discrimination methods for the classification of tumors using gene expression data. J. Am. Stat. Assoc. **97**(457), 77–87 (2002)
6. Guyon, I., Elisseeff, A.: An introduction to variable and feature selection. J. Mach. Learn Res. **3**, 1157–1182 (2003)
7. Han, Y., Yu, L.: A variance reduction framework for stable feature selection. Stat. Anal. Data Min. **5**, 428–445 (2012)
8. Kalousis, A., Prados, J., Hilario, M.: Stability of feature selection algorithms: a study on high-dimensional spaces. Knowl. Inf. Syst. **12**(1), 95–116 (2006)
9. Kira, K., Rendell, L.A.: The feature selection problem: traditional methods and a new algorithm, pp. 129–134. AAAI Press and MIT Press, Cambridge (1992)
10. Křížek, P., Kittler, J., Hlaváč, V.: Improving stability of feature selection methods. In: Kropatsch, W.G., Kampel, M., Hanbury, A. (eds.) CAIP. Lecture Notes in Computer Science, vol. 4673, pp. 929–936. Springer, Berlin (2007)
11. Kuncheva, L.I.: A stability index for feature selection. In: IASTED International Conference on Artificial Intelligence and Applications, Innsbruck, Austria. ACTA Press, Anaheim, pp. 390–395 (2007)
12. Newman, D.J., Hettich, S., Blake, C.L., Merz, C.J.: UCI repository of machine learning databases, vol. 55. Department of Information and Computer Science, University of California, Irvine (1998). http://www.ics.uci.edu/mlearn/MLRepository.html
13. Raudys, A., Baumgartner, R., Somorjai, R: On understanding and assessing feature selection bias. LNCS, vol. 3581, pp. 468–472. Springer, Berlin (2005)
14. Saeys, Y., Abeel, T., Peer, Y.: Robust feature selection using ensemble feature selection techniques. ECML-PKDD, pp. 313–325. Springer, Berlin (2008)

15. Singhi, S.K., Liu, H.: Feature subset selection bias for classification learning. In: Cohen, W.W., Moore, A. (eds.) ICML, vol. 148, pp. 849–856 (2006)
16. Somol, P., Novovičová, J.: Evaluating stability and comparing output of feature selectors that optimize feature subset cardinality. IEEE Trans. Pattern Anal. Mach. Intell. **32**(11), 1921–1939 (2010)

Towards a Parallel Computationally Efficient Approach to Scaling Up Data Stream Classification

Mark Tennant, Frederic Stahl, Giuseppe Di Fatta
and João Bártolo Gomes

Abstract Advances in hardware technologies allow to capture and process data in real-time and the resulting high throughput data streams require novel data mining approaches. The research area of Data Stream Mining (DSM) is developing data mining algorithms that allow us to analyse these continuous streams of data in real-time. The creation and real-time adaption of classification models from data streams is one of the most challenging DSM tasks. Current classifiers for streaming data address this problem by using incremental learning algorithms. However, even so these algorithms are fast, they are challenged by high velocity data streams, where data instances are incoming at a fast rate. This is problematic if the applications desire that there is no or only a very little delay between changes in the patterns of the stream and absorption of these patterns by the classifier. Problems of scalability to Big Data of traditional data mining algorithms for static (non streaming) datasets have been addressed through the development of parallel classifiers. However, there is very little work on the parallelisation of data stream classification techniques. In this paper we investigate K-Nearest Neighbours (KNN) as the basis for a real-time adaptive and parallel methodology for scalable data stream classification tasks.

M. Tennant (✉) · F. Stahl · G. Di Fatta
School of Systems Engineering, University of Reading,
Whiteknights, Reading RG6 6AY, UK
e-mail: m.tennant@pgr.reading.ac.uk

F. Stahl
e-mail: f.t.Stahl@reading.ac.uk

G. Di Fatta
e-mail: g.difatta@reading.ac.uk

J.B. Gomes
Institute for Infocomm Research (I2R), A*STAR, 1 Fusionopolis Way Connexis,
138632 Singapore, Singapore
e-mail: bartologjp@i2r.a-star.edu.sg

© Springer International Publishing Switzerland 2014
M. Bramer and M. Petridis (eds.), *Research and Development
in Intelligent Systems XXXI*, DOI 10.1007/978-3-319-12069-0_4

51

1 Introduction

The problem of dealing with 'Big Data' for 'classification' is that classical methods of classification are not suitable with respect to the new scale, speed and the variety (possible unstructured format) of 'Big Data'. Traditional data mining methods for classification of static data take several passes through the training data in order to generate the classification model, which is then applied on previously unseen data instances. Streaming models differ from this learning procedure of Train and Test to a system that continuously needs to be evaluated and updated. As the data is often either too fast to process in depth, or too vast to store, data stream classifiers must be naturally incremental, adaptive and responsive to single exposures to data instances. The continuous task of re-learning and adaptation aims to tackle the problem of concept drift [1] (changes of the patterns encoded in the streams over time). An ideal data stream classifier should incorporate certain features [2]: the classifier must limit its size (memory footprint) as streams are theoretically infinitely long; the time taken to process each instance is short and constant so as not to create a bottleneck; each data instance is only seen once by the classifier; the classification model created incrementally should be equivalent to a 'batch' learner given the same training data; and the classifier must be able to handle concept drift. Data streams come in all forms as technologies merge and become more interconnected. Classic applications are: sensor networks; Internet traffic management and web log analysis [3]; TCP/IP packet monitoring [4]; and intrusion detection [5]. However, capturing, storing and processing these data streams is not feasible, as the data stream is potentially infinite. Systems that could analyse these very fast and unbounded data streams in real-time are of great importance to applications such as the detection of credit card fraud [6, 7] or network intrusion. For many data mining problems parallelisation can be utilised to increase the scalability. It is a way for classifiers to increase the speed of both model creation and usage, notable developments are for example the tree and rule based parallel classifiers [8–10]. Working with data streams limits the processing time available for classifications (both testing and training), to the small window of time in between the arrival of instances. Parallelisation of data stream mining algorithms offers a potential way to create faster solutions that can process a much larger amount of data instances in this small time window and thus can scale up these algorithms to high velocity data streams. One of the currently fastest streaming decision tree based classifiers VFDT (Very Fast Decision Tree) [11] is simple, incremental and has great performance. Unfortunately they are not inherently scalable and lack the ability to be efficiently parallelised. The problem with distributing complex streaming classifier models (such as decision trees) over a cluster, is that it reduces their ability to adapt to concept drift and creates new problems such as load imbalance and time delays. KNN is typically not suited to data stream mining without adaptation (such as employing KD-Trees, P-Trees, L-Trees, MicroClusters) [12], as they incur a relatively high real-time processing cost, proportional to their training data size. In this paper we propose KNN as a basis for the creation of a parallel data stream classifier. The motivation for using KNN is because KNN is inherently parallelisable, for example [13] developed a parallel KNN using the MapReduce parallel programming paradigm [14]. This has been demonstrated in the past for KNN on static data [15], but not yet on data streams. Versions of KNN for data streams exist, such as [16], but to our knowledge there are no parallel approaches for KNN on data streams.

This paper investigates the feasibility of developing a parallel data stream classifier based on KNN. For this we build a simple serial (non-parallel) KNN classifier for data streams that is able to process and adapt on streaming data in real-time. We then show that this basic real-time KNN classifier is competitive (in terms of classification accuracy) compared with other existing well established data stream classifiers, yet very slow. Next we propose a parallel version of this real-time KNN classifier and show empirically that it achieves the same level of classification accuracy compared with its serial version, but has the potential to process data much faster. This parallel real-time KNN method is not only expected to compete well with the current best classification algorithms in terms of accuracy and adaptivity to concept drifts, but also in terms of its scalability to future increases in 'Big Data' in terms of volume and velocity [17].

The remainder of this paper is organised as follows: Sect. 2 proposes to use KNN as the basis for a data stream classifier that is parallelisable and provides a comparative analysis with other data stream classifiers. We go on to highlight its weakness (processing time), and propose to counter this with parallisation. Section 3 then introduces a parallel version of the KNN based data stream classifier. Section 4 discusses ongoing work and concluding remarks.

2 Adapting KNN for Data Stream Classification

This sections outlines a basic KNN algorithm for data streams and evaluates this real-time KNN in terms of its predictive performance compared with other currently leading data stream classifiers. Motivated by the results a parallel version of real-time KNN is proposed in Sect. 3.

2.1 A Simple Real-Time KNN Based Classifier for Data Streams

When dealing with data streams of infinite length it is apparent that the size of the training set needs to be managed, simply adding more and more items into the training set over time is not feasible for long-term performance. Too many items in the training set will have a negative effect on the performance of the classifier in terms of CPU-time, memory consumption but also accuracy as old potentially obsolete concepts may be used for prediction. Too few items will negatively impact the classifiers accuracy as well, as new emerging concepts may only be covered partially. Loosely speaking, the classifier needs to be able to adapt to 'concept drifts'.

In this section we give a brief overview on the principle of KNN and discuss the *sliding window* approach as a means to manage the training pool and sampling from the data stream. The classifier maintains a training set of previously classified instances. New instances that require classification are evaluated against each instance in the training set using a distance measure indicating how 'similar' the new (test) instance is to each training instance. Once all training items have been evaluated against the new instance the training instances are re-ordered by their 'distance' evaluation. The top 'K' training instances (the Nearest Neighbours) are used as individual votes with their classification labels. The classification label with the majority vote is assigned to the new instance.

Each training item must be evaluated against the new instances to see how closely they resemble each other. This resemblance is quantified by the squared Euclidean distance.

$$Squared\ Euclidean\ Distance = \sum_{i=1}^{N}(X_i - \bar{X}_i)^2$$

where N is the number of attributes, X is the training record and $\bar{X}$ is the test record.

Categorical attributes are re-quantified to give an equal distance of 1 unit between any different attribute values. More categorical attribute tailored distance metrics could be used, however, for showing that KNN is competitive with other existing data stream classifiers this simple approach is sufficient. A few 'Windowing' techniques exist to manage the number and relevance of training items in a training set.

A popular, yet simple technique is the 'Sliding Window' as depicted in Fig. 1, where a buffer of predetermined size is created to hold the training instances according to the First In First Out (FIFO) principle w.r.t the timestamps of the instances. Therefore, as a new instance arrives it replaces the oldest instance in the sliding window. As the sliding window size (number of items it contains) is fixed it

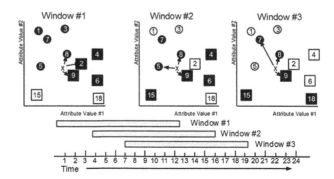

Fig. 1 Sliding windows with KNN. The shapes represent the data instances' classes (*circle* and *square*) and the number inside the shape is the time stamp. *Shaded instances* (*shapes*) are within the current sliding window and available for KNN

naturally keeps the training focused upon the recently seen instances without the need of additional management or complex statistical evaluations. Possibly more for KNN tailored Windowing approaches could be developed, however, Sect. 2's purpose is to show KNN's general suitability for data stream classification rather than optimising windowing. The reader is referred to [3] for a more comprehensive review of windowing techniques and change detectors.

2.2 Experimental Analysis of Real-Time KNN

This Section evaluates the illustrated simple real-time KNN approach (Sect. 2.1), empirically regarding its potential adaptation as a parallel data stream classifier. Multiple sliding window sizes were used to also evaluate the effects of a larger training set size upon KNN classifiers. As KNN classifiers only utilise training instances as the basis for classification, we would expect that more training instances potentially produce better results. The tradeoff we expect is that the classification times become much larger.

2.2.1 Experimental Setup

To evaluate the accuracy and the speed versus the training sizes, multiple sliding window sizes were created (50, 100, 250, 500, 1,000, 2,000 and 5,000 instances) and evaluated. The KNN classifier was set to a fixed K-value of 3 and utilised a 'Sliding Window' approach. Each simulation was run 6 times. Both, the Mean accuracy of the complete streams and their corresponding Mean processing times were recorded for evaluation. Our KNN Classifier is implemented using the MOA framework [18]. The MOA framework is an open source environment for evaluating data stream classification, clustering and regression algorithms and comparing them against one-another. The framework incorporates the most widely used algorithms for comparison, such as Hoeffding Trees [11] and Naive Bayes classifiers.

In this paper we utilise three of MOA's artificial stream generators with concept drift. Multiple stream generators were created from which new instances are taken from for training and testing. Initially all instances were taken from one data stream generator. As the concept drifts, overlap occurs, increasingly more instances were taken from a second data stream generator, finally, after the concept drift finished, all instances were taken from the second generator. Each stream generator was set to create 35,000 instances sequentially for classification. Three individual tests were performed on each of the streams: *sudden concept drift*, where the concept in the stream switches to a different one immediately after the 10,000th instance, there is no overlap of instances from both generators; *gradual concept drift*, where two concepts coexist over a period of 1,000 instances from the 10,000th instance onwards, instances from both data streams are taken over a period of 1,000 instances; *recurring concept drift*, which are essentially two sudden concept drifts, where the first concept is replaced by the second one from the 10,000th instance onwards and then the second concept is again replaced by the first concept from the 12,000th instance onwards. The MOA stream generators were used to generate unique data instances at run time. Each instance generated, conformed to the underlying concept of the generator at the time of creation. Each instance was tested upon the classifier to log the classifier's performance before being used for training: this is also know as Prequential testing (Test then Train) [19].

The **STAGGER** data stream [20] consists of three attributes, each valuing one of three values. The stream generator classes each instance according to the attributes randomly assigned and the 'function' selected at run-time. Only two possible class labels exist (True and False). Depending on the function preset by the user the generator labels the instances (True or False). The default value of False is used unless combinations of attribute values are met to produce True classifications, dependant on the function selected. For function 1 an attribute pair of 'Red' and 'Small' gives a True value. Function 2 requires an attribute value of 'Green' or 'Circle', while function 3 needs a 'Medium' or 'Large' value. We have arbitrarily chosen function 1 for the initial concept and function 3 for the concept change.

The **SEA** data stream [21] contains three continuous attributes and two class labels. The stream can be set to 1 of 4 function parameters. The function selected determines the value of a sum threshold of the first 2 continuous attribute values (Threshold Values: 8, 9, 7, 9.5). A class label of True is given only if the Threshold level is surpassed, otherwise class label False is given. We have arbitrarily chosen function 1 for the initial concept and function 3 for the concept change.

The **Random Tree Generator** [11] creates a random tree with split points using the attributes associated with the stream. Each tree leaf is labelled with class labels and then all subsequent randomly generated instances are labelled according to their traversal of the tree to a leaf node. In our experiments the random tree(s) comprise five continuous attributes and three distinct class labels. A drift is achieved by simply generating a different random tree.

2.2.2 Results and Interpretation

All streams were randomly generated by the MOA artificial stream generators with a finite number of instances (35,000). The classification accuracies are listed in Table 1 for our real-time KNN approach, Naive Bayes and Hoeffding Tree, where the accuracy is the mean accuracy achieved over the whole 35,000 instances.

The categorical synthetic data stream (STAGGER) on all KNN evaluations was observed to perform best with a training set size of 250 (Table 1). This is probably due to the random nature of the generated stream and an overall imbalance of classes in the training instances.

The SEA dataset [21] represents an interesting problem. On first glance the Hoeffding Tree produces a good tree quickly with good classification results. Upon further investigation of the Hoeffding Tree we found that the underlying tree structure is larger than expected. For the simple task of summing two attributes for

Table 1 Simple real-time KNN classification accuracy and overall processing time needed for the 35,000 instances

Concept	Learner (training size)	STAGGER % (s)	SEA % (s)	Random Tree % (s)
Sudden	Naive Bays	83.24(0.08)	94.15(0.11)	69.87(0.12)
	Hoeffding Tree	96.65(0.20)	95.59(0.36)	78.65(0.36)
	KNN(50)	96.53(0.22)	90.35(0.25)	68.53(0.21)
	KNN(100)	98.14(0.38)	92.56(0.37)	71.86(0.35)
	KNN(250)	**99.25(0.90)**	94.66(0.94)	75.03(0.96)
	KNN(500)	98.87(1.76)	95.70(1.97)	77.32(2.03)
	KNN(1000)	97.76(3.55)	96.37(4.33)	78.75(4.46)
	KNN(2000)	95.58(6.70)	96.80 (9.05)	**79.84(9.25)**
	KNN(5000)	90.65(15.65)	**97.12(23.53)**	79.18(24.00)
Gradual	Naive Bays	83.24(0.08)	94.15(0.11)	69.86(0.11)
	Hoeffding Tree	96.65(0.21)	95.57(0.33)	78.36(0.33)
	KNN(50)	96.35(0.18)	90.33(0.19)	67.97(0.18)
	KNN(100)	98.14(0.36)	92.51(0.34)	71.32(0.35)
	KNN(250)	**99.25(0.86)**	94.63(0.92)	74.56(0.95)
	KNN(500)	98.87(1.71)	95.71(1.96)	77.01(2.00)
	KNN(1000)	97.76(3.48)	96.35(4.34)	78.48(4.39)
	KNN(2000)	95.58(6.66)	96.80(9.02)	**79.75(9.20)**
	KNN(5000)	90.11(15.93)	**97.12(23.42)**	79.13(23.90)
Recurring	Naive Bays	83.62(0.08)	94.20(0.11)	71.07(0.16)
	Hoeffding Tree	88.00 (0.2)	95.37(0.27)	77.60(0.31)
	KNN(50)	95.10(0.18)	90.31(0.16)	68.98(0.18)
	KNN(100)	97.17(0.35)	92.55(0.34)	72.13(0.34)
	KNN(250)	**98.45(0.85)**	94.64(0.94)	75.04(0.95)
	KNN(500)	98.01(1.69)	95.67(1.96)	77.17(2.03)
	KNN(1000)	96.61(3.40)	96.37(4.30)	78.38(4.41)
	KNN(2000)	93.53(6.57)	96.78(9.08)	**79.14(9.25)**
	KNN(5000)	85.63(15.55)	**97.00(23.50)**	77.82(24.14)

A sudden (or gradual) concept drift occurs at instance 10,000, the gradual concept drift lasts until instance 11,000. A recurring concept drift appears 12,000 instances. The run-time s is listed in seconds

classification (True if over a set threshold or False if under) the tree structure contained over 40 decision tree branches, many of which were a split decision based upon the 3rd attribute; an attribute that is known to have no relation to the classification decision within the stream generator.

The increase in accuracy of the KNN classifier with respect to continuous attributes is promising. It shows that the real-time KNN approach is competitive with standard popular data stream classifiers (Hoeffding Trees and Naive Bayes), at the cost of processing time. In general the results of SEA and Random Tree suggest that probably

Fig. 2 Simple real-time
KNN processing time with
respect to training set size:
sudden concept drift

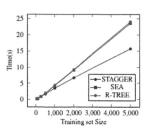

Fig. 3 Simple real-time
KNN processing time with
respect to training set size:
gradual concept drift

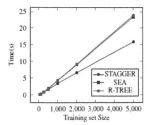

Fig. 4 Simple real-time
KNN processing time with
respect to training set size:
recurring concept drift

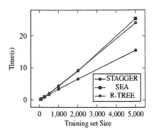

even higher accuracies could have been achieved with a larger sliding window (on the expense of processing time).

As expected, Figs. 2, 3 and 4 show that as the sliding window size increases the time taken to process all 35,000 instances increases linearly. This is because test instances have to be compared with a larger number of training instances, which involves more distance calculations. In each graph the STAGGER data stream is quicker to process due to its categorical variables.

We expect a rise in computational cost due to the training size increase. However, KNN is inherently parallel as each training item can be independently evaluated. Motivated by the analysis results of this section, Sect. 3 discusses how real-time KNN can be parallelised.

3 Parallel Data Stream Classifier Based Upon KNN

As the process of calculating the distance of a test instance to a training instance is an isolated function, with no other information required other than these two instances; it is a data parallel problem. We propose to utilise the open source MapReduce framework (Hadoop) to create a distributed real-time KNN classifier.

3.1 The MapReduce Parallelisation Paradigm

Apache Hadoop [22] is a programming framework and implementation for processing large data, based on Google's MapReduce framework. The MapReduce framework can be simplified into 2 logical steps (Mapping and Reducing). Mapping, refers to the 'bulk' processing that needs to be parallelised. The function to be processed (Map function) is sent to each node (worker) in the cluster, along with a subset of the data to be processed. Each Mapper works independently upon its subset of the data and reports its findings to the Reducer. The Reducer collates all the information from each Mapper and merges it into a logical order (using a Reduce function) before further processing and/or reporting final results. Apache Spark [23] is an open source implementation of MapReduce tailored for streaming data, which we utilise in this research.

3.2 Adaptive Parallel Real-Time KNN for Data Streams

To show parallel real-time KNN's classification performance we have created a set of KNN classifiers in the form of Mappers, each employing a sliding window of size:

$$\frac{size\ of\ the\ total\ sliding\ window}{number\ of\ Mappers}$$

Each Mapper holds a unique subset of the training data and an instance of our serial real-time KNN classifier as Map function (described in Sect. 2). Training upon the cluster involves the additional step of splitting the training data evenly across the cluster. Each Mapper keeps its local training data in its own local 'Sliding Window' to manage its size and recency. As long as additional training data is evenly distributed across the cluster each Mapper will only hold the most relevant (latest) instances. Each Mapper does not have any direct access or communication with other Mappers, or their data. All data input and output is facilitated through the MapReduce paradigm.

The complexity of KNN is approximately $O(N)$ where N is the number of training instances. An overview of the cluster layout and process path can be seen in Fig. 5. By distributing the training set out over different Mappers the costly distance evaluations for each of the training items can be calculated in parallel. As shown in Algorithm 1,

Fig. 5 Parallel real-time
KNN classifier architecture

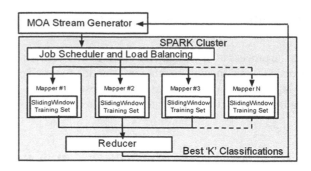

each Mapper only posts the locally 'K' required best classification/distance metrics pairs. Thus reducing the internal communication of the cluster down to a minimum level, as there is no expensive communication required between the Mappers in order to find the overall 'K' best candidates. This results in $m * K$ instances to be forwarded to the Reducer (if m is the number of Mappers), amongst which the 'K' best candidates are contained. The Reducer phase ensures that the globally best 'K' instances' class labels are returned to the 'MOA Stream Generator'. This ensures that always the globally 'K' best candidates are returned. Please note that this approach ensures that the parallel real-time KNN will always return the same 'K' best candidates as the serial real-time KNN would, irrespectively of how many Mappers have been used.

Algorithm 1: Parallel real-time KNN Process

Data: Test Instance
Result: Closest Neighbour(s) Class Labels
Broadcast(Test Instance);
foreach *Parallel Mapper KNN Classifier* **do**
    foreach *Training Instance in the Sliding Window* **do**
      | Evaluate Test Instance(TrainingInstance[i]);
    end
    Sort Evaluations by Distance();
    Send K Best Evaluations to Reducer();
end
Reducer;
foreach *MapperResultSet* **do**
    | MergeSort(MapperResults);
end
ReturnKBestEvaluations();

The scalability of this approach can be broken down into several steps that consume time for both, the maintenance of a sliding window (training data) and performing classifications: communication of instances to the computer cluster T_{COMIN}; adding and removing instances to the sliding window T_{WIN}; process instances (distance calculations) T_{DIST}; communication of the K locally best instances from each Mapper to the Reducer T_{COMOUT}; rank the $K \cdot m$ instances and determine the classification based on the majority vote of the K top ranked instances on the Reducer

T_{RANK}. Thus the total online training and testing time consumed can be summed up as follows:

$$T_{TOTAL} = T_{COMIN} + T_{WIN} + T_{DIST} + T_{COMOUT} + T_{RANK}$$

whereas we currently parallelised the computationally most expensive steps T_{WIN} and T_{DIST}. The times T_{COMOUT} and T_{RANK} are negligibly small, T_{COMOUT} only requires the communication of only $K \cdot m$ instances to the Reducer on a classification request; and T_{RANK} essentially performs a sort on only the $K \cdot m$ instances also only on a classification request. T_{COMIN} is currently not parallelised but required in order to get the training and test data into the cluster. Thus the total time needed is:

$$T_{TOTAL} = T_{COMIN} + \frac{T_{WIN}}{m} + \frac{T_{DIST}}{m} + T_{COMOUT} + T_{RANK}.$$

3.3 Initial Experimental Analysis of Parallel KNN Prototype

3.3.1 Experimental Setup

The experimental setup is the same used for the evaluation of the serial real-time KNN in Sect. 2.2.1, this allows comparing parallel real-time KNN against its serial version. The purpose of this preliminary evaluation is to show that our parallel approach is able to reproduce the same accuracy as serial real-time KNN. The evaluation in this Section is based on a real parallel implementation, however, with only one physical computer node. The empirical scalability analysis is subject to ongoing work.

3.3.2 Results and Interpretation

As the results in Table 2 show, our parallel real-time KNN is just as accurate as the serial real-time KNN classifier (see Table 1), but has the ability to be scaled up to faster data streams through the use of a computer cluster. Please note that both, the serial and parallel versions of real-time KNN are deterministic on the same data; and that the parallel version always produces exactly the same classifications as the single real-time KNN classifier does. However, the parallel version is expected to be faster. The small discrepancies in accuracy between the single real-time KNN classifier and the parallel version exist because MOA's stream generators are non-deterministic.

On average using the same classification metrics and settings on a single machine the SPARK framework adds approximately 40 s to the task of classification regardless of the training sets sizes as can be seen in Table 2. This shows that the times taken for classifications do not follow a linear increase as we would expect. This may be partly due to using MOA and SPARK frameworks together. The MOA framework generates single instances for classification while SPARK works more efficiently with

Table 2 Parallel real-time KNN classification accuracy for 35,000 instances

Concept	Learner (training size)	STAGGER % (s)	SEA % (s)	Random tree % (s)
Sudden	SPARK-NN(50)	96.87(39.39)	90.53(41.70)	67.45(40.86)
	SPARK-NN(100)	98.75(39.03)	92.65(39.97)	70.28(41.37)
	SPARK-NN(250)	**99.26(40.48)**	94.37(40.05)	73.85(42.03)
	SPARK-NN(500)	98.84(40.36)	95.49(40.06)	75.70(40.15)
	SPARK-NN(1000)	97.73 (40.12)	96.22(37.69)	77.56(38.97)
	SPARK-NN(2000)	95.53(38.98)	96.34(39.67)	**78.56(40.79)**
	SPARK-NN(5000)	76.02(44.77)	**96.89(39.89)**	78.46(40.45)
Gradual	SPARK-NN(50)	95.88(41.23)	90.56(41.56)	66.88(41.35)
	SPARK-NN(100)	97.88(41.97)	92.64(42.34)	69.77(41.37)
	SPARK-NN(250)	**98.70(41.52)**	94.30(41.72)	73.35(42.84)
	SPARK-NN(500)	98.46(41.26)	95.40(41.87)	75.23(41.35)
	SPARK-NN(1000)	97.62(41.16)	96.16(42.25)	77.23(42.34)
	SPARK-NN(2000)	95.54(41.17)	96.71(41.82)	**78.40(41.54)**
	SPARK-NN(5000)	90.52(42.80)	**96.83(42.69)**	78.35(42.86)
Recurring	SPARK-NN(50)	95.68(40.91)	90.49(40.43)	67.76(39.81)
	SPARK-NN(100)	97.76(39.29)	92.60(40.01)	70.38(40.06)
	SPARK-NN(250)	**98.47(40.57)**	94.43(39.93)	73.84(41.33)
	SPARK-NN(500)	97.96(40.17)	95.47(39.94)	75.54(41.76)
	SPARK-NN(1000)	96.75(38.87)	96.19(40.22)	77.22(39.75)
	SPARK-NN(2000)	93.47(39.90)	96.67(40.53)	**77.93(40.35)**
	SPARK-NN(5000)	85.58(40.41)	**96.70(41.25)**	77.09(41.47)

(sudden concept drift occurs at instance 10,000. Secondary concept drift for recurring occurred at 12,000 instances)

batch broadcasts. The internal communications and data handling of the MapReduce jobs inside the SPARK framework are therefore overly large. As mentioned above, we have used MOA in this implementation in order to evaluate the accuracy of our approach, but it will be removed in the future in order to allow an evaluation regarding parallel real-time KNN's scalability to fast data streams on a real cluster. An interesting feature has been observed with the continuous data sets (Random Tree and SEA). When the training data is set to 1,000 instances, a small reduction in overall time can be seen in the *sudden* and *recurring* results. We believe this is due to the data propagation through the SPARK framework and internal data handling.

Our further works will aim to improve the performance of the classifier. As we have highlighted that our parallel real-time stream classifier performs with similar accuracy as our single real-time classifier we can remove the feedback loop to MAO for testing (see Fig. 5) the MOA framework. This should increase the SPARK performance as multiple classification broadcasts and evaluations can be batch transmitted throughout the cluster.

4 Conclusions and Ongoing Work

In this research we developed a data stream classifier that is highly parallelisable and hence can be applied to high velocity data streams. So far there are very few approaches of parallel data stream classification algorithms. In our algorithm we have proposed to use KNN as a basis, due to the fact that it is inherently parallel. We have introduced real-time KNN as a simple non-parallel data stream classifier and compared it to existing popular non-parallel data stream classifiers. The results showed that our simple real-time KNN is competitive in terms of the average mean accuracy; however, as expected, on the expense of processing time. Yet, processing time is crucial in many high velocity real-time applications. Motivated by real-time KNN's competitiveness in terms of its mean accuracy we further proposed a parallel version of real-time KNN that is expected to increase the scalability of the classifier to much larger high velocity data streams compared with existing data stream classification algorithms. A prototype implementation of this parallel real-time KNN classifier has been evaluated in terms of its accuracy and it has been found that it achieves a very similar level of accuracy compared with the serial real-time KNN.

We are currently migrating our implementation to a dedicated server cluster. We have created a SPARK cluster of 64 nodes capable of running our real-time parallel KNN classifier. We aim to show that the SPARK overhead can be mitigated with large enough data streams of high velocity. We propose to use both, artificial stream generators and real stream data for parallel classification.

Also a new data structure for real-time parallel KNN, based on 'Micro-Clusters' similar to those used in the CluStream [24] data stream clustering algorithm, is currently in development. The Micro-Clusters are intended to be used in real-time statistical summaries of the training instances in terms of their attributes values and time stamps. This is expected to make real-time KNN as well as its parallel version computationally more efficient in terms of processing times, as well as memory consumption. The second objective of using these Micro-Clusters is to improve the accuracy and speed of adaptation to concept drifts, as Micro-Clusters present a more dynamic way of dealing with the recency of data instances compared with a simple sliding window. Currently the size of the sliding window is set by the user; however, the micro clusters would allow us to take the actual recency of a Micro-Cluster into account as a weight for the distance calculation to test instances. Furthermore Micro-Clusters present a memory efficient way of storing obsolete concepts that could be reactivated quickly in the case an old concept suddenly recurring, which in turn is expected to improve the speed of adaptation in the case of recurring concepts. KNN queries can also be accelerated by using efficient data structures such as multidimensional binary search trees (KD-Trees) [25]. KD-Trees are very efficient and effective for many problems in Computer Graphics and Image Processing where the space dimensionality is intrinsically low. Binary Space Partitioning (BSP) trees are a general technique for recursively dividing a multidimensional space into convex sets by means of hyperplanes. Every non-leaf node of a BSP-tree defines a splitting hyperplane that divides the space into two subspaces. KD-Trees are a particular case where each hyperplane is perpendicular to an axis. BSP-trees can be built by means of approximate Hierarchical Clustering [26] and have been shown to be better suited to improving the efficiency of KNN queries even in high dimensional spaces.

References

1. Gaber, M., Zaslavsky, A., Krishnaswamy, S.: A survey of classification methods in data streams. In: Data Streams, pp. 39–59 (2007)
2. Domingos, P., Hulten, G.: A general framework for mining massive data streams. J. Comput. Graph. Stat. **12**, 945–949 (2003)
3. Gama, J.: Knowledge Discovery from Data Streams. Chapman and Hall/CRC, Boca Raton (2010)
4. Bujlow, T., Riaz, T., Pedersen, J.M.: A method for classification of network traffic based on c5.0 machine learning algorithm. In: 2012 International Conference on Computing, Networking and Communications (ICNC), pp. 237–241 (2012)
5. Jadhav, A., Jadhav, A., Jadhav, P., Kulkarni, P.: A novel approach for the design of network intrusion detection system(nids). In: 2013 International Conference on Sensor Network Security Technology and Privacy Communication System (SNS PCS), pp. 22–27 (2013)
6. Behdad, M., French, T.: Online learning classifiers in dynamic environments with incomplete feedback. In: IEEE Congress on Evolutionary Computation, Cancùn, Mexico (2013)
7. Salazar, A., Safont, G., Soriano, A., Vergara, L.: Automatic credit card fraud detection based on non-linear signal processing. In: 2012 IEEE International Carnahan Conference on Security Technology (ICCST), pp. 207–212 (2012)
8. Joshi, M.V., Karypis, G., Kumar, V.: Scalparc: a new scalable and efficient parallel classification algorithm for mining large datasets. In: Parallel Processing Symposium, pp. 573–579 (1998)
9. Shafer, J., Agrawal, R., Mehta, M.: Sprint: a scalable parallel classifier for data mining. In: Proceedings of the 22nd VLDB Conference (1996)
10. Stahl, F., Bramer, M.: Computationally efficient induction of classification rules with the PMCRI and J-PMCRI frameworks. Knowl. Based Syst. **35**, 49–63 (2012)
11. Domingos, P., Hulten, G.: Mining high-speed data streams. KDD, pp. 71–80 (2000)
12. Zhang, P., Gao, B.J., Zhu, X., Guo, L.: Enabling fast lazy learning for data streams. In: 2011 IEEE 11th International Conference on Data Mining (ICDM), pp. 932–941 (2011)
13. Zhang, C., Li, F., Jestes, J.: Efficient parallel kNN joins for large data in MapReduce. In: Proceedings of the 15th International Conference on Extending Database Technology, EDBT 12, pp. 38–49. ACM, New York (2012)
14. Dean, J., Ghemawat, S.: MapReduce: simplified data processing on large clusters. In: Proceedings of the 6th Symposium on Operating Systems Design and Implementation (2004)
15. Liang, S., Wang, C., Liu, Y., Jian, L.: CUKNN: A parallel implementation of K-nearest neighbor on CUDA-enabled GPU. In: 2009 IEEE Youth Conference on Information, Computing and Telecommunication, YC-ICT '09, pp. 415–418 (2009)
16. Dilectin, H.D., Mercy, R.B.V.: Classification and dynamic class detection of real time data for tsunami warning system. In: 2012 International Conference on Recent Advances in Computing and Software Systems (RACSS), pp. 124–129 (2012)
17. Gantz, J., Reinsel, D.: Extracting value from chaos. IDC iview, pp. 1–12 (2011)
18. Massive online analysis (http://moa.cms.waikato.ac.nz) (2014)
19. Dawid, A.: Stastical theory the prequential approach. Royal Stat. Soc. **147**, 278–292 (1984)
20. Schlimmer, J.C., Granger, R.: Beyond incremental processing: tracking concept drift. In: Proceedings of the Fifth National Conference on Artificial Intelligence, vol. 1, pp. 502–507 (1986)
21. Street, W.N., Kim, Y.S.: A streaming ensemble algorithm (sea) for large-scale classication. In: Proceedings of the Seventh ACM SIGKDD International Conference on Knowledge Discovery and Data Mining, pp. 377–382 (2001)

22. Hadoop, http://hadoop.apache.org/ (2014)
23. Spark: Lightning fast cluster computing (http://spark.apache.org) (2014)
24. Aggarwal, C., Han, J., Wang, J.,Yu P.: A framework for clustering evolving data streams. In: Proceedings of the 29th VLDB Conference. Berlin, Germany (2003)
25. Bentley, J.L.: Multidimensional binary search trees used for associative searching. Commun. ACM **18**(9), 509–517 (1975)
26. Pettinger, D., Di Fatta, G.: Space partitioning for scalable k-means. In: The Ninth IEEE International Conference on Machine Learning and Applications (ICMLA 2010), pp. 319–324, Washington DC, USA (2010)

Machine Learning

Following the Trail of Source Languages in Literary Translations

Carmen Klaussner, Gerard Lynch and Carl Vogel

Abstract We build on past research in distinguishing English translations from originally English text, and in guessing the source language where the text is deemed to be a translation. We replicate an extant method in relation to both a reconstruction of the original data set and a fresh data set compiled on an analogous basis. We extend this with an analysis of the features that emerge from the combined data set. Finally, we report on an inverse use of the method, not as guessing the source language of a translated text, but as a tool in quality estimation, marking a text as requiring inspection if it is guessed to be a translation, rather than a text composed originally in the language analysed. We obtain c. 80 % accuracy, comparable to results of earlier work in literary source language guessing—this supports the claim of the method's validity in identifying salient features of source language interference.

1 Introduction

The present study replicates and extends research reported on by Lynch and Vogel [13], which sought to identify salient features in texts of translated English that discriminate them from texts originally composed in English. That study focused on an English literary text corpus, containing translations into English from French, German and Russian, as well as non-translated English, and achieved c. 80 % accuracy. The feature set included 'document level' statistics, in addition to token-level distinctive features. By comparison, van Halteren [18] achieved 86 % accuracy in a similar task with speeches in the EUROPARL corpus, using token-level features.

C. Klaussner (✉)
CNGL, Trinity College, Dublin, Ireland
e-mail: klaussnc@tcd.ie

G. Lynch
CeADAR, University College Dublin, Dublin, Ireland
e-mail: gerard.lynch@ucd.ie

C. Vogel (✉)
Centre for Computing and Language Studies, Trinity College, Dublin, Ireland
e-mail: vogel@tcd.ie

© Springer International Publishing Switzerland 2014
M. Bramer and M. Petridis (eds.), *Research and Development in Intelligent Systems XXXI*, DOI 10.1007/978-3-319-12069-0_5

The task in the literary discrimination problem considered by Lynch and Vogel [13] was to correctly identify the source language (L1) of a text, be it a translation or a text composed originally in English. They explored the classification efficacy of different feature types—document-level statistics, such as average word length or readability metrics, aggregated over the whole text and other features are individuated as n-grams of tokens used in the text, or abstracted as part-of-speech (POS) sequences.[1] The features identified can be regarded as tangible influences of the source language on the target language texts. Text selection was restricted to those originating in the late 19th to early 20th century as made available through the Project Gutenberg archive. To limit the effect of individual author (or translator) style, each author and translator was only represented once in the corpus, the aim being the detection of features of source-language translation effects rather than *fingerprints* of individual authors or translators. In the present replication of the task, we test the method's overall validity. For this purpose our data set is independent with respect to internal overlaps of translators and authors with distinct texts from those used in earlier work. We deem demonstration of replicability and extensibility to be an important part of scientific endeavour, and therefore we report our findings here.

Thus, Sect. 2 gives an insight into more intrinsic causes of translation effects and recent work that studied their manifestations in translations. Section 3 describes the current data set and methodology for the experiment. In Sect. 4, we present our experiments and results on the detection of source languages. Finally, Sect. 5 closes with an in-depth analysis of the prediction properties of identified highly discriminatory attributes and compares to those identified on the basis of the previous study's data set. For the purpose of comparison, we further combine both data sets and subsequently analyse highly characteristic features. As a final exercise in quality estimation, we use salient features to see whether they will lead us to texts that contain irregularities in terms of *normal* English style.

2 Motivation and Previous Research

The task of predicting the source language of a text by identifying elements that indirectly originated in the source language presupposes *translation effects*, in this case influence of the source language on the target language. Since translators traditionally translate into their native language, an interesting question to consider is how differences between translated and original language might arise. The act of translation requires fidelity to the source text while maintaining adequacy of representation in the target text [19], and given these two forces pulling in opposite directions, it might be expected that one of these factors will prevail.

Translation studies brought forth the idea of *translation universals*, properties inherent to the act of translation, such as *explicitation* (the tendency to render content

---

[1] Lexical items that could identify works by topic or theme were discarded to retain and use only features widely applicable.

that is expressed implicitly in the source language more explicitly the target language) or *simplification* (the act of simplifying content in the form of, for instance a less complex sentence structure) that were originally introduced by Blum-Kulka [3]. However, the concept of translation universals remains a controversial issue, partly connected to the difficulty of proving it and thus is constantly challenged by emerging research, as recently by Becher [2] and by Volansky et al. [19].

The property of *interference*, which is by nature a phenomenon specific to language pairs, is simpler to investigate, as irregularities in the target language (L2) can generally be traced back to properties of the source language (L1). Most of these influences are likely to be more abstract, and thus difficult to prevent, as well as being highly dependent on the particular translation pair and therefore not necessarily pertaining to universal translation effects. For instance, a study into bidirectional transfer has shown that one is influenced by a language's conceptual properties regardless of whether it was learned as a first or second language [4].[2]

Another issue is related to register interference for pairs of languages that have a similar syntactic realisation of certain items, but which do not share the same register or code relation, where two items in L2 of different register point to one lexical and register representation in L1. For example, in German *"Ich bin"* can translate to both *"I am"* and *"I'm"* in English. Whereas the German non-contracted version can also appear in colloquial speech, in contrast *"I am"* appears mostly in written text and would be rather marked elsewhere. However, a translator would need to be able to make this distinction (in an abstract way) and decide based on the context rather than only the lexical realisation and similarity to the target language construction. Practically, this is likely to result in more instances of *"I am"* in the target language. Also, interference may arise due to lack of language competence of the translator, for instance with respect to comprehending metaphors or idiomatic expressions to a degree that would allow rephrasing them appropriately given the cultural context of the target language. Failure of this sort may result in word-by-word translations that are structurally and lexically unusual in the target language.

Text classification methods trying to identify specific items of translated text likely originating in the source language have only recently been adopted. Work by Baroni and Bernardini [1] spearheaded the usage of machine learning techniques to separate translated from non-translated text, in this case based on a large Italian newspaper corpus comprising both original texts and translations. A variety of feature types were explored, the most distinguishing being a combination of lemma and mixed unigrams, bigrams and POS trigrams resulting in ~86% classification accuracy. It showed that mandatory linguistic features in the source language that are optional in the target language are more likely to be overrepresented in a translation from L1 to L2. In work on the EUROPARL corpus, van Halteren [18] was able to achieve 87.2–96.7% accuracy in guessing the source language of speeches, taking into account the signalling strength of individual sequences

---

[2] Consequently, both L1 and L2 speakers of Hebrew and English rated word pairs in English closer when they shared a translation in Hebrew, e.g. *tool* and *dish* both translate into *kli*. English monolinguals did not perceive this relatedness of the same word pair.

being present for texts originating in particular languages as well as degrees of over- or under-use. A number of studies explored translation direction and *translationese*, focused on machine translation. Kurokawa et al. [9] investigated translation direction in the parallel Canadian Hansard FR-EN corpora and discovered that using FR-EN corpora for a FR-EN machine translation system resulted in the same accuracy (BLEU score) as a much larger corpus of EN-FR. Lembersky et al. [10] extended this using both the EUROPARL and the Canadian Hansard corpus, focusing in the first instance on the language model used in machine translation, not the phrase tables themselves as done by Kurokawa et al. [9]. Using the *perplexity* metric on the EUROPARL data, they find that a language model made up of language translated from the source language of the text to be translated had the highest perplexity score, with the worst perplexity score derived from the original English data. A mixed model containing text from all four languages plus original English performed second best. They claim this validates the *convergence* universal of translation: a mixed *translationese* model was still a better fit than original English; *translationese* is more similar to other *translationese* from different source languages than original English. Lembersky et al. [11] focused on the phrase tables themselves like Kurokawa et al. [9], replicating the earlier results, delving further into the explanation of why translation direction matters. Using the information-theoretic metrics of *entropy* and *cross entropy*, they find that phrase tables trained in the correct direction contain more unique source phrases and less translations per source phrase, which results in a lower entropy and cross entropy score on the data.

Koppel and Ordan [8] investigated dialects of *translationese*, which concerned the different subsets of translations in a target language from numerous source languages. Their work used the frequencies of the top three hundred most frequent words and SVM classifiers and obtained 90 % classification accuracy between translations into English from Hebrew, Korean and Greek from the International Herald Tribune (IHT). They replicated and extended the work on EUROPARL from van Halteren [18] and reported that results on corpora of same-language translations and original English are generally better than training a *translationese* detection system on one corpus (EUROPARL) and testing on another (IHT), which returned poor results (50–60 %), showing that features of *translationese* vary not only by source language, but also genre. Ilisei et al. [7] examined translation universals of simplification using document-level metrics only, and reported high accuracy (80–90 %) on two separate corpora of medical and technical translations from Spanish to English. Distinguishing features included lexical richness, the ratio of closed-class to open-class words and the ratio of numerals, adjective and finite verbs, each to total words. This work motivated the inclusion of 'document-level' features in the original study on source language detection by Lynch and Vogel [13], which is continued here. Similar methods are used in Lynch [12] for the task of distinguishing authorial style in a corpus of literary translations of three Russian authors by the same translator, distinguishing reliably between authors using document-statistics alone. Forsyth and Lam [5] also investigates authorial style in parallel translations of Van Gogh's letters using clustering techniques and finds that authorial discriminability is stronger than translatorial discriminability.

Volansky et al. [19] tested translation hypotheses, such as *simplification, explicitation* and *interference*, using supervised machine learning in English with linguistically informed features in a text classification setting. They tentatively conclude that claims to universal translation effects should be reconsidered, as they not only rely on genre and register, but also vary greatly depending on language pairings. *Interference* yields the highest performing and most straightforwardly interpretable distinguishing features ("source language shining through"). In the present study, we attempt to identify salient features of interference that originated in the source language, but are not translator-specific issues; i.e. features detected should be general enough to appear across different translations of the same source language and thus signal more intrinsic transfer issues for that particular language pair.

3 Methods

First, we replicated the study of Lynch and Vogel [13], finding results that agree with the earlier work.[3] Below we describe our methods in extending this method to a new set of literary translations. We used the same restrictions introduced by the earlier study. The data selection process and preprocessing are described in Sect. 3.1; feature sets, in Sect. 3.2; classification methods, in Sect. 3.3.

3.1 Corpus Selection and Preparation

In order to have a comparable data sample with respect to the first analysis, the present set was selected according to the same criteria of being available in machine-readable format and being in the public domain.[4] Each author should have a unique text and no translators should be repeated in the case of translations. Additionally, all texts should be of sufficient length, meaning at least 200 kB in size.

Since this is the second experiment using this configuration, to be able to generalise better about the model's validity, there should be little or if possible no overlap in authors/translators with the first data set. While being able to meet all other requirements, the last one proved to be more difficult with respect to the Russian corpus, as there were fewer texts available (not theoretically, but practically, since new books are still in the process of being added) and after discarding all those not suited for other reasons, there remained one overlap with the first data set for the translator, *Constance Garnett*. Generally, one of the objectives was also to not have too large a

---

[3] See previous and replicated results in Table 8 and 9 in the online appendix to this paper: https://www.scss.tcd.ie/~vogel/SGAI2014/KlaussnerLynchVogel-Appendix-SGAI2014.pdf.

[4] All were taken from *Project Gutenberg*: http://www.gutenberg.org/–last verified Nov. 2013.

time-frame, i.e. have all novels written and translated within 100 years, so as to limit differences in style with respect to diachronic effects. It would also have been more desirable for the corpus to contain only novels, whereas in the current set, there is one collection of short stories in the Russian set, namely *"The wife and other stories"* rather than a continuous text. The difference in genre may affect the distribution of proper nouns, since there are likely to be more *different* ones.

In the original study, there were five novels per source language, while for ours, one more novel per source language corpus is added, as shown in Table 1. We retain the

Table 1 Text corpora for current source language detection experiment

Title	Author	Source	Pub.	Translator	T. Pub.
Vanity Fair	William M. Thackeray	English	1848	n/a	n/a
Wives and Daughters	Elizabeth Gaskell	English	1866	n/a	n/a
The Moonstone	Wilkie Collins	English	1868	n/a	n/a
Sylvie and Bruno	Lewis Carroll	English	1889	n/a	n/a
The Return of Sherlock Holmes	Arthur Conan Doyle	English	1904	n/a	n/a
House of Mirth	Edith Wharton	English	1905	n/a	n/a
Only a girl: or, A Physician for the Soul	Wilhelmine von Hillern	German	1869	A.L. Wister	1870
Villa Eden	Berthold Auerbach	German	1869	Charles C. Shackford	1871
The Northern Light	E. Werner	German	1890	D.M. Lowrey	1891
Cleopatra—Complete	Georg Ebers	German	1893	Mary J. Safford	1894
Dame Care	Hermann Sudermann	German	1887	Bertha Overbeck	1891
Royal Highness	Thomas Mann	German	1909	A. Cecil Curtis	1916
The King of the Mountains	Edmond About	French	1856	Mrs. C.A. Kingsbury	1897
Against The Grain	Joris-Karl Huysmans	French	1884	John Howard	1924
The Dream	Emile Zola	French	1888	Eliza E. Chase	1893
Pierre and Jean	Guy de Maupassant	French	1888	Clara Bell	1890
Arsne Lupin versus Herlock Sholmes	Maurice Leblanc	French	1908	George Moorehead	1910
The Gods are Athirst	Anatole France	French	1912	Mrs. Wilfrid Jackson	1913
Dead Souls	Nicholas Gogol	Russian	1840	D.J. Hogarth	1846
The Wife, and Other Stories	Anton Chekhov	Russian	1892	Constance Garnett	1918
The Daughter of the Commandant	Aleksandr S. Pushkin	Russian	1836	Mrs. Milne Home	1891
The Precipice	Ivan Goncharov	Russian	1869	Unknown/M. Bryant	1916
A Russian Gentleman	Sergei T. Aksakov	Russian	1858	J.D. Duff	1917
Satan's Diary	Leonid Andreyev	Russian	1919	Herman Bernstein	1920

same file sizes and partition scheme, whereby 200 kB of text are randomly extracted from each novel and divided into pieces of 10 kB each, resulting in 120 text chunks per source language. Limited availability of texts translated from the chosen languages, as well as our constraints on novel size and non-overlap with the previous experiment rendered a significantly larger data set infeasible.

3.2 Feature Sets

We consider the same set of document-level features as Lynch and Vogel [13], shown in Table 2, the inclusion of which had originally been motivated by the exploration of Ilisei et al. [7] into features other than n-grams. In addition, we selected both part-of-speech (POS) trigrams (including all punctuation) and word unigrams. The previous study used POS bigrams but not POS trigrams. Another alteration in our system is the weighting scheme. In the previous study lexical and POS features were binarized: a feature was observed as either present or not regardless of the exact frequency magnitude. We adopt relative frequency weighting, which in this case is reduced to taking raw frequency counts, since all text chunks have about the same size and there is no need to normalise over document length. Binarized weighting

Table 2 Document level features

No	Feature	Description
1	Avgsent	Average sentence length
2	Avgwordlength	Average word length
3	ARI	Readability metric
4	CLI	Readability metric
No	Feature	Ratio description
5	Typetoken	Word types : total words
6	Numratio	Numerals : total words
7	Fverbratio	Finite verbs : total words
8	Prepratio	Prepositions : total words
9	Conjratio	Conjunctions : total words
10	Infoload	Open-class words : total words
11	Dmarkratio	Discourse markers : total words
12	Nounratio	Nouns : total words
13	Grammlex	Open-class words : closed-class words
14	Simplecomplex	Simple sentences : complex sentences
15	Pnounratio	Pronouns : total words
16	Lexrichness	Lemmas : total words
17	Simpletotal	Simple sentences : total sentences
18	Complextotal	Complex sentences : total sentences

was abandoned because of a considerable drop in performance on the current data set, while tests using relative frequency were successful on both data sets.

3.3 Classification and Selection Methods

For processing of all features, we use R scripts dealing separately with word unigrams, POS n-grams and document-level features. POS tagging of the text is done using the R **koRpus** package for POS tagging (that uses TreeTagger POS tagger [14, 15]). For both POS trigrams and word unigrams, we only retain the most frequent ones to dispose of highly novel-specific proper nouns and terms, as well as thus ensuring that we build a very general classifier.

The process of feature selection and classification remains the same as in the previous study. In order to preselect characteristic features from both the word unigram and POS trigram list, we rank each group separately according to classification power using the χ^2 metric, where a higher value indicates a greater divergence over different classes. We retain the most discriminatory 100 POS trigrams and 50 word unigrams. All 18 document-level features, shown in Table 2, enter the final selection.

For machine-learning, we also used Weka [6], specifically the Naive Bayes classifier, the SMO classifier, which is an implementation of a Support Vector Machine (SVM) classifier and the Simple Logistic classifier.

4 Experiments

4.1 Method

For benefit of comparison, experiments are repeated using the same general settings as in the previous study by testing document-level features on their own and having two different configurations for selecting the highest discriminatory features from the preselected features (out of the two n-gram feature groups). In the 1st setting, we classify on only the 18 document-level statistics. For the 2nd setting, we select the best 50 features out of the preselected 100 best POS trigrams, 50 best unigrams and all document-level features, and for the 3rd configuration, rather than 50, we retain only the best 30 features according to χ^2 for the final model. The training and test sets are constructed by randomly extracting $\sim20\%$ from the full set, resulting in a division of 390 instances for training and 90 instances for evaluation randomly spread over all four classes, which increases the proportion of test items in the total set by 3% compared to what was used previously. Using only the training data, we execute ten-fold cross validation; we apply separate evaluation against the test set.

4.2 Results

Table 3 shows the results for all experiments, using both test set and cross-validation for evaluation. In relation to the first study, results for document-level features in run 2–7 are occasionally both higher and lower. Similarly for the different configurations in runs 8–13 and runs 14–19, results move in the same range between ~60 and ~85 % accuracy. Our highest value, also run 13 is slightly lower, but differences in terms of data set and training and test division probably account for that. Table 4 shows the first 20 features selected for run 8–13 (the full list is shown in Table 11 of the online appendix—see fn. 3). In comparison to the earlier study, it is notable that the χ^2 value for the most discriminatory features is with 53.13 less than three times as low as the best in the previous study (χ^2 of 191.1184). Another general difference is the composition of the best 50 features. Although the same number of each feature type was considered, our set has a higher number of POS features and fewer unigrams (13), whereas in the previous study, 32 features out of 50 were word unigrams. This difference might be due to our discarding rather infrequent unigrams and thus not considering the complete set of features for preselection. However, it could also be due to POS trigrams (including punctuation) being better discriminators than POS bigrams, at least on the current data set. For a clearer analysis, we need

Table 3 Results for model evaluation based on three different sets of features

Run	Training	Test	Classifier	Features	Accuracy (%)
1	Full	10-f-cv	Baseline	n/a	25.0
2	Full	Test	Naive Bayes	18 doc-level	51.8
3	Full	Test	SVM (SMO)	18 doc-level	63.0
4	Full	Test	SimpLog	18 doc-level	67.9
5	Full	10-f-cv	Naive Bayes	18 doc-level	56.2
6	Full	10-f-cv	SVM (SMO)	18 doc-level	60.2
7	Full	10-f-cv	SimpLog	18 doc-level	65.3
8	Full	Test	Naive Bayes	Top50(100POStri+18doc+50wuni)	73.8
9	Full	Test	SVM (SMO)	Top50(100POStri+18doc+50wuni)	79.7
10	Full	Test	SimpLog	Top50(100POStri+18doc+50wuni)	76.9
11	Full	10-f-cv	Naive Bayes	Top50(100POStri+18doc+50wuni)	77.2
12	Full	10-f-cv	SVM (SMO)	Top50(100POStri+18doc+50wuni)	82.7
13	Full	10-f-cv	SimpLog	Top50(100POStri+18doc+50wuni)	84.5
14	Full	Test	Naive Bayes	Top30(100POStri+18doc+50wuni)	59.4
15	Full	Test	SVM (SMO)	Top30(100POStri+18doc+50wuni)	76.2
16	Full	Test	SimpLog	Top30(100POStri+18doc+50wuni)	71.1
17	Full	10-f-cv	Naive Bayes	Top30(100POStri+18doc+50wuni)	69.7
18	Full	10-f-cv	SVM (SMO)	Top30(100POStri+18doc+50wuni)	77.7
19	Full	10-f-cv	SimpLog	Top30(100POStri+18doc+50wuni)	74.5

Table 4 Top 20 ranked features selected for run 8–13

Top 20 features	χ^2
it's	53.13
NP NP POS	52.35
IN NP NP	50.12
, VBN IN	46.79
, DT NN	43.26
, DT JJ	43.68
don't	43.56
got	43.46
DT NNS IN	43.40
and	42.63
NNS , DT	42.31
VBP JJ PP	41.70
suppose	40.38
NNS IN DT	40.35
i'll	40.11
IN NN ,	39.40
NN , VBN	39.23
NN , DT	38.79
Typetoken	38.77
NP NP SENT	38.54

to examine the features identified on the previous data set to determine consistent features over both data sets and therefore the ones that are more prevalent for a particular class.

5 Analysing Discriminatory Features

In this section, we explore salient features of this study, and the features obtained when we combine the two data sets, with respect to what they predict for the source languages. We close with an exercise in quality estimation using salient features.

5.1 Feature Prediction

To determine what features predict in terms of the four classes, we analyse those prominent 50 features that emerged from this study as being best in discriminating

Table 5 Feature proportions over entire source language corpus

Ranked features	% English	% French	% German	% Russian
it's	**61**	*3*	*19*	*17*
NP NP POS	**65**	*9*	*17*	*9*
IN NP NP	**48**	*17*	*16*	*18*
, VBN IN	*15*	**46**	*19*	20
, DT NN	*17*	**37**	22	24
, DT JJ	*16*	**40**	21	23
don't	**42**	*9*	*16*	**33**
got	**48**	*10*	*17*	25
DT NNS IN	*19*	**34**	25	22
and	*23*	*23*	25	**29**
NNS , DT	*13*	**48**	*17*	23
VBP JJ PP	**64**	*6*	16	14
suppose	**59**	*10*	*15*	16
NNS IN DT	*21*	**35**	23	*21*
i'll	**58**	*0*	27	*14*
IN NN ,	*20*	**29**	24	**27**
NN , VBN	*15*	**46**	*19*	20
NN , DT	*14*	**40**	22	24

Bold features are those significantly more frequent and Italic features and those significantly less frequent than expected with respect to the feature type distribution over that source language

between classes. If a feature is frequent in a particular translation, but not in the English corpus, this might signal potential source language interference. We again extracted a sample and analysed each feature in all four source languages by taking the raw counts of the feature in the language and normalising it by the total count for its feature type in the language.[5] Table 5 shows the proportion of use for each source language in relation to the other languages for 18 of the 50 best features (Table 10 in the online appendix contains the complete list—see fn. 3).[6] Most interesting are those where translations deviate vastly from the English source corpus, e.g. IT'S and I'LL—most uses of these items are in the English sample (61 and 58 %, respectively) and least are in the French source corpus (3 and 0 %, resp.). Differences in syntactic sequences can be seen for ⟨NP NP POS⟩,[7] a frequent sequence in English which appears less frequently in translations, or ⟨NNS , DT⟩, which is less common in originally composed English than in translations, especially from French.

---

[5] A POS trigram, e.g. ⟨NP NP NP⟩, is relativized to the total number of witnessed POS trigrams.

[6] All proportions are rounded to the nearest integer.

[7] In this context, ⟨POS⟩ refers to "possessive ending" rather than part-of-speech (POS).

5.1.1 Comparison Between Data Sets

To compare the two data sets ("A" for previous data and "B" for current) on equal terms, we consider the features emerging from only our system. Shared items among the top 50 features selected from each are indicated in Table 6. That they are the most robust features over both sets indicates that they have strongest predictive power. However, there might be differences in what exactly they are predicting between the data sets, and thus we need to look the distributions of the individual features in each set separately. Table 6 shows the raw counts for each feature and the total count of observed feature type (unigrams/POS trigrams). For each feature, the table also shows the percentage of its occurrences within texts of each language.[8] Consider ⟨DT NNS IN⟩ and ⟨, DT NN⟩, both relatively consistent in use between data sets. The use in English, German and Russian as source language is lower, and in French, higher than would expected if the feature did not depend on source language (with significant Pearson residuals, $p < 0.001$). Thus, these features are potentially reliable *interference* signals. However, other features have distributions that differ significantly between data sets. Consider IT'S, for example: the proportion in B for English is much greater than in A (61 % vs. 28 %), where the "Russian" sample has 57 % of the instances of IT'S. A χ^2 comparison of corresponding tables of counts shows significant difference ($p < 0.001$), with significant Pearson residuals indicating that the frequency of IT'S for English in set B and Russian in set A lies well over expected counts for each. Even though this feature is discriminatory for both sets, in A it is indicative of translation from Russian and in B is diagnostic of English original texts. Thus, finding discriminatory features may be easier than finding true signals of source language interference. That candidate interference signals are not reliably diagnostic across data sets is revealed by replication studies like this.

5.1.2 Combined Data Set

As a final exercise, we combine both data sets into one, by discarding one novel per class in data set B in order to create a balanced set with respect to the two different sets, i.e. have 5 novels per class per data set, thus 400 instances for each data set and 800 in total. We apply the settings from the most successful runs/configurations, 11–13, by preselecting 168 features out of 50 word unigrams, 100 POS trigrams and 18 document-level features and then again selecting the best 50 out of those using χ^2. The results although slightly lower than what was achieved earlier are still well in the range of the previous one (c. 75–82 %). In Table 7 the "Data" column indicates if the feature was prevalent in data set **A**, **B** or **S** if it was shared.

---

[8] This is relevant, since there are four more data samples in set B.

Table 6 Comparison between prediction direction of shared features of both data sets by looking at proportion in % and raw counts # of shared items in both sets separately

Features	Data set A								Data set B							
	Eng		Fr		Ger		Ru		Eng		Fr		Ger		Ru	
	%	(#)	%	(#)	%	(#)	%	(#)	%	(#)	%	(#)	%	(#)	%	(#)
it's	28	(82)	6	(16)	9	(27)	**57**	(163)	**61**	(218)	3	(10)	19	(68)	17	(59)
don't	**43**	(386)	5	(46)	23	(205)	30	(270)	**42**	(286)	9	(58)	16	(109)	33	(221)
, DT NN	20	(302)	**38**	(556)	22	(330)	20	(291)	17	(320)	**37**	(697)	22	(402)	24	(441)
DT NNS IN	23	(432)	**35**	(663)	22	(422)	20	(368)	19	(437)	**34**	(780)	25	(563)	22	(515)
which	21	(582)	**36**	(1,011)	24	(680)	18	(505)	24	(1,051)	30	(1,286)	**28**	(1,222)	18	(771)
NP NP NP	21	(163)	**50**	(383)	19	(144)	11	(82)	**39**	(280)	24	(171)	29	(206)	9	(62)
that's	33	(96)	5	(14)	11	(31)	**52**	(153)	**48**	(105)	3	(7)	32	(71)	17	(38)
the	23	(9,655)	**30**	(12,262)	25	(10,419)	22	(9,176)	24	(12,296)	**27**	(13,869)	**26**	(13,131)	22	(11,286)
and	**27**	(6,363)	20	(4,866)	**27**	(6,473)	26	(6,092)	23	(6,526)	23	(6,466)	25	(6,944)	**29**	(8,179)
NN : SENT	12	(6)	2	(1)	0	(0)	**85**	(41)	3	(5)	29	(51)	16	(28)	**53**	(93)
Type totals																
Unigram	147,912		146,380		150,362		147,637		174,428		172,429		177,486		173,300	
POS trigram	228,923		223,340		230,062		226,778		268,539		260,359		262,624		264,828	

The left and right table contain proportions of data set A and B respectively. Bold and Italic are used as in Table 5

Table 7 18 most discriminatory features in the combined data set

Sequence	χ^2	Data	Sequence	χ^2	Data	Sequence	χ^2	Data
NP NP POS	41.27	B	and	34.55	S	NN , DT	33.39	B
don't	40.69	S	SENT NN VBZ	34.07	A	CD NNS ,	32.89	A
IN NP NP	38.11	B	SENT " NN	33.90	A	SENT NNS VBP	32.07	A
, VBN IN	38.01	B	suppose	33.90	B	Conjratio	31.98	S
SENT CC PP	36.30	A	Nounratio	33.67	A	Avgwordlength	31.93	A
, DT NN	34.92	S	which	33.43	S	though	31.90	A

5.2 An Exercise in Quality Estimation

We have identified features in translated texts likely to have originated in the source language. However, since these were selected on the basis of a statistical optimisation, it is not guaranteed that texts containing these features would be judged equally anomalous by native speakers of English. Here, we examine whether highly salient features of our model index sentences that a human would strongly perceive as either original English or translated English. We selected a number of salient features for each source language and sentences in which they occur from the data set.

The gold standard notes whether an item was a translated text or originally composed in English. We obtained two human judgements on 17 sentences selected across our corpus.[9] The sentences were selected manually to maximise features distinctive for the source languages. The kappa statistic computed between the raters' judgements is 0.406, suggesting moderate agreement. On 7 out of 17 instances, both raters agreed with the gold standard and on another 5, at least one was correct.[10]

Some sentences were identified by both annotators as translations, as for instance example (1), which has feature type ⟨, CC RB⟩ of *German* English. There is some disagreement between other ratings of translated pieces, but interestingly, not only the translated sentences introduced confusion: example (2) is an instance of original English (tagged to include ⟨IN NP NP⟩), but was judged translated by both raters.

(1) ... whereupon the Professor [...] had then expressed his thanks with glad surprise, **and indeed** emotion, for the kind interest the Prince had expressed in his lectures.
(2) The young couple had a house **near Berkeley Square** and a small villa at Roehampton, among the banking colony there.

This evidence constitutes anecdotal support for the claim that the method is correctly discriminative: items that are high in distinctive features that would be identified by the system slip by human scrutiny in 29 % of cases, but agree with human judgement in 70 % of cases. While this requires more elaborate testing and rigorous methodology

---

[9] The assessors are native English speakers among co-authors of this paper.

[10] On 5 items, both raters agreed with each other but not with the gold standard: in 3 of those the source was English, and both raters judged the item odd; in the other 2, the source was not English.

[16, 17], the results tentatively indicate that highly discriminatory features could be useful in quality estimation of translations.

6 Conclusion and Future Directions

We have replicated and extended past work in identifying linguistic features that discriminate the source language of literary texts presented in English, perhaps through translation. Our replication of work by Lynch and Vogel [13], using the same source data, method and feature types,[11] yielded results which were a slight improvement on the original regarding the task of source language identification. Our reproduction of the experimental design on a fresh data set confirmed some signals of source language *interference* and showed others to be unreliable between data sets. This speaks to the need for still more extensive studies of this sort. We conclude that this method of source language detection on unseen texts of English from the literary genre is robust, and has potential for use in quality assessment for English prose. This work contributes to the emergent practice within the digital humanities of applying automatic text classification techniques to explore hypotheses.

Acknowledgments This research is conducted partly within the Centre for Global Intelligent Content (www.cngl.ie), funded by Science Foundation Ireland (12/CE/I2267).

References

1. Baroni, M., Bernardini, S.: A new approach to the study of translationese: machine-learning the difference between original and translated text. Literary Linguist. Comput. **21**(3), 259–274 (2006)
2. Becher, V.: Abandoning the notion of 'translation-inherent' explicitation. Against a dogma of translation studies. Across Lang. Cultures **11**, 1–28 (2010)
3. Blum-Kulka, S. "Shifts of cohesion and coherence in translation". In: Interlingual and intercultural communication: Discourse and cognition in translation and second language acquisition studies 17 (1986), p. 35
4. Degani, T., Prior, A., Tokowicz, N.: Bidirectional transfer: the effect of sharing a translation. J. Cogn. Psychol. **23**(1), 18–28 (2011)
5. Forsyth, R.S., Lam, P.W.: Found in translation: to what extent is authorial discriminability preserved by translators? Literary Linguist. Comput. **29**(2), 199–217 (2014)
6. Hall, M., et al.: The WEKA data mining software: an update. ACM SIGKDD explor. news. **11**(1), 10–18 (2009)
7. Ilisei, I., et al.: Identification of translationese: a machine learning approach. In: Computational linguistics and intelligent text processing. pp. 503–511. Springer (2010)
8. Koppel, M., Ordan, N.: Translationese and its dialects. In: Proceedings of the 49th Annual Meeting of the Association for Computational Linguistics: Human Language Technologies, Vol. 1, pp. 1318–1326 (2011)

---

[11] Our current study employed POS trigrams, the original work used POS bigrams but not trigrams.

9. Kurokawa, D., Goutte, C., Isabelle, P.: Automatic detection of translated text and its impact on machine translation. In: Proceedings of the XII MT Summit, Ottawa, Ontario, Canada. AMTA. 2009
10. Lembersky, G., Ordan, N., Wintner, S.: Language models for machine translation: original versus translated texts. In: Proceedings of the 2011 Conference on Empirical Methods in Natural Language Processing, pp. 363–374 (2011)
11. Lembersky, G., Ordan, N., Wintner, S.: Adapting translation models to translationese improves SMT. In: Proceedings of the 13th EACL Conference, Avignon, France, p. 255 (2012)
12. Lynch, G.: A supervised learning approach towards profiling the preservation of authorial style in literary translations. In: Proceedings 25th COLING, pp. 376–386 (2014)
13. Lynch, G., Vogel, C.: Towards the automatic detection of the source language of a literary translation. In: Proceedings 24th COLING (Posters), pp. 775–784 (2012)
14. Michalke, M. koRpus: An R Package for Text Analysis. (Version 0.04-40), last verified: 11(08), 2014, 2013. http://reaktanz.de/?c=hacking&s=koRpus
15. Schmid, H.: Probabilistic part-of-speech tagging using decision trees. In: Proceedings of International Conference on New Methods in Language Processing, vol. 12, pp. 44–49. Manchester, UK (1994)
16. Schütze, C.: The Empirical Base of Linguistics: Grammaticality Judgements and Linguistic Methodology. University of Chicago Press, Chicago (1996)
17. Schütze, C.: Thinking about what we are asking speakers to do. In: Kepsar, S., Reis, M., Mouton De Gruyter. (eds.) Linguistic Evidence: Empirical, Theoretical and Computational Perspectives. pp. 457–485 (2005)
18. van Halteren, H.: Source language markers in EUROPARL translations. In: Proceedings 22nd COLING, pp. 937–944 (2008)
19. Volansky, V., Ordan, N., Wintner, S.: On the features of translationese. In: Literary and Linguistic Computing (2013)

Reluctant Reinforcement Learning

Chris Jones and Malcolm Crowe

Abstract This paper presents an approach to Reinforcement Learning that seems to work very well in changing environments. The experiments are based on an unmanned vehicle problem where the vehicle is equipped with navigation cameras and uses a multilayer perceptron (MLP). The route can change and obstacles can be added without warning. In the steady state, no learning takes place, but the system maintains a small cache of recent inputs and rewards. When a negative reward occurs, learning restarts, based not on the immediate situation but on the memory that has generated the greatest error, and the updated strategy is quickly reviewed using the cache of recent memories within an accelerated learning phase. In the resulting Reluctant Learning algorithm the multiple use of a small quantity of previous experiences to validate updates to the strategy moves the MLP towards convergence and finds a balance between exploration of improvements to strategy and exploitation of previous learning.

1 Introduction

Reinforcement Learning (RL) is an area of artificial intelligence that is inspired by behavioural psychology. An example of RL in the natural world is the way in which a child learns to move its arms or legs, using trial and error. A young child will learn to gain control of these limbs by interacting with its environment and using its senses as feedback.

Given the example above it could be conceived that RL would be the ideal methodology to deal with real world problems. However, there remain several fundamental challenges that have thwarted the wide-spread application of RL methodology to real world problems. Real world problems normally have large or infinite state and action spaces (Curse of Dimensionality) [1], and a non-static environment that requires

C. Jones (✉)
New College Lanarkshire, Scotland, UK
e-mail: chris.jones@nclan.ac.uk

M. Crowe
University of the West of Scotland, Scotland, UK

© Springer International Publishing Switzerland 2014
M. Bramer and M. Petridis (eds.), *Research and Development in Intelligent Systems XXXI*, DOI 10.1007/978-3-319-12069-0_6

continuous or online learning. As a result, any unsupervised learning process takes many episodes, requires large quantities of data, and in many applications do not converge at all. After all, a young child takes years of learning to reach the dexterity of an adult.

This paper proposes a new algorithm called "Reluctant Learning", for a multi-layer perceptron (MLP) learning system. We reached this successful model through experimenting with caching of various sorts, and to our surprise the Reluctant Learning algorithm can be interpreted in human terms. Our MLP learns only on a negative reward above a threshold (pain). It then reviews recent episodes, and learns not from the very latest event (the crash) but the recent episode that it found most surprising (the swerve). Also to continue the analogy, if a driver swerved and was involved in a near miss, once the car continued down the road the learning would be linked to the swerve and the actions and reward related to that, not on the current road.

2 A Learning Scenario

Computer games simulations can be used to provide large state/action spaces, combined with continuous unsupervised action and rewards. As a first example, we set the algorithm to learn to drive a car, learning only from a number of simple visual sensors (5 × 5 pixel cameras). Figure 1 shows the first track and a car travelling along the road.

The learning system does not know the configuration or colours used in the 25-pixel camera images. Clearly more cameras result in more data to process and more inputs. The dots shown in Figs. 1 and 2 indicate an example set of 31 sensors (27 cameras, and the 4 wheels).

Fig. 1 Track one

Fig. 2 Position of sensors

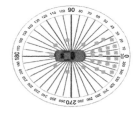

The car is rewarded for staying on the road. In this example the road is grey, so if all pixels in a camera are grey there will be a score of 1 (and similarly for the wheels). The colour blue (water) scores 0 and other colours are 0.3 (off road). The reward value is made up of 4 components: wheel sensors, close proximity cameras, distant cameras and current velocity.

There are 3 possible actions: straight on, steer left, steer right. This affects the direction of the car depending on the speed, the direction of the wheels and the road surface. The car will steadily accelerate up to the maximum speed for the road surface, but if the surface changes (e.g. water, off road) it will decelerate to the new maximum speed.

Once the system has learned this simple track, we change the track to something more complex (Figs. 1, 3 and 4), to see how the learning system adapts to the modified situation. The first track (Fig. 1) is simple with 4 similar corners (all left turning). The second track (Fig. 3) adds the complexity of right hand corners and longer sweeping corners. The final track (Fig. 4) has the increased complexity of water that slows the car more.

All the tracks have the same boundaries and if the car attempts to leave the boundary the system will place them back on the track at the default start position. In all experiments in this paper the car completes 100 laps of each track (300 in total). The MLP has no warning of the track change and values are not reset. The track changes are intended to simulate a more dynamic real world environment. The changing environment means that the system must always be learning or monitoring its environment. The standard algorithm often has a training/validation/testing configuration. In real world models or simulations like this that is not possible and it must continually learn and update the MLP.

Fig. 3 Track two

Fig. 4 Track three

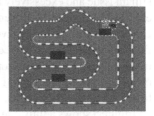

3 Reinforcement Framework

Almost all RL applications can be described as Markovian decision processes (MDP). An MDP is described by a set of states (S), a set of actions (A), and a cost function $S \times A \rightarrow R$. A Stochastic function is also required to allow exploration. The goal is to find an optimal policy $\pi : S \rightarrow A$, which allows the state to assume the Action.

For very simple systems, Q-Tables use an action-value function that calculates the expected outcome of performing a specific action. For a successful learning function the agent must involve three phases; memorisation, exploration and updating. The Q-Table Learning update rule is given by where S is the state, a is the action, and S' is the state resulted. α is the learning rate and γ is a discount factor.

$$Q_{R+1}(s, a) := (1 - \alpha)Q(S, a) + \alpha(c(s, a) + \gamma_b^{\min}Q_k(s', b))$$

The optimal Q-function will converge when there is a finite state and action spaces. The update is executed for each sampled input online.

This method has been successfully proven in small finite state/action spaces, but real world examples often require an enormous amount of test data and processing power to ensure that enough of the feature space is sampled.

The Neural Network approach disperses with the table based representation and uses an error function. The error function aims to measure the difference between generated Q-value and the target value. A common method to update the error value is backpropagation, this adjusts the weights within the neural network to minimise the error. Similar to the Q-Table, this is updated after each input.

4 MLP for Reinforcement Learning

The shortcomings associated with the multilayer perceptron (MLP) have been widely reported in the literature since the 1980s. The primary shortcoming associated with the MLP using backpropagation (BP) is the uncertainty of finding a global minimum. The second shortcoming is the slow rate of convergence. Several modifications [2–5] of the original BP algorithm have already been suggested however performance is still an issue and excessively large training datasets are required to obtain acceptable error rates. Nevertheless, even with these shortcomings the MLP has many favourable characteristics, such as universal approximation [6]. It is also a popular tool for financial modelling and predictions [7] and robotic control systems, [8] amongst other usages. The standard topology of a fully connected feed forward MLP consists of three layers: input layer, hidden layer and output layer. All layers are connected to the next layer by weights.

Figure 5 shows a fully connected feed-forward MLP with an additional bias input to all layers other than the inputs. This is how most MLPs are constructed. The bias normally has a constant output, which is commonly set to 1. The nodes have an activation function which acts on the activations of each node. This research uses the

Fig. 5 Standard MLP
network

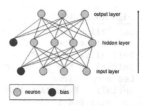

sigmoid activation function, however Tanh functions are common. Each activation
is made up of a sum of all the values of all the connections going into that node.
As the name feed-forward suggests, input data travels through the MLP in a single
direction, from input to hidden (times the number of hidden layers) and finally to the
output layer.

The basic mechanism behind most MLP learning is BP. The BP takes place during
the training cycle. BP has a simple process for comparing the output of the MLP
and the expected output or target. The difference is used within the BP algorithm to
update the weights; this slowly moves the output closer to the target value.

The standard algorithm used within this scenario is shown in Algorithm 1, the
process is relatively simple for every action a reward is calculated and this reward is
used to update the MLP.

Algorithm 1 Standard Learning with Backpropagation

```
Initialise Sensors Array (31); ActionCount = 3; // Number of actions
Initialise RandomActionRate = 90 (10% Random); Initialise LearnRate = 0.3;
Initialise MLP (31 Inputs, 20 Hidden, 3 outputs); Learning = True;
Do Until End {
        Inputs = ReadSensors;
        Inputs -> MLP -> Result;
        NextAction = Select_Actions (RandomActionRate);
        Complete Action (MoveCar(NextAction));
        Reward = Calculate_Current_Reward (NextAction);
        Target = GetTarget(Result,NextAction,Reward);
        Error = Calculate_Error(Reward, NextAction,Target);
        If (Learning = true){
                MLP.Learn(Target, Result, LearnRate);
}
```

---

5 The Reluctant Learning Algorithm

The Reluctant method works in two phases, the first phase initialises the memory
locations and then an ongoing phase that updates the MLP, and controls a variable
"Learn Rate" and the rate that Random Actions are selected. It uses the data within
the memory to assess its performance. If the MLP is performing well then the system
will use decay value to reduce both the learn rate and the chances of a random action
being selected. Once the chances of performing a random action have been removed
the system will opt to move into exploitation mode and stop learning. Algorithm 2
is an example of how this would be integrated into a standard algorithm, additions
are highlighted in grey. The new methods are explained in the next section.

```
Algorithm 2 Reluctant Learning Algorithm
Initialise Sensors Array (31); ActionCount = 3;
Initialise RandomActionRate = 90; // 10% Random
Initialise LearnRate = 0.3;
Initialise MLP (31 Inputs, 20 Hidden, 3 outputs);
Learning = True;
Intialise ReluctantLearning(50,200);
Do Until End {
        Inputs = ReadSensors;
        Inputs -> MLP -> Result;
        NextAction = SelectActions (RandomActionRate);
        Complete_Action (MoveCar(NextAction))
        Reward = Calculate_Current_Reward (NextAction);
        Target = GetTarget(Result,NextAction,Reward);
        Error = Calculate_Error(Reward, NextAction,Target);
        If (Learning = true){
                MLP=ReluctantLearning.CacheLearning(
                New NewEntry(Target,Result, NextAction,Inputs), MLP);
                MLP = ReluctantLearning.UpdateCache(MLP);
        }
        Learning = ReluctantLearning.UpdateLearnRate (Learning, Error);
        NewLearnRate = ReluctantLearning.GetLearnRate();
        RandomAction = ReluctantLearning.GetInvestigationRate();
}
```

Reluctant learning requires a pool of previous experiences prior to moving into the monitoring and learning mode. The pool of experiences are created in phase one of the learning process. The second phase controls the learning rate and probability of the action being chosen at random. If the MLP is performing well it can also stop the learning process and exploit its knowledge. For performance reasons the number of previous experiences has to be limited; all examples within the paper are limited to 50 experiences in phase one (accelerated learning) and 200 (50*4) in phase two. These memories are within a dynamic table and are referred to as Cached Experiences within this paper. Depending on processing power these numbers can be increased but for this example increasing has limited improvement in performance.

• Phase One

The Initialisation of the reluctant learning method creates a number of variables that enables the system to monitor the learning. *LowError*: The Lowest error value from within the cache; *HighError*: The Highest error value from within the cache; *NewEntry*: the most recent inputs, target and reward; and *CachedExperiences*: the dynamic array of past input and reward.

• The *CacheLearning* function

There are three purposes to the *CacheLearning* function:

1. To add the Current experience into the cache (Inputs, reward, target, etc.).
2. Complete initial Accelerated Learning
3. Find the lowest and highest error value and the corresponding position within the cache list.

Algorithm 3 CacheLearning Algorithm (NewEntry , MLP)

```
CachedExperiences.Add(NewEntry);
    foreach CachedExperiences {
        NewInput = CachedExperiences.GetInput();
        Action = CachedExperiences.GetAction();
                SingleTarget = CachedExperiences.GetTarget();
                result = MLP.feedForward(NewInput);
                NewTarget = result;
                NewTarget[Action] = SingleTarget;
        if (AcceleratedLearning > 0)    {
                        MLP.Learn(NewTarget, result, NewLearnRate);
                }
                ErrorLevel = MLP.LearnError(NewTarget, result, iAction);
                if (ErrorLevel < LowError)    {
                        LowError = ErrorLevel;
                        LowPos = a;
                }
            if (ErrorLevel > HighError){
                        HighError = ErrorLevel;
                        HighPos = a;
                }
}
    if (AcceleratedLearning > 0) { AcceleratedLearning --; }
return MLP;
```

The Accelerated Learning element of algorithm involves an initial period of executing a BP for every entry within the cache each time an action is undertaken. This results in a large number of BP execution compared with the standard algorithm. The Accelerated learning element is very short; in all these examples it only lasts for 500 actions. However in that short period over 25 k BP will have been executed; by comparison the standard algorithm will have only executed 500. Once the Accelerated period has finished, the code simply adds the most recent action to the cache and finds the highest and lowest error value.

• Phase Two

This phase involves the updating of the cache in *UpdateCache* and the update of the Learning Rate and Random Action Rate in UpdateLearnRate. This phase will implement the learning and monitoring. It starts once *CachedExperiences* has a minimum of two entries.

• The *UpdateCache* function

There are three purposes to the *UpdateCache* function:

1. Set a Target Level for the learning
2. Execute the BP as required for the Experience that generated the highest error value.
3. To control the size of the Cached Experiences.

Key variables are the *HighPos* and *LowPos* these are set within *CacheLearning* function and hold the relative position of high and low error within Cached Experiences array.

Algorithm 4 UpdateCache(MLP)

```
if (CachedExperiences.Count > 2)
{
        NewInput = CachedExperiences[HighPos].GetInput();
        Action = CachedExperiences[HighPos].GetAction();
        SingleTarget = CachedExperiences[HighPos].GetTarget();
        SetLevel = HighError *0.9;
        Do( a maximum of 50 times){
         result = MLP.feedForward(NewInput);
         NewTarget = (double[])result.Clone();
         NewTarget[Action] = SingleTarget;
         MLP.Learn(NewTarget, result, NewLearnRate);
         ErrorLevel = MLP.LearnError(NewTarget, result, iAction);
        } while (ErrorLevel > SetLevel);
        if (count > CacheSize){
                CachedExperiences.RemoveAt(LowPos);
        }
}
return MLP;
```

The *UpdateCache* function only starts to execute when two or more experiences are stored within the cache. It then uses the *HighPos* variable created within *CachLearning* function to select the experience that had the highest error value. It then sets a target value using the high error value (*HighError = HighError* × 0.9). The system then performs the BP update within a loop. This is unusual as normally the system would carry out only one BP update per action. Due to performance issues the system is limited to 50 per action, however in the scenarios the average number of loops is far lower. This recursive nature can be interpreted in human terms; when a difficult task or some unexpected outcome occurs humans tend to concentrate on that task. The limit of 50 loops stops the system becoming fixated on a single target and neglecting the next action. The final action is to check the number of experiences within the cache, if the number is over the limit; in the scenario the limit for experiences is 50 in phase one and 200 in phase two. The system will remove the experience that generated the lowest error value. The lowest error value maybe the latest or an experience from many actions ago. It is possible that an experience may stay within the cache indefinitely, although this is not likely.

The unusual element to Reluctant Learning is that many current actions will be added to the cache and immediately removed.

- The *UpdateLearnRate* function

There are three purposes to the *UpdateLearnRate* function:

4. To check if the last error value is more than the Highest error within the cache
5. Reduce the Learn Rate and the Random Rate using a Decay value
6. Reset the Learner Rate and Random rate and turn on Learning if environment change detected.

```
Algorithm 5 UpdateLearnRate(Learning, TotalError)
Learning = true;
if (HighError < TotalError){
        if (Learning & InvestigationRate >= InvestigationRateReset){
                NewLearnRate = NewLearnRateLimit;
                InvestigationRate = InvestigationRateReset;
        }
        Else {
                NewLearnRate = NewLearnRateLimit;
                InvestigationRate = InvestigationRateReset;
                CachedExperiences.Add(LastEntry);
                CachedExperiences.RemoveAt(0);
        }
}
else{
        NewLearnRate = Math.Abs(NewLearnRate - (NewLearnRate * DecayRate));
InvestigationRate = Math.Abs(InvestigationRate + (InvestigationRate/10 *
DecayRate));
}
if (InvestigationRate > 99.5){
        InvestigationRate = 100;
        NewLearnRate = 0;
        Learning = false;
}
return Learning;
```

Algorithm 5 start by comparing the Highest Error value generated from the cache and the error value from the current action. If the current error value is more than the Highest Error value then this results in the 2 possible actions. The action depends on what mode the reluctant system is in. If the system is in learning mode then the learn rate and the random rate will be reset to the initial system setting. If the system is in monitoring mode (not learning) then it will reset the values, add the most recent entry to the cached experiences, remove the experience at position 0 and turn on learning mode. If the value of the current action generates an error values less than the highest error then the system will simply use the decaying value and reduce the learn rate and the random rate (chances of the system selecting a random action). If the chances of the system selecting a random action is less than 0.5 % the system will move into monitoring mode and stop learning.

6 Results

The Scenario has previously been discussed, it is an unmanned car driving around three changing tracks. The tracks change but there the experiment continues using the same MLP and data. The track changes are not to test how it can learn each different track, but to see how it copes with change. The car movement uses a basic level of physics, so the faster the car the more unstable and difficult it is to control, especially around corners. In each scenario the car was limited to four different speeds. Within the virtual world the car movement is measured in pixels, however, as it is difficult to quantify the speed within a game, the comparison is done in percentages. The standard speed is set as at a base percentage of 100 %, the faster Mid Speed is 125 % and the high speed is set at 150 % from the base rate. A slower speed has also been included at 75 %. At all speeds when turning, the speed reduces by 20 % of the top speed. Driving off the road reduces Maximum speed to 15 % of the base speed, so to maximise speed and complete the most laps the car must stay on the road and drive as straight as possible.

The number of hidden nodes within a MLP is key to generating efficient and effective outputs. All of the runs within this scenario are using 20 single layer hidden nodes. The scenario has been run with much smaller hidden layers, however the 3rd and more complex track has proven to be difficult and many runs fail to achieve the required 300 laps. Interestingly the reluctant learning algorithm performed well with small and large hidden layers. With a large hidden layer the standard algorithm struggled to converge on the simple 1st track. In contrast, the reluctant learning algorithm always managed to converge even with overly large hidden layers over 50. More research into large and multi-layer hidden nodes is required.

Three performance indicators (PI) have been selected; first the number of moves required, secondly the number of updates (BP) to the MLP executed and third the number of times the car crashed out of the game space and the start position reset. The first PI is effected by the speed and accuracy of the driving, so a car moving at full speed and following the track accurately will require less moves to complete the full 300 laps. The second PI is based on the number of BP updates executed. The standard algorithm will therefore have the same number of moves as BP executions. The new method can select not to learn or even increase the level of learning for more challenging situations. The third PI shows the number of time the car got lost and moved out of range (of the screen). All runs are based on completing 300 laps (100 laps with each track). If the car takes a short cut across the grass and misses an element of the track the lap will not count.

Table 1 shows the results for both the standard algorithm and the new reluctant learning algorithm at the 4 speeds discussed previously. It shows that the reluctant algorithm completed more laps at every speed and also required less updates to the MLP (the exception of this is at 150 % however the standard algorithm was unable to complete 300 laps within one million moves).

There is no significant difference between performances at low speeds, as the car can easily turn and manoeuvre around the track; this is shown in Fig. 6. The most obvious difference is how quickly reluctant learning starts lapping in comparison

Table 1 Totals for moves, BP update and out of bounds for each algorithm at various speeds

Speed	75 %	75 %	100 %	100 %	125 %	125 %	150 %	150 %
Algorithm	Standard	Reluctant	Standard	Reluctant	Standard	Reluctant	Standard	Reluctant
Move count	719,637	576,864	574,549	423,409	842,867	410,738	1,000,000[a]	754,826
BP update	719,637	549,637	574,549	232,769	842,867	514,873	1,000,000[a]	1,397,523
Out of bounds	16	1	26	1	55	7	62[a]	24

[a]The 150 % standard run was unable to complete the final track

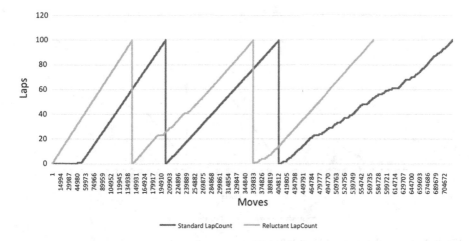

Fig. 6 Laps completed (3 × 100) at 75 % speed

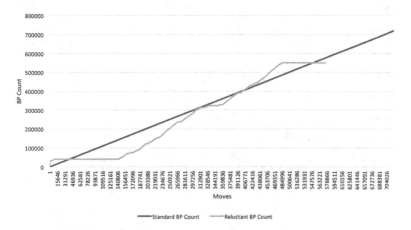

Fig. 7 Back propagation (BP) comparison at 75 % speed

to the standard algorithm. The reluctant algorithm requires around 20 % less moves and 24 % less BP updates. It also only requires resetting once, whereas the standard algorithm requires resetting 16 times in this scenario.

However it is interesting that the number of BP updates is lower in the reluctant learning method. On the simplest track it stops learning relatively early and only starts increasing the number of BP executions when driving around the second and third track. At several stages the reluctant algorithm executes far more BP, but once the track is learned it stop learning this can be seen in Fig. 7.

With the car moving at 100 % speed the standard algorithm learns to manoeuvre the car around the track slower than at 75 % speed and only after several thousand actions does it finally manage to start lapping consistently. It then performs adequately and

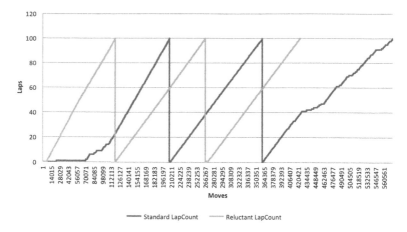

Fig. 8 Laps completed (3 × 100) at 100 % speed

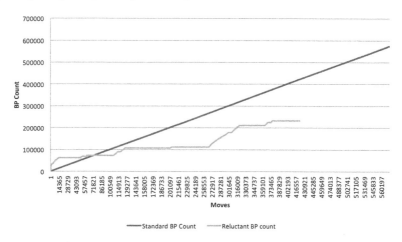

Fig. 9 Back propagation (BP) comparison at 100 % speed

matches the reluctant learning model on track 1 and 2, however it struggles with the
more complex track 3; this is shown in Fig. 8. The Reluctant algorithm out performs
the standard algorithm by around 26 % over all 300 Laps.

Figure 9 shows the different ways that the reluctant algorithms approaches learning, turning the learning on and off. The reluctant algorithm requires half the number
of learning (BP) at this speed. The car position has to be reset an average of 26 times,
whereas the reluctant algorithm requires only 1 reset.

As the car speed increases it becomes harder to control and the requirement for
faster more efficient learning is greater. Figure 10 shows how the reluctant algorithm
performance increases at high speed while the standard algorithm is now struggling to
adapt especially with the more complex tracks. The reluctant algorithm has completed

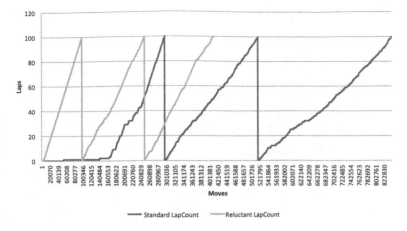

Fig. 10 Laps completed (3 × 100) at 125 % speed

tracks 1 and 2, while the standard algorithm is still working on track 1. The number of average resets required has already risen sharply to 55 for the standard algorithm while the reluctant algorithm only raises to 7. As the standard algorithm is spending time off the track the reluctant algorithm actually completes 300 laps twice as fast as the standard.

Due to the standard algorithm spending twice the amount of time on the track, it is going to require more learning. However the interesting point here is that during the 3rd track the reluctant shows a rise in the number of BP executions; this is shown in Fig. 11.

At this new increased speed both algorithms find it difficult to drive around the track with an increase is all reset rates.

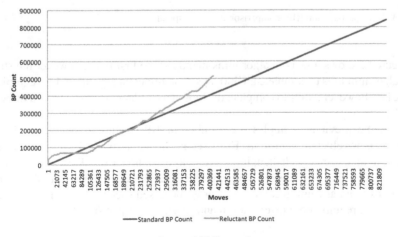

Fig. 11 Back propagation (BP) comparison at 125 % speed

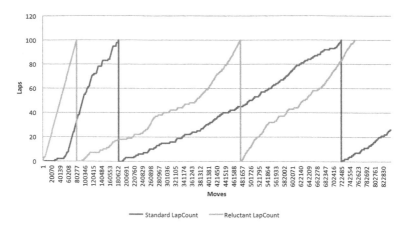

Fig. 12   Laps completed (3 × 100) at 150% speed

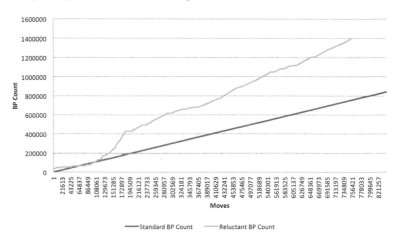

Fig. 13   Back propagation (BP) comparison at 150% speed

Figure 12 shows how both algorithms managed the first track, however the standard algorithm fails to complete all 300 laps. The reluctant algorithm does complete all the laps however it fails to complete the laps at a similar rate to the previous slower speeds.

In all the previous examples the number of BP executions have been lower using the new reluctant algorithm. However in this experiment the high speed has resulted in an extremely challenging drive, and the number of BP executions is far higher. Figure 13 shows how the reluctant algorithm has increases the number of BP executions in an attempt to control the car at increased speed. This ability to detect difficult task require more investigation and to see if it is just specific to this type of domain.

7 Conclusion

This paper introduces a new concept called "Reluctant Learning" and has demonstrated some key improvements in performance using the unmanned car scenario. The key concept is moving away from learning from the most recent action and using the error value to target learning where it is most required. This new concept now needs to be verified using a different scenario, but could lead to a more efficient use of MLP within reinforcement learning.

References

1. Bellman, R.: Dynamic Programming. Princeton University Press, Princeton (1957)
2. Andina, D., Jevtic, A.: Improved multilayer perceptron design by weighted learning. In: IEEE International Symposium on Industrial Electronics (ISIE 2007), pp. 3424–3429 (2007)
3. Becker, S., Cun, Y.L.: Improving the convergence of back-propagation learning with second order methods. In: Touretzky, D.S., Hinton, G.E., Sejnowski, T.J. (eds.) Proceedings of the 1988 Connectionist Models Summer School, pp. 29–37. Morgan Kauf-mann, San Francisco, CA (1989)
4. Bhattacharya, U., Parui, S.K.: Self-adaptive learning rates in backpropagation algorithm improve its function approximation performance. In: Proceedings of IEEE International Conference on Neural Networks, vol. 5, pp. 2784–2788 (1995)
5. Xun, C., Tyagi, K., Manry, M.T.: Training multilayer perceptron by using optimal input normalization. In: IEEE International Conference on Fuzzy Systems (FUZZ 2011), pp. 2771–2778 (2011)
6. Jones, L.K.: Constructive approximations for neural networks by sigmoidal functions. Proc. IEEE **78**(10), 1586–1589 (1990)
7. Patra, J.C., Thanh, N.C., Meher, P.K.: Computationally efficient flannbased intelligent stock price prediction system. In: International Joint Conference on Neural Networks (IJCNN 2009), pp. 2431–2438 (2009)
8. Pavon, N., Sanchez, O., Ollero, A.: MLP neural networks to control a mobile robot using multiple goal points. In: Proceedings of World Automation Congress, vol. 15, pp. 395–398 (2004)

Preference and Sentiment Guided Social Recommendations with Temporal Dynamics

Xavier Ferrer, Yoke Yie Chen, Nirmalie Wiratunga and Enric Plaza

Abstract Capturing users' preference that change over time is a great challenge in recommendation systems. What makes a product feature interesting now may become the accepted standard in the future. Social recommender systems that harness knowledge from user expertise and interactions to provide recommendation have great potential in capturing such trending information. In this paper, we model our recommender system using sentiment rich user generated product reviews and temporal information. Specifically we integrate these two resources to formalise a novel aspect-based sentiment ranking that captures temporal distribution of aspect sentiments and so the preferences of the users over time. We demonstrate the utility of our proposed model by conducting a comparative analysis on data extracted from Amazon.com and Cnet. We show that considering the temporal preferences of users leads to better recommendation and that user preferences change over time.

1 Introduction

Recommender systems traditionally provide users with a list of recommended items based on users preferences. The huge success of these systems in the retail sector demands innovative and improved recommendation algorithms. The dawn of the social web has created opportunities for new recommendation algorithms to utilise knowledge from such resources and so the emergence of social recommender

X. Ferrer (✉) · E. Plaza
Artificial Intelligence Research Institute (IIIA), Spanish Scientific Research Council (CSIC),
Campus UAB, Bellaterra, Spain
e-mail: xferrer@iiia.csic.es

E. Plaza
e-mail: enric@iiia.csic.es

Y.Y. Chen · N. Wiratunga
IDEAS Research Institute, Robert Gordon University, Aberdeen AB25 1HG, Scotland
e-mail: y.y.chen@rgu.ac.uk

N. Wiratunga
e-mail: n.wiratunga@rgu.ac.uk

© Springer International Publishing Switzerland 2014
M. Bramer and M. Petridis (eds.), *Research and Development
in Intelligent Systems XXXI*, DOI 10.1007/978-3-319-12069-0_7

101

system. These systems harness knowledge from user generated reviews to generate better recommendations by incorporating sentiment expressed in opinions to bias the recommendation list [5]. Similarly preference knowledge and temporal dynamics have also separately been applied to influence recommendations [7, 18].

Purchase choices are based on comparison of artefacts; which implicitly or explicitly involve comparison of characteristics or aspects of these artefacts. In particular a user's purchase decision hints at the aspects that are likely to have influenced their decision and as such be deemed more important. Additionally it is also not unusual to expect that the criteria used for this comparison may also change with time. For example, in the domain of Cameras, the LCD display may have been an important aspect users were interested in the past but now this is given in almost every camera and so is likely to be an aspect of contention.

In recent work [1] we explored how preference knowledge can be captured and exploited within a social recommendation application. Our findings suggested that preference knowledge allows us to extract important aspects from reviews, in terms of those that are likely to have influenced the users' purchase decision. However, would recency of reviews have an impact on aspect weights? How far back in time must we go before extracted weights improve the recommendations? Our main focus in this paper is to study temporal and preference context for social recommendations with a view to integrate these contexts with aspect-based sentiment analysis.

Our contribution is three-fold: firstly we demonstrate how sentiment distribution analysis can impact the quality of recommendations; and secondly show how a preference-based algorithm can be incorporated to derive rankings on the basis of preference relationships; and finally provide a formalism to combine sentiment and temporal information. Our results confirm that incorporating temporal information in aspect-based sentiment analysis is comparable to preference knowledge.

The rest of the paper is organized as follows: In Sect. 2 we present the background research related to this work. Next in Sect. 3 we describe aspect preference over time and how preference graphs can be generated by using a case study from Amazon.com. The process of aspect extraction and weight learning for sentiment analysis is presented in Sect. 4. Finally, our evaluation results are presented in Sect. 5 followed by conclusions in Sect. 6.

2 Related Works

Social recommenders recognise the important role of sentiment analysis of user reviews [5]. Instead of relying on user logs and sessions to model user preference [12], in this paper we infer aspect preferences from comparing the sentiment-rich content generated by users. However, extracting sentiment from natural language constructs is a challenge. Lexicons are often used to ascertain the polarity (positive or negative) and strength of sentiment expressed at word-level (e.g. SentiWordNet [6]). However sophisticated methods are needed to aggregate these scores at the sentence, paragraph and document level to account for negation and other forms of sentiment modifiers [15]. Increasingly aggregation is organised at the aspect level, since the distribution of a user's sentiment is typically mixed and expressed over the aspects of the artefact (e.g. I love the *colour* but not too keen on *size*). Hu and Liu [8] propose an association mining driven approach to identify frequent nouns or noun phrases as aspects. Thereafter sentences are grouped by these aspects and sentiment scores assigned to each aspect group [13]. Whilst there are many other statistical approaches to frequent noun extraction [16]; others argue that identifying semantic relationship in text provides significant improvements in aspect extraction [14]. Here we explore how semantic based extraction can be augmented by frequency counts.

Temporal dynamics is a crucial dimension for both content and collaborative approaches. Initial work on concept drift was applied to classification tasks with focus on optimising the learning time window [9]. More recently with recommender systems, temporal knowledge in the form of rules was used to predict purchase behaviour over time [2]. Similarly the temporal influence on changes in user ratings has been observed on the Netflix movies [10], and short and long-term preference changes on products [19]. The association of importance weights to aspects according time is not new [4]. Here they generate aspect weights that are a function of time. Whilst our work also acknowledges the need for time-aware aspect weight learning, we exploit knowledge from both user review histories and preferences for this task.

3 Social Recommendation Model

An overview of our proposed process appears in Fig. 1. The final outcome is a recommendation of products (hitherto referred to as artefacts) that are retrieved and ranked, with respect to a given query product. Central to this ranking are aspect weights, which are derived from two knowledge sources: sentiment rich user generated product reviews and preferences from purchased summary statistics. Generally preference knowledge is captured in a graph according to purchase behavior and reviews depending on recency will influence both weight extraction and ranking algorithms. Here, we are interested in exploring how aspects can be weighted. Accordingly alternative aspect weight functions, AW_i, will be explored by taking into account the time and preference contexts of aspects. These in turn will influence the sentiment scores assigned to each extracted aspect, AS_i. Therefore the final ranking of products is based on an aspect weighted sentiment score aggregation.

3.1 Time context

New and improved product aspects grow over time. While there are product aspects that are continuously improving, others stabilise when a majority of the products possess them. We can observe such trends on a sample of data collected from Amazon between 2008 and April 2014 (see Table 1). Here we summarise the statistics of

Fig. 1 Social recommendations with temporal dynamics

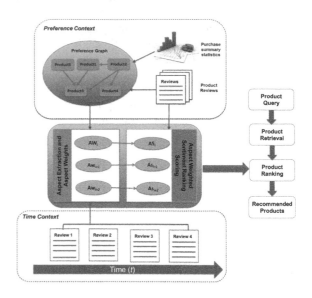

Table 1 Top 10 aspects and #aspects ordered by frequency between years 2008 until April 2014

TOP 10 ASPECTS

2008	2009	2010	2011	2012	2013	2014 until April
Resolution	Photographer	Photographer	Photographer	Photographer	Photographer	*Picture*
Photographer	*Picture*	*Picture*	*Picture*	*Picture*	*Picture*	Feature
Software	Feature	Feature	Feature	Feature	Feature	Photographer
Feature	Battery	Setting	Battery	Battery	Setting	Battery
Picture	Setting	Battery	Setting	Setting	Battery	Setting
Battery	Reason	Result	Result	Result	Result	Photography
Setting	*Resolution*	Photography	Photography	Photography	Photography	Image
Reason	Result	Reason	Video	Video	Image	Result
Noise	Photography	Image	Capability	Image	Video	Time
Result	Software	Video	Image	Time	Time	Quality

#ASPECTS

315	557	672	721	866	934	896

aspect frequency and show the top 10 most frequently mentioned aspects. In 2008, *resolution* was most frequent, however, this aspect's importance diminished during the following years finally disappearing from the top 10 in 2010. On the other hand, aspects like *picture* or *battery* remain in the top 10 list. *Resolution* might have been an important aspect for a camera in 2008, but it is no longer a determinant as the majority of the SLR cameras are now equipped with high resolution. On the contrary, users keep seeking for better pictures or longer battery life in SLR cameras.

Another interesting observation is that the number of different aspects grows with time (see Table 1). This is not suprising as manufacturers introduce new product features every year (e.g. HD video, creative image function, etc.) to attract customers. This situation also explains the top aspect weights presented in Table 1 decreasing over time: a higher number of aspects per year means a lower aspect average frequency. Accordingly such situations challenge existing recommendation systems, calling for adaptive extraction and weighting algorithms that can leverage temporal information from product reviews for product recommendation.

3.2 Preference Context

Like time, the product preference behaviour of users also paints a picture about what aspects are likely to be important when making purchase decisions. For instance if we know which product was preferred over which other product then by comparing the product aspect differences we can infer a degree of aspect importance. To do this we need to establish preference knowledge and thereafter quantify product aspect differences on the basis of sentiment.

We acquire preference knowledge from preference graph generated from viewed and purchased product pairs. The weight of an aspect is derived by comparing the sentiment difference between node pairs in the graph. A preference relation between a pair of products denotes the preference of one product over the other through the analysis of viewed and purchased product relationships. A preference graph, $G = (\mathcal{P}, \mathcal{E})$, is generated from such product pairs (see Fig. 2). The set of nodes, $p_i \in \mathcal{P}$, represent products, and the set of directed edges, $\mathcal{E}$, are preference relations, $p_j \succ p_i$, such that a directed edge from product p_i to p_j with $i \neq j$ represents that, for some users, p_j is preferred over product p_i. For any p_i, we use $\mathcal{E}^i$ to denote in-coming and $\mathcal{E}_i$ for outgoing product sets.

Figure 2 illustrate a preference graph generated from a sample of Amazon data on *Digital SLR Camera*. The number of reviews/questions for a product is shown below each product node. It is not surprising that such products appear in Amazon's *Best Seller* ranking (e.g. *B003ZYF3LO* is amongst Amazon's top 10 list). In our recent

Fig. 2 Preference graph for
Amazon SLR Camera

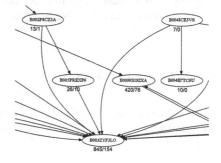

work [1], we observed that the higher the number of incoming edges (quantity) from preferred products (quality), the more preferred, is for product, p_i. However we also observed that while our assumption is true with most studied products, it is not always the case that a product with higher number of incoming edges will also have a higher rank in Amazon's *Best Seller*. This motivates the need to leverage further dimensions of knowledge sources such as sentiment from product reviews.

4 Aspect Weighted Sentiment-Based Ranking

User generated product reviews contain user opinion in the form of positive and negative sentiment. Strength of sentiment expresses the intensity with which an opinion is stated with reference to a product [17]. We exploit this information as a means to rank our products, such that products ranked higher denote higher positive sentiment. *ProdSenti* of a product, p_i, given a set of related reviews $\mathscr{R}^i$, a weighted summation of sentiment expressed at the aspect level is computed as follows:

$$ProdSenti(p_i) = \frac{\sum_{j=1}^{|\mathscr{A}^i|} AW(a_j, t) * AS(p_i, a_j)}{|\mathscr{A}^i|} \tag{1}$$

$$AS(p_i, a_j) = \frac{\sum_{m=1}^{|\mathscr{R}_j^i|} SentiScore(r_m)}{|\mathscr{R}_j^i|} * (1 - Gini) \tag{2}$$

where $\mathscr{R}_j^i$ is a set of reviews for product p_i related to aspect a_j, $r_m \in \mathscr{R}_j^i$ and AW is a function of a_j's weight over time t. *SentiScores* are generated by the SMARTSA system [15]. The negative and positive strength is expressed as a value in the range $[-1:1]$. It allows the sentiment of product, p_i, to be associated with individual aspects $a_j \in \mathscr{A}^i$ where $\mathscr{A}^i \subseteq \mathscr{A}$. The aspects of a product are extracted by using Algorithm 1. These sentiment instantiated values for product aspects allows the comparison of product pairs in terms of sentiment. We exploit this further to derive aspects weights based on preference and sentiment knowledge. We use Gini index [20] to acknowledge higher sentiment scores to an aspect when there is consensus on the distribution of the sentiment and otherwise penalised accordingly.

4.1 Aspect Extraction

Grammatical extraction rules [14] are used to identify a set of candidate aspect phrases from sentences. These rules operate on dependency relations in parsed sentences.[1] Rules involving negation are not included because our SMARTSA system already takes this into consideration when generating sentiment scores.

Algorithm 1 : Aspect Selection by Dependency Patterns

1: **INPUT:** S = sentences
2: **for all** s_j **in** S **do**
3:    $g = grammaticalRelations(s_j)$
4:    $candidateAspects = \{\}$
5:    **for** $dp_i \in DP$ and $1 \leq i \leq 4$ **do**
6:      **if** $g.matches(dp_i) \wedge g.contains(nn)$ **then**
7:        $candidateAspects. \leftarrow g.apply(\{dp_5, dp_6\})$ ▷ Apply rules dp_5, dp_6
8:      **end if**
9:    **end for**
10:   $aspects \leftarrow candidateAspects.select(N, NN)$ ▷ select nouns and compound nouns
11: **end for**
12: $filterByFrequency(aspects)$ ▷ ignore low frequency aspects
13: **return** $aspects$

Consider the sample sentence *"The camera lens is good"*. According to Algorithm 1 applies to rule three: $cop(good, is) + nsubj(good, lens) \rightarrow \langle lens, good \rangle$. Next, if a Noun Compound Modifier (nn) exists in the sentence, rules five and six apply; and in this example rule five applies resulting in the following candidate aspects: $\langle lens, good \rangle + nn(lens, camera) \rightarrow \langle camera\ lens, good \rangle$. In this way given a set of reviews a set of candidate phrases are extracted. For each candidate, non noun (N) words are eliminated. Thereafter frequency of each candidate is calculated according to its N and NN phrase; retaining only those candidates above a frequency cut-off. See [1, 14] for the detailed definition of grammatical relations and its application.

4.2 Time-Dependent Aspect Weight

The first of our weighting schemes assumes that an aspect is deemed important when it is frequently mentioned by the authors of the reviews. This allows us to monitor aspect trends in straight forward manner as reviews can be fairly easily grouped into seasonal time windows. Based on this principle, an aspect weight is derived by the relative aspect frequency at time window t.

$$AW(a_j, t) = \frac{Freq(a_j, t)}{\sum_{a_i \in \mathscr{A}} Freq(a_i, t)} \qquad (3)$$

where *Freq* returns the frequency of an aspect a in a given time window t into which reviews have been grouped. Frequency here is the number of times a term appears in a

---

[1] Sentences can be parsed using the Stanford Dependency parser [11].

Table 2 Time-dependent *megapixel* aspect weight in 2008–2011

Year	Freq (*megapixel, year*)	$\sum_{a_i \in \mathscr{A}}$ Freq (a_i, *year*)	AW (*megapixel, year*)
2008	434	1.4×10^6	3×10^{-5}
2011	504	4.5×10^6	1.1×10^{-5}

specified group of reviews. Table 2 shows an example of aspect weight calculations in two different time windows i.e. 2008 and 2011. We observe that although the aspect frequency of *megapixel* has increased overtime, its relative weight compared to all other aspects decreased significantly. Therefore, we suggest that the importance of the aspect *megapixel* has dropped over the years. Whilst frequency of aspects over time allows us to infer global trends about aspect usage, it does so without considering the relationship between aspects from preferred products. Therefore an alternative approach is to compare aspects shared between preferred products.

4.3 Preference Aspect Weight

A product purchase choice is a preference made on the basis of one or more aspects. The notion of aspect importance arises when the same set of aspects contribute to similar purchase decisions. Using this principle, aspects weights are derived by comparing the aspect sentiment score differences between purchased and viewed product pairs in which $(p_x, p_y) \in \{(p_x, p_y)\}_{x,y=1 \wedge x \neq y}^d$

$$AW(a_j) = \frac{\sum_{x=1}^{|\mathscr{P}|} \sum_{y=1}^{|\mathscr{P}|} \delta(a_j, p_x, p_y)}{|d \in \mathscr{E}|} \qquad (4)$$

where either $p_x \succ p_y$ or $p_y \succ p_x$ or both and d is the set of product preference pairs containing aspect a_j. The preference difference between any pairs of products is computed as:

$$\delta(a_j, p_x, p_y) = |L_{\min}(\mathscr{A}, \mathscr{E})| + \delta'(a_j, p_x, p_y) \qquad (5)$$

$$\delta'(a_j, p_x, p_y) = AS(a_j, p_x) - AS(a_j, p_y) \qquad (6)$$

Here $|L_{\min}(\mathscr{A}, \mathscr{E})|$ is the least minimum preference difference scores obtained over all aspects and product preference pairs. $AS(a_j, p_x)$ is the sentiment associated to aspect a_j of product p_x. Higher values of δ contribute to higher weights. Since δ' computation can results in negative scores, we use $|L_{\min}(\mathscr{A}, \mathscr{E})|$ to bound the lowest value to zero. Thereafter we normalise $\delta(a_j, p_x, p_y)$ such that it is in range [0, 1].

5 Evaluation

In this section we evaluate our proposed integration of time and preference guided aspect weight extraction applied to product recommendation. We first conduct a pilot study to evaluate the quality of aspects extracted by our algorithm with the state-of-the-art. Thereafter, we evaluate how well the recommendation system works in practice on Amazon and Cnet data using two derived benchmark rankings.

5.1 Comparative Pilot Study—Aspect Extraction Analysis

We use a public dataset on product reviews containing manually marked-up product aspects [3, 8]. For this study we use phone category products with at least hundred reviews. Precision, recall and F measure is used to compare manually labeled aspects with extracted ones. We advocate higher precision because this would mean we are able to identify true aspects of a product and this will lead to better learning of users preference. Therefore, $F_{0.5}$ measure was used in this study. We compare our proposed approach with the following alternative extraction algorithms:

- FQITEMSETS uses Apriori algorithm to identify candidate aspects that are then pruned using a frequency cut-off threshold [8].
- FQPOS uses Part-of-Speech (POS) extraction patterns that are then pruned using sentiment informed frequency cut-off threshold [5].
- DPRULES uses the dependency extraction rules in [14].
- FQDPRULES same as DPRULES but prunes candidate aspects using a frequency cut-off (see Algorithm 1).

Precision of all frequency based extraction approaches are significantly better compared to DPRULES (see Table 3). As expected best results are achieved with FQDPRULES when deep NLP semantics is combined with frequency pruning. Here we observe a 49 and 7 % improvement in precision over FQITEMSETS and FQPOS respectively. We also observe that FQDPRULES has the highest $F_{0.5}$ value (0.54). Recall trends suggests that FQDPRULES must have many false negatives and so missed extraction opportunities compared to DPRULES. However, in the context of our study a lower precision is more damaging as it is likely to introduce aspect sparsity problems which have detrimental effect on sentiment difference computations. Therefore, we

Table 3 Preliminary results for aspect extraction approaches

Approach	Precision	Recall	$F_{0.5}$ measure
FQITEMSETS	0.51	0.24	0.41
FQPOS	0.71	0.11	0.22
DPRULES	0.28	0.66	0.32
FQDPRULES	**0.76**	0.25	0.54

confirm that FQDPRULES provides a better result in the context of our study. On the basis of these results we use FQDPRULES to extract product aspects for sentiment analysis in social recommender experiments in the next sections.

5.2 Amazon Dataset and Ranking Strategies

We crawled 2,264 Amazon products during April 2014. From this we use the *Digital SLR Camera* category containing more than 20,000 user generated reviews. Since we are not focusing on the cold-start problem, newer products and those without many user reviews are removed. Here we use 1st January 2008 and less than 15 reviews as the pruning factor for products. Finally, any synonymous products are united leaving us data for 80 products. The FQDPRULES algorithm extracted 981 unique aspects and on average 128 different aspects for each product. Importantly more than 50 % of the products shared at least 70 different aspects, while 30 % shared more than 90 aspects on average. The fact that there are many shared aspects is reassuring for product comparison when applying Eq. 1.

The retrieval set of a query product consists of products that share a similar number of aspects. This retrieval set is ranked using the following sentiment-based recommendation strategies considering only the k top shared aspects between the retrieved and the query product. The comparative weighting schemes used in our experiments are as follows:

- BASE: recommend using aspect sentiment analysis without considering aspect weights presented in Eq. 2;
- *PrefW*: same as BASE but with the additional preference aspect weighting component from Eq. 4;
- *TimeW$_t$*: same as *PrefW* but considering the time context for aspect weighting component (instead of preference) presented in Eq. 3.

It is worth noting that Amazon only provides the current snap shot of preferences. Therefore we are unable to explore the impact of time on preference-based aspect weight extraction. We will present the *TimeW$_t$* strategy considering all reviews created between three different time windows: 2008–2014, 2011–2014 and 2014.

5.3 Evaluation Metrics

In the absence of ground truth data and individual user specific purchase trails, we generated two different benchmark lists according to the following dimensions:

- *Popular*: Derived from Amazon's reviews, questions and timeline data. Products are ranked based on its popularity computed by means of Eq. 7.

$$Popular(p) = \frac{nReviews + nQuestions}{days\_online} \qquad (7)$$

where *nReviews* and *nQuestions* refer to the number of reviews and questions of a product respectively, and *days_online* is the number of days the product has been on Amazon's website. We found that this formula has some correlation with the top 100 Amazon Best Seller ranking (*Spearman correlation* of -0.4381). Unlike Amazon's benchmark this allows us to experiment with query products that may not necessarily be in the Amazon top 100.[2] Using a *leave-one-out* methodology, the average gain in rank position of recommended products over the left-out query product is computed relative to a benchmark product ranking.

$$\%RankGain = \frac{\sum_{i=1}^{n=3} benchmark(P_q) - benchmark(P_i)}{n * |\mathscr{P} - 1|} \qquad (8)$$

where n is the size of the retrieved set and *benchmark* returns the position on the benchmark list. The greater the gain over the query product the better.

• *Precision*: Derived from Cnet.com expert recommendations for DSLRs cameras.[3] This groundtruth is divided in three subcategories (*entry-level DSLRs*, *midrange DSLRs* and *professional DSLRs*), each containing a list of cameras recommended by Cnet experts.

$$Precision = \sum_{i=1}^{n} \frac{TopN_{cat} \cap Cnet_{cat}}{n} \qquad (9)$$

In the absence of a defined Cnet ranking per category, we use a leave-one-out methodology to evaluate the capacity of our strategies to recommend the expert-selected cameras in each category. We compute the precision by means of Eq. 9, where $TopN_{cat}$ is the list of the top n recommended products for category *cat*, and $Cnet_{cat}$ is the list of Cnet expert recommended cameras for that category.

5.4 Results—Amazon

Here we present results from our exploration of aspects trends in terms of weights over time followed by a comparison of the two weighting schemes.

---

[2] http://www.amazon.co.uk/Best-Sellers-Electronics-Digital-SLR-Cameras.

[3] http://www.cnet.com/uk/topics/cameras/best-digital-cameras/.

5.4.1 Importance of Time on Aspect Weighting

Figure 3 shows the weight of the aspects *megapixel*, *autofocus* and *battery* computed by using strategy $TimeW_t$ for years between 2008 and 2014. We observe that *megapixel* was an important aspect in *2008* with a frequency weight of close to 0.009. However, its importance decreased dramatically during the following years, reducing its weight up to five times in *2011*. In contrast, with *autofocus*, we see an increasing trend. A different trend can be observed for *battery* in Fig. 3. Here it is interesting to note that the aspect weight is maintained over the years. Whilst there is a negligible loss in the raw score this is explained by the difference in number of unique aspects in the time period. For example, in 2008 we found approximately 250 aspects whilst in 2014 this had increased to 900.

In Fig. 4, we use $TimeW_t$ with $t = 2008–2014$, $2011–2014$ and 2014 to rank the recommendations for increasing number of shared aspects (k) on benchmark *Popular*. In general, we observe that weights learned using $TimeW_{2014}$ perform up to a 15 % better for $k = 30$ when recommending actual products. $TimeW_{2011}$ falls close to the recommendations made by $TimeW_{2014}$, being the weights learned by $TimeW_{2008}$ the ones that perform worst. These results indicate that considering the most recent time frame for computing the aspect weights improves the recommendations made by the system, and that aspect frequency over time is a good indicator of what users care most when considering cameras.

5.4.2 Time Versus Preference Weighting

In Fig. 5 we compare the three strategies, $TimeW_{2014}$, *PrefW* and BASE using the *Popular* benchmark. We include the strategy agnostic of aspects weights, BASE,

Fig. 3 Aspects weight over time (in years)

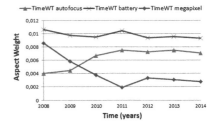

Fig. 4 *ProdSenti* on *Popular* benchmark

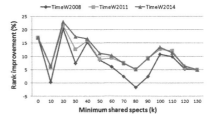

Fig. 5 Comparison of
different strategies

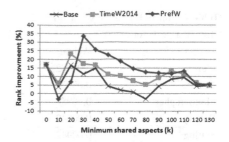

in order to compare the impact that weights have on the recommendations while considering preferences and time weighted aspects. As we can observe, $TimeW_{2014}$ and $PrefW$ strategies outperform BASE by more than 10 % on average. On the other hand, we observe that $PrefW$ outperforms $TimeW_{2014}$ for all values of k comprised between 30 and 100. This suggests that preference weights used by $PrefW$ are able to recommend better products for the *Popular* benchmark since they represent the most recent snapshot of the current users preferences. We also observe that there seems to be a sweet spot in terms of the number of shared aspects (k), with $PrefW$ achieving best results with $k = 30$, and a gradual decrease thereafter. However, the rank improvement obtained by considering time in $TimeW_{2014}$ should not be ignored as it performs 15 % better than $PrefW$ with smaller subsets of shared aspects (e.g. $k = 20$) and obtains a similar rank improvement with increased numbers of shared aspects (e.g. $k \geq 100$). Close examination of $TimeW_{2014}$'s performance suggests that the retrieval set consisted of a high number of products (85 %) with $k = 20$ shared aspects. This is in contrast to its poorer performance with higher values of k. This suggests that with less frequently occurring aspects the frequency based weight computation of $TimeW_{2014}$ is likely to be less reliable compared to $PrefW$.

Table 4 presents the top 10 aspects extracted by means of $TimeW_{2014}$ and $PrefW$ and its correspondent weight. Here the lists and weights of the top aspects obtained by the two strategies are different except for aspects *picture*, *setting* and *photography*. Although occupy different ranking positions and are weighted differently, both strategies seem to agree on their importance. Also, the weight distribution of each strategies is different. For example, $TimeW_{2014}$ gives almost five times more weight to *picture* (0.00982) than $PrefW$ does to its top aspect *shutter* (0.00229). We also notice that there are several semantically related aspects that appear in the top 10: *image* represents a similar concept to *picture* and *photography*; similarly *system* and *settings*. It is likely that mapping such related aspects to a common concept is able to generate more accurate aspect weights.

Table 4 Top 10 for $TimeW_{2014}$ and $PrefW$

Top 10 aspects for $TimeW_{2014}$		Top 10 aspects for $PrefW$	
Aspect	Weight	Aspect	Weight
Picture	0.00982	Shutter	0.00229
Feature	0.00974	Photography	0.00181
Photographer	0.00956	Point	0.00179
Battery	0.00948	System	0.00176
Setting	0.00942	Video	0.00166
Photography	0.00877	Setting	0.00165
Image	0.00857	Picture	0.00158
Result	0.00847	Advantage	0.00152
Time	0.00843	Sensor	0.00150
Quality	0.00842	Manual	0.00150

5.5 Results—Cnet

Next we compare our recommendations using weights extracted from Amazon sources against expert advice. For this purpose we use Cnet expert advice on DSLR camera recommendations. We divided all cameras from our dataset in three subsets, corresponding to *entry-level*, *midrange* and *professional* categories presented on the website, based on price (see Table 5). Table 5 also shows the number of products included in every subset and the number of Cnet products used as the gold standard.

Table 6 shows the average precision of $TimeW_{2014}$, $PrefW$ and BASE for every Cnet category and different recommendation set sizes ($n = 1$ and $n = 3$) computed by means of Eq. 9. We included a strategy that randomly recommends products, *Random*, to facilitate understanding of the results. As we observe, strategies that are aware of aspect sentiments are able to improve precision in every category. In *Popular* dataset, *PrefW* is the strategy that performs better: for the *Cnet Entry-level* subset, it is able to recommend a Cnet top product 37 % of the time for $n = 1$ and a 24 % for $n = 3$ on average. These results are promising considering that the probability of recommending a top Cnet product amongst the entire subset in this category is 20 % and 7 % respectively. Precision results for all three strategies are much higher when applied to smaller Cnet datasets; *Mid-range* and *Professional*,

Table 5 Cnet dataset classification for Entry-level, Mid-range and Professional

	Entry-level	Mid-range	Professional
Price range ($)	0–1k	1–2.2k	2.2–7k
#$products_{cat}$	60	14	6
#$products_{cat} \cap Cnet_{cat}$	7	8	3

Table 6 Precision for different retrieved set sizes n in Cnet

Strategy	Entry-level		Mid-range		Professional	
	$n = 1$	$n = 3$	$n = 1$	$n = 3$	$n = 1$	$n = 3$
Random	0.203	0.077	0.461	0.376	0.2	0.277
BASE	0.220	0.146	0.769	0.423	0.4	0.377
$TimeW_{2014}$	0.254	0.128	**0.846**	0.444	**0.8**	0.4
$PrefW$	**0.372**	**0.243**	**0.846**	**0.461**	0.6	**0.422**

achieving a precision of close to 85 and 80 % for $n = 1$ respectively and doubling the precision of the random recommender. Overall, since our system recommendations closely match with Cnet expert's recommendations, we conclude that the aspect weights learned from Amazon are likely to correspond with criteria that the Cnet expert's might have implicitly used. However, we cannot accurately verify this without manually demanding user trial. Nevertheless it is interesting that consensus knowledge discovered from social media seem to closely echo views of the domain experts.

6 Conclusions

Tracking users preference over time raises unique challenges for recommendation systems. Every product potentially goes through a series of changes which typically involves functional improvements resulting in a broader range of aspects which in turn will be echoed in changes to user preferences. Our previous findings suggested that preference knowledge allows us to identify aspects that are important to users but lacks the capability to trace aspect importance trends.

In this paper, we formalise a novel aspect-based sentiment ranking that utilise both time and preference contexts. The benefits are demonstrated in a realistic recommendation setting using benchmarks generated from Amazon and Cnet. We show that monitoring aspect frequency in product reviews allows to capture changes to aspect importance over time. Importantly, we confirm that time context can be conveniently exploited by using the recent time frame to improve recommendations.

We compare the performance of aspect-based sentiment ranking in the context of time and preference. We observed that both contexts perform well in different number of shared aspects, further work is needed to study the benefit of integrating both contexts in ranking algorithm. Our results show that similar aspects were mentioned using different terms, further work is needed to study how sparsity problems might impact the difference calculations. Finally, it would be interesting to integrate trending information within the aspect weight computation to infer their importance.

Acknowledgments This research has been partially supported by AGAUR Scholarship (2013FI-B 00034), ACIA mobility scholarship and Project Cognitio TIN2012-38450-C03-03.

References

1. Chen, Y., Ferrer, X., Wiratunga, N., Plaza, E.: Sentiment and preference guided social recommendation. In: International Conference on Case-Based Reasoning. Accepted in ICCBR '14 (2014)
2. Cho, Y., Cho, Y., Kim, S.: Mining changes in customer buying behavior for collaborative recommendations. Expert Syst. Appl. **28**, 359–369 (2005)
3. Ding, X., Liu, B., Yu, P.: A holistic lexicon-based approach to opinion mining. In: Proceedings of International Conference on Web Search and Data Mining (2008)
4. Ding, Y., Li, X.: Time weight collaborative filtering. In: Proceedings of International Conference on Information and Knowledge Management, CIKM '05, pp. 485–492 (2005)
5. Dong, R., Schaal, M., O'Mahony, M., McCarthy, K., Smyth, B.: Opinionated product recommendation. In: International Conference on Case-Based Reasoning (2013)
6. Esuli, A., Sebastiani, F.: Sentiwordnet: A publicly available lexical resource for opinion mining. In: Proceedings of Language Resources and Evaluation Conference, pp. 417–422 (2006)
7. Hong, W., Li, L., Li, T.: Product recommendation with temporal dynamics. Expert Syst. Appl. **39**, 12398–12406 (2012)
8. Hu, M., Liu, B.: Mining and summarising customer reviews. In: Proceedings of ACM SIGKDD International Conference on Knowledge Discovery and Data Mining, pp. 168–177 (2004)
9. Kolter, J., Maloof, M.: Dynamic weighted majority: A new ensemble method for tracking concept drift. In: International Conference on Data Mining, pp. 123–130 (2003)
10. Koren, Y.: Collaborative filtering with temporal dynamics. Commun. Assoc. Comput. Mach. **53**, 89–97 (2010)
11. Marneffe, M., MacCartney, B., Manning, C.: Generating typed dependency parses from phrase structure parses. In: Proceedings of Language Resources and Evaluation Conference (2006)
12. McCarthy, K., Salem, Y., Smyth, B.: Experience-based critiquing: reusing experiences to improve conversational recommendation. In: International Conference on Case-Based Reasoning (2010)
13. Moghaddam, S., Ester, M.: Opinion digger: An unsupervised opinion miner from unstructured product reviews. In: Proceedings of International Conference on Information and Knowledge Management (2010)
14. Moghaddam, S., Ester, M.: On the design of lda models for aspect-based opinion mining. In: Proceedings of International Conference on Information and Knowledge Management (2012)
15. Muhammad, A., Wiratunga, N., Lothian, R., Glassey, R.: Contextual sentiment analysis in social media using high-coverage lexicon. In: Research and Development in Intelligent. Springer, Berlin (2013)
16. Popescu, A., Etzioni, O.: Extracting product features and opinions from reviews. In: Natural Language Processing and Text Mining, pp. 9–28. Springer, London (2007)
17. Turney, P.: Thumbs up or thumbs down?: semantic orientation applied to unsupervised classification of reviews. In: Proceedings of Annual Meeting for Computational Linguistics (2002)
18. Vasudevan, S., Chakraborti, S.: Mining user trails in critiquing based recommenders. In: Proceedings of International Conference on World Wide Web Companion, pp. 777–780 (2014)
19. Xiang, L., Yuan, Q., Zhao, S., Chen, L., Zhang, X., Yang, Q., Sun, J.: Temporal recommendation on graphs via long- and short-term preference fusion. In: Proceedings of International Conference on Knowledge Discovery and Data Mining, pp. 723–732 (2010)
20. Yitzhaki, S.: Relative deprivation and the Gini coefficient. Q. J. Econ., 321–324 (1979)

Agents, Ontologies
and Genetic Programming

Query Failure Explanation in Inconsistent Knowledge Bases: A Dialogical Approach

Abdallah Arioua, Nouredine Tamani, Madalina Croitoru
and Patrice Buche

Abstract In the EcoBioCap project (www.ecobiocap.eu) about the next generation of packaging, a decision support system has been built that uses argumentation to deal with stakeholder preferences. However, when testing the tool the domain experts did not always understand the output of the system. The approach developed in this paper is the first step to the construction of a decision support system endowed with an explanation module. We place ourselves in the equivalent setting of inconsistent Ontology-Based Data Access (OBDA) and addresses the problem of explaining Boolean Conjunctive Query (BCQ) failure. Our proposal relies on an interactive and argumentative approach where the processes of explanation takes the form of a dialogue between the User and the Reasoner. We exploit the equivalence between argumentation and inconsistency tolerant semantics to prove that the Reasoner can always provide an answer for user's questions.

1 Introduction

In the popular ONTOLOGY-BASED DATA ACCESS setting the domain knowledge is represented by an ontology facilitating query answering over existing data [18]. In practical systems involving large amounts of data and multiple data sources, data

A. Arioua
IATE, INRA, UM2, 2 Place Pierre Viala, 34060 Montpellier, France
e-mail: abdallaharioua@gmail.com

N. Tamani
INRIA GraphIK, LIRMM, 161 rue Ada, 34095 Montpellier, France
e-mail: tamani@lirmm.fr; ntamani@gmail.com

M. Croitoru (✉)
INRIA GraphIK, LIRMM, UM2, 161 rue Ada, 34095 Montpellier, France
e-mail: croitoru@lirmm.fr

P. Buche
IATE, CIRAD-INRA-Supagro-UM2, 2 Place Pierre Viala, 34060 Montpellier, France
e-mail: patrice.buche@supagro.inra.fr

© Springer International Publishing Switzerland 2014
M. Bramer and M. Petridis (eds.), *Research and Development
in Intelligent Systems XXXI*, DOI 10.1007/978-3-319-12069-0_8

119

inconsistency with respect to the ontology is difficult to prevent. Many inconsistency-tolerant semantics [4, 5, 15, 16] have been proposed that rely on the notion of data repairs i.e. subsets of maximally consistent data with respect to the ontology. Different query answering strategies over these repairs (called semantics) are investigated in the literature. For instance, computing answers that hold in every repair corresponds to AR-semantics, computing answers that hold under the intersection closed repairs corresponds to ICR-semantics, etc.

In this paper we will consider inconsistent knowledge bases and focus on the ICR-semantics (**I**ntersection of **C**losed **R**epair). The particular choice of the ICR-semantics for OBDA is due to its interesting productivity properties as shown in [4]. Under the ICR-semantics, maximal (in terms of set inclusion) consistent subsets (called repairs) of the knowledge base are constructed. Querying the knowledge base is done on the intersection of repairs after ontology closure.

Query answering under these semantics may not be intuitively straightforward and can lead to loss of user's trust, satisfaction and may affect the system's usability [17]. Moreover, as argued by Calvanese et al. in [10] *explanation facilities* should not just account for user's "Why Q?" question (why a query holds under a given inconsistency-tolerant semantics) but also for question like "Why not Q ?" (why a query does not hold under a given inconsistency-tolerant semantics), which is the research problem addressed in this paper. Precisely: *"Given an inconsistent knowledge base, denoted by KB, and a boolean conjunctive query Q, why Q is not entailed from KB under the ICR-semantics?"*.

To address this issue we use argumentation as an approach for explanation. We consider the logical instantiation of Dung's [13] abstract argumentation framework for OBDA in [12] and we exploit the equivalence result shown by the authors between the ICR-semantics and sceptical acceptance under preferred semantics to guarantee the existence of an explanation for any failed query. The explanation takes the form of a dialogue between the *User* and the *Reasoner* with the purpose of explaining the query failure. At each level of the dialogue we use (an introduced) language-based primitives such as *clarification* and *deepening* to refine the answer.

This paper improves over and extends the work presented in [1]. Its contribution lies in the following points: first, it puts forward the explanation power of argumentation in the benefit of Semantic Web. Second, it improves OBDA system's usability and enhances its user-friendliness. To the best of our knowledge, we are the first to propose query failure explanation in the context of OBDA for **inconsistent knowledge bases** by means of **argumentation**. Our approach differs from [6, 10] in handling query failure since we consider an inconsistent setting within OBDA. In addition, the work presented in [14] is neither applied to an OBDA context nor to the Datalog $+/-$ language.

The paper is organized as follows. In Sect. 2, we recall the scope of the paper and formally define the addressed problem using an illustrating example. In Sects. 3 and 4, we present the language and the corresponding argumentation framework. In Sect. 5, we introduce the formalization of the proposed explanation and show several properties. Finally, Sect. 6 concludes the paper.

2 Paper in a Nutshell

Let us introduce the motivation and the formal context of our work.

The Context. Several semantics that allow answering queries in the presence of inconsistency have been studied in the literature and here we only focus on the ICR-semantics.

The Problem. Consider a knowledge base about university staff and students which contains inconsistent knowledge. This inconsistency is handled by ICR-semantics. The *User* might be interested in knowing why the knowledge base *does not entail* the query Q: "Luca is a student". Such query failure might occur, for instance, after the *User* does not find in the answer set the expected output (e.g. the *User* asked for all the students and Luca did not appear in the answer list). Another possibility is when the *User* asks for the existence of students in the knowledge base and the system answers that no students are present in the knowledge base.

Observe that the answer δ (e.g. Luca in the example above) is a negative answer for a conjunctive query Q (e.g. get me all the students in the example above) if and only if the boolean conjunctive query $Q(\delta)$ (e.g. student(Luca) in the example above) has failed. Hence, in this paper we concentrate only on explaining the failure of a boolean conjunctive query. Let us formally introduce the problem of *Query Failure Explanation* in inconsistent knowledge bases.

Definition 1 (*Query Failure Explanation Problem $\mathscr{P}$*) Let $\mathscr{K}$ be an inconsistent knowledge base, Q a Boolean Conjunctive Query such that $\mathscr{K} \not\models_{ICR} Q$. We call $\mathscr{P} = \langle \mathscr{K}, Q \rangle$ a Query Failure Explanation Problem (QFEP).

The Guidelines of the Solution. To address the *Query Failure Explanation Problem*, we use a logical instantiation of Dung's [13] abstract argumentation framework for OBDA in [12] which ensures that the argumentation framework used throughout the paper respects the rationality postulates [11]. The arguments are then used for query failure explanation as follows. First, we introduce two notions over the arguments: clarifying and deepening. The former notion serves at unfolding the knowledge used in the argument, and the latter is used to explicit the reason for an attack between two arguments. We then describe a dialectical system of explanation custom-tailored for inconsistent knowledge base query failure. The *User* and the *Reasoner* will take turns in a dialogue with the final aim that the *User* understands why a query is not entailed under the ICR-semantics by the knowledge base.

3 Formal Settings

In this section, we introduce: (1) OBDA setting and representation language, (2) the inconsistency-tolerant semantics.

3.1 OBDA Setting

There are two major approaches to represent an ontology for the OBDA problem: Description Logics (such as $\mathscr{EL}$ [2] and DL-Lite [9] families) and rule-based languages (such as Datalog $+/-$ [7] language, a generalization of Datalog that allows for existentially quantified variables in rules heads). Despite Datalog $+/-$ undecidability when answering conjunctive queries, different decidable fragments of Datalog $+/-$ are studied in the literature [3]. These fragments generalize the aforementioned Description Logics families and overcome their limitations by allowing any predicate arity as well as cyclic structures. Here we follow the second method and use a general rule-based setting knowledge representation language equivalent to the Datalog $+/-$ language.

3.2 Language Specification

We consider *the positive existential* conjunctive fragment of first-order logic, denoted by FOL($\wedge$, $\exists$), which is composed of formulas built with the connectors ($\wedge$, $\rightarrow$) and the quantifiers ($\exists$, $\forall$). We consider first-order vocabularies with constants but no other function symbol. A term t is a constant or a variable, different constants represent different values (unique name assumption), an atomic formula (or atom) is of the form $p(t_1, ..., t_n)$ where p is an n-ary predicate, and $t_1, ..., t_n$ are terms. A *ground* atom is an atom with no variables. A variable in a formula is free if it is not in the scope of a quantifier. A formula is closed if it has no free variables. We denote by $\mathbf{X}$ (with a bold font) a sequence of variables $X_1, ..., X_k$ with $k \geq 1$. A *conjunct* $C[\mathbf{X}]$ is a finite conjunction of atoms, where $\mathbf{X}$ is the sequence of variables occurring in C. Given an atom or a set of atoms A, $vars(A)$, $consts(A)$ and $terms(A)$ denote its set of variables, constants and terms, respectively.

An *existential rule* (or simply a rule) is a first-order formula of the form $r = \forall \mathbf{X} \, \forall \mathbf{Y} \, (H[\mathbf{X}, \mathbf{Y}]) \rightarrow \exists \mathbf{Z} \, C[\mathbf{Z}, \mathbf{Y}]$, with $vars(H) = \mathbf{X} \cup \mathbf{Y}$, and $vars(C) = \mathbf{Z} \cup \mathbf{Y}$ where H and C are *conjuncts* called the hypothesis and conclusion of r respectively. We denote by $r = (H, C)$ a contracted form of a rule r. An existential rule with an empty hypothesis is called a *fact*. A fact is an existentially closed (with no free variable) conjunct. i.e. $\exists x (teacher(x) \wedge employee(x))$.

We recall that a *homomorphism* π from set of atoms A_1 to set of atoms A_2 is a substitution of $vars(A_1)$ by $terms(A_2)$ such that $\pi(A_1) \subseteq A_2$. Given two facts f and

f'. We have $f \models f'$ if and only if there is a homomorphism from f' to f, where $\models$ is the first-order semantic entailment.

A *rule* $r = (H, C)$ is *applicable* to a set of facts F if and only if there exists $F' \subseteq F$ such that there is a homomorphism π from H to the conjunction of elements of F'. For instance, the rule $\forall x(teacher(x) \rightarrow employee(x))$ is applicable to the set $\{teacher(Tom), cute(Tom)\}$ because there is a homomorphism from $teacher(x)$ to $teacher(Tom)$. If a rule r is applicable to a set of facts F, its application according to π produces a set $F \cup \{\pi(C)\}$. The new set $F \cup \{\pi(C)\}$, denoted also by $r(F)$, is called *immediate derivation* of F by r. In our example the produced set (immediate derivation) is $\{teacher(Tom), employee(Tom), cute(Tom)\}$.

A *negative constraint* (or constraint) is a first-order formula $n = \forall \mathbf{X}\, H[\mathbf{X}] \rightarrow \perp$ where $H[\mathbf{X}]$ is a conjunct called hypothesis of n and $\mathbf{X}$ is a sequence of variables appearing in the hypothesis. For example, $n = \forall x \forall y \forall z(supervises(x, y) \wedge work\_in(x, z) \wedge directs(y, z)) \rightarrow \perp$, means that it is impossible for x to supervise y if x works in department z and y directs z.

The Knowledge Base $\mathcal{K} = (\mathcal{F}, \mathcal{R}, \mathcal{N})$ consists of a finite set of facts $\mathcal{F}$, a finite set of existential rules $\mathcal{R}$ and a finite set of negative constrains $\mathcal{N}$.

Example 1 The following example is inspired from [8]. In an enterprise, employees work in departments and use offices which are located in departments, some employees direct departments, and supervise other employees. In addition, a supervised employee cannot be a manager. A director of a given department cannot be supervised by an employee of the same department, and an employee cannot work in more than one department. The following sets of (existential) rules $\mathcal{R}$ and negative constraints $\mathcal{N}$ model the corresponding ontology:

$$\mathcal{R} = \begin{cases} \forall x \forall y\,(works\_in(x, y) \rightarrow emp(x)) & (r_1) \\ \forall x \forall y\,(directs(x, y) \rightarrow emp(x)) & (r_2) \\ \forall x \forall y\,(directs(x, y) \wedge works\_in(x, y) \rightarrow manager(x)) & (r_3) \\ \forall x \forall y \forall z\; locate\_office(y, z) \wedge uses\_office(x, y) \rightarrow works\_in(x, z) & (r_4) \end{cases}$$

$$\mathcal{N} = \begin{cases} \forall x \forall y\,(supervises(x, y) \wedge manager(y)) \rightarrow \perp & (n_1) \\ \forall x \forall y \forall z\,(supervises(x, y) \wedge works\_in(x, z) \wedge directs(y, z)) \rightarrow \perp & (n_2) \\ \forall x \forall y \forall z\,(works\_in(x, y) \wedge works\_in(x, z)) \rightarrow \perp & (n_3) \end{cases}$$

Let us suppose the following set of facts $\mathcal{F}$ that represent explicit knowledge:

$$\mathcal{F} = \begin{cases} directs(John, d_1) & (f_1) & directs(Tom, d_1) & (f_2) \\ directs(Tom, d_2) & (f_3) & supervises(Tom, John) & (f_4) \\ works\_in(John, d_1) & (f_5) & works\_in(Tom, d_1) & (f_6) \\ works\_in(Carlo, Statistics) & (f_7) & works\_in(Luca, Statistics) & (f_8) \\ works\_in(Jane, Statistics) & (f_9) & works\_in(Linda, Statistics) & (f_{10}) \\ uses\_office(Linda, o_1) & (f_{11}) & locate\_office(o_1, Accounting) & (f_{12}) \end{cases}$$

$\mathcal{R}$-**derivation.** Let $F \subseteq \mathcal{F}$ be a set of facts and $\mathcal{R}$ be a set of rules. An $\mathcal{R}$-derivation of F in $\mathcal{K}$ is a finite sequence $\langle F_0, ..., F_n \rangle$ of sets of facts such that $F_0 = F$, and for all $i \in \{0, ..., n-1\}$ there is a rule $r_i = (H_i, C_i) \in \mathcal{R}$ and a homomorphism π_i from H_i to F_i such that $F_{i+1} = F_i \cup \{\pi(C_i)\}$. For a set of facts $F \subseteq \mathcal{F}$ and a query Q and a set of rules $\mathcal{R}$, we say $F, \mathcal{R} \models Q$ if and only if there exists an $\mathcal{R}$-derivation $\langle F_0, ..., F_n \rangle$ such that $F_n \models Q$.

Closure. Given a set of facts $F \subseteq \mathcal{F}$ and a set of rules $\mathcal{R}$, the closure of F with respect to $\mathcal{R}$, denoted by $\mathrm{Cl}_{\mathcal{R}}(F)$, is defined as the smallest set (with respect to $\subseteq$) which contains F and is closed under $\mathcal{R}$-derivation.

Finally, we say that a set of facts $F \subseteq \mathcal{F}$ and a set of rules $\mathcal{R}$ *entail* a fact f (and we write $F, \mathcal{R} \models f$) if and only if the closure of F by all the rules entails f (i.e. $\mathrm{Cl}_{\mathcal{R}}(F) \models f$).

A conjunctive query (CQ) has the form $Q(\mathbf{X}) = \exists \mathbf{Y} \Phi[\mathbf{X}, \mathbf{Y}]$ where $\Phi[\mathbf{X}, \mathbf{Y}]$ is a conjunct such that $\mathbf{X}$ and $\mathbf{Y}$ are variables appearing in Φ. A boolean CQ (BCQ) is a CQ of the form $Q()$ with an answer yes or no, e.g. $Q = \exists x\, emp(x)$. We refer to a BCQ with an answer *no* as **failed query**, whereas a query with the answer *yes* as **accepted query**. Unless stated otherwise, we refer to a BC query as a query.

3.3 Inconsistency-Tolerant Semantics

Given a knowledge base $\mathcal{K} = (\mathcal{F}, \mathcal{R}, \mathcal{N})$, a set $F \subseteq \mathcal{F}$ is said to be *inconsistent* if and only if there exists a constraint $n \in \mathcal{N}$ such that $F \models H_n$, where H_n is the hypothesis of the constraint n. A set of facts is consistent if and only if it is not inconsistent. A set $F \subseteq \mathcal{F}$ is $\mathcal{R}$-*inconsistent* if and only if there exists a constraint $n \in \mathcal{N}$ such that $\mathrm{Cl}_{\mathcal{R}}(F) \models H_n$. A set of facts is said to be $\mathcal{R}$-*inconsistent* if and only if it is not $\mathcal{R}$-consistent. A knowledge base $(\mathcal{F}, \mathcal{R}, \mathcal{N})$ is said to be *inconsistent* if and only if $\mathcal{F}$ is $\mathcal{R}$-inconsistent.

Notice that (like in classical logic), if a knowledge base $\mathcal{K} = (\mathcal{F}, \mathcal{R}, \mathcal{N})$ is inconsistent, then everything is entailed from it. A common solution [4, 15] is to construct maximal (with respect to set inclusion) consistent subsets of $\mathcal{K}$. Such subsets are called *repairs* and denoted by $\mathcal{R}epair(\mathcal{K})$. Once the repairs are computed, different semantics can be used for query answering over the knowledge base. In this paper we focus on (**I**ntersection of **C**losed **R**epairs semantics) [4].

Definition 2 (*ICR-semantics*) Let $\mathcal{K} = (\mathcal{F}, \mathcal{R}, \mathcal{N})$ be a knowledge base and let Q be a query. Q is ICR-entailed from $\mathcal{K}$, written $\mathcal{K} \models_{ICR} Q$ if and only if:

$$\bigcap\nolimits_{\mathcal{A} \in \mathcal{R}epair(\mathcal{K})} \mathrm{Cl}_{\mathcal{R}}(\mathcal{A}) \models Q.$$

Example 2 The knowledge base in Example 1 is inconsistent because the set of facts $\{f_1, f_4, f_6\} \subseteq \mathcal{F}$ is inconsistent since it violates the negative constraint n_2. To be able to reason in presence of inconsistency one has to construct first the repairs and intersect their closure. The following is of the repairs:
$\mathcal{A}_1 = \{directs(John, d_1), supervises(Tom, John), works\_in(Linda, Statistic),$
$uses\_office(Linda, o_1), directs(Tom, d_1), directs(Tom, d_2), works\_in(Carlo,$

Statistic), *works_in*(*Jane, Statistic*), *works_in*(*Luca, Statistic*), *emp*(*John*),
emp(*Tom*), *emp*(*Carlo*), *emp*(*Luca*), *emp*(*Jane*), *emp*(*Linda*)}.
The intersection of all closed repairs is:
$\bigcap_{A \in \mathcal{R}epair(\mathcal{K}))} \text{Cl}_{\mathcal{R}}(A) = \{directs(Tom, d_1), directs(Tom, d_2), works\_in(Carlo,$
Statistics), *works_in*(*Luca, Statistics*), *works_in*(*Jane, Statistics*), *emp*(*Carlo*),
emp(*Jane*), *emp*(*Luca*), *emp*(*Tom*), *emp*(*John*), *emp*(*Linda*)}.

Note that in the intersection of all closed repairs there is *works_in*(*Luca, Statistics*).
That means that *works_in*(*Luca, Statistics*) is ICR-entailed from the knowledge base.
Whereas, *works_in*(*Linda, Statistics*) is not ICR-entailed since the facts about Linda
are conflicting (because she works also for the department of Accounting). We can
conclude that the ICR-semantics is prudent and operates according to the principle
"when in doubt, throw it out".

4 Argumentation Framework, Deepening and Clarification

In what follows we present the definition of argumentation framework in the context
of rule-based languages. We use the definition of argument of [12] and extend it to
the notions of deepened and clarified arguments.

4.1 Rule-Based Dung Argumentation Framework

As defined in [12], given a knowledge base $\mathcal{K} = (\mathcal{F}, \mathcal{R}, \mathcal{N})$, the *corresponding
argumentation framework* $\mathcal{A}\mathcal{F}_{\mathcal{K}}$ is a pair (Arg, Att) where Arg is the set of
arguments that can be constructed from $\mathcal{F}$ and Att is an asymmetric binary relation
called *attack* defined over Arg × Arg. An argument is defined as follows.

Definition 3 (*Argument* [12]) Given a knowledge base $\mathcal{K} = (\mathcal{F}, \mathcal{R}, \mathcal{N})$, an argument a is a tuple $a = (F_0, F_1, \ldots, F_n, C)$ where:

- $(F_0, \ldots, F_n)$ is an $\mathcal{R}$-derivation of F_0 in $\mathcal{K}$,
- C is an atom, a conjunction of atoms, the existential closure of an atom or the
  existential closure of a conjunction of atoms such that $F_n \models C$.
  F_0 is the support of the argument a (denoted Supp(a)) and C is its conclusion
  (denoted Conc(a)).

Example 3 (Argument) The following argument indicates that John is an employee
because he directs department d_1:

$a = (\{directs(John, d_1)\}, \{directs(John, d_1), emp(John)\}, emp(John)).$

Definition 4 (*Attack* [12]) The attack between two arguments expresses the conflict
between their conclusion and support. An argument a attacks an argument b iff there
exists a fact $f \in$ Supp(b) such that {Conc(a), f} is $\mathcal{R}$-inconsistent.

Example 4 (Attack) Consider the argument a of Example 3, the following argument $b = (\{supervises(Tom, John), works\_in(Tom, d_1)\}, supervises(Tom, John) \land works\_in(Tom, d_1))$ attacks a, because $\{supervises(Tom, John) \land works\_in(Tom, d_1), directs(John, d_1)\}$ is $\mathscr{R}$-inconsistent since it violates the constraint n_2.

4.1.1 Admissibility, Semantics and Extensions

Let $\mathscr{K} = (\mathscr{F}, \mathscr{R}, \mathscr{N})$ be a knowledge base and $\mathscr{A}\mathscr{F}_{\mathscr{K}}$ its *corresponding argumentation framework*. Let $\mathscr{E} \subseteq \texttt{Arg}$ be a set of arguments. We say that $\mathscr{E}$ is *conflict free* if and only if there exist no arguments $a, b \in \mathscr{E}$ such that $(a, b) \in \texttt{Att}$. $\mathscr{E}$ *defends* an argument a if and only if for every argument $b \in \texttt{Arg}$, if we have $(b, a) \in \texttt{Att}$ then there exists $c \in \mathscr{E}$ such that $(c, b) \in \texttt{Att}$. $\mathscr{E}$ is *admissible* if and only if it is conflict free and defends all its arguments. $\mathscr{E}$ is a *preferred extension* if and only if it is maximal (with respect to set inclusion) admissible set (please see [13] for other types of semantics).

We denote by $\texttt{Ext}(\mathscr{A}\mathscr{F}_{\mathscr{K}})$ the set of all extensions of $\mathscr{A}\mathscr{F}_{\mathscr{K}}$ under one Dung's semantics. An argument is sceptically accepted if it is in all extensions, credulously accepted if it is in at least one extension and not accepted if it is not in any extension.

4.1.2 Equivalence Between ICR and Preferred Semantics

Let $\mathscr{K} = (\mathscr{F}, \mathscr{R}, \mathscr{N})$ be a knowledge base and $\mathscr{A}\mathscr{F}_{\mathscr{K}}$ the corresponding argumentation framework. A query Q is sceptically accepted under preferred semantics if and only if $(\bigcap_{\mathscr{E} \in \texttt{Ext}(\mathscr{A}\mathscr{F}_{\mathscr{K}})} \texttt{Concs}(\mathscr{E})) \vDash Q$, such that $\texttt{Concs}(\mathscr{E}) = \bigcup_{a \in \mathscr{E}} \texttt{Conc}(a)$. The results of [12] show the equivalence between sceptically acceptance under preferred semantics and ICR-entailment:

Theorem 1 (Semantics equivalence [12]) *Let $\mathscr{K} = (\mathscr{F}, \mathscr{R}, \mathscr{N})$ be a knowledge base, let $\mathscr{A}\mathscr{F}_{\mathscr{K}}$ the corresponding argumentation framework, Q a query. $\mathscr{K} \vDash_{ICR} Q$ if and only if Q sceptically accepted under preferred semantics.*

4.2 Deepening and Clarifying

In the following we show some functionalities that give the *User* the possibility to manipulate arguments to gain clarity, namely: deepening and clarification.

Deepening aims at showing the reason why an argument attacks another. In our knowledge base the attack is justified by the violation of a constraint. Put differently, an argument attacks another argument if the conclusion of the former and the hypothesis of the latter are mutually exclusive. Thus deepening amounts to explain the attack between two arguments by showing the violated constraint.

Definition 5 (*Deepening* $\mathbb{D}$) Given two arguments $a, b \in \texttt{Arg}$. The mapping *deepening* denoted by $\mathbb{D}$ is a total function from the set $\texttt{Att}$ to $2^{\mathscr{N}}$ defined as follows:
$\mathbb{D}(b, a) = \{nc |$ 1. $nc \in \mathscr{N}$ and,

$\quad\quad\quad$ 2. $\exists f \in \texttt{Supp}(a)$ such that $\texttt{Cl}_{\mathscr{R}}(\{Conc(b), f\}) \models H_{nc}$.
$\}$, where H_{nc} is the hypothesis of the constraint nc.

Example 5 (Deepening) Consider the argument a of Example 3, the argument $b = (\{supervises(Tom, John), works\_in(Tom, d_1)\}, supervises(Tom, John) \wedge works\_in (Tom, d_1))$ attacks a. Deepening is $\mathbb{D}(b, a) = \{\forall x \forall y \forall z \, (supervises(x, y) \wedge work\_in(x, z) \wedge directs(y, z)) \rightarrow \bot\}$ which explains why the argument b attacks a.

The information carried by the argument would be more useful if the structure exhibits the line of reasoning that leads to the conclusion. We call this *clarifying* the argument.

Definition 6 (*Clarifying* $\mathbb{C}$) Given an argument $a \in \texttt{Arg}$. The mapping *clarification* denoted by $\mathbb{C}$ is a total function from the set $\texttt{Arg}$ to $2^{\mathscr{R}}$ such that:
$\mathbb{C}(a = \langle F_0, ..., F_n, C \rangle) = \{r | r \in \mathscr{R}$ such that r is applicable to F_i and the application of r on F_i yields F_{i+1} for all $i \in \{0, ..., n-1\}\}$.

Definition 7 (*Clarified Argument*) Given an argument $a \in \texttt{Arg}$. The corresponding *clarified* argument C_a is a 3-tuple $\langle \texttt{Supp}(a), \mathbb{C}(a), \texttt{Conc}(a) \rangle$ such that $\mathbb{C}(a) \subseteq \mathscr{R}$ are the rules used to derive the conclusion $\texttt{Conc}(a)$.

Example 6 (Clarification count. Example 3) A clarified version of the argument a is $C_a = (\{diretcs(John, d_1)\}, \{\forall x \forall d \, directs(x, d) \rightarrow emp(x)\}, emp(John))\}$ such that $\texttt{Supp}(a) = \{directs(John, d_1)\}$, $\mathbb{C}(a) = \{\forall x \forall d \, directs(x, d) \rightarrow emp(x)\}$ and $\texttt{Conc}(a) = emp(John)$.

Relying on these notions, in the next section, we provide an argumentation-based explanation called *dialectical-explanation* based on Walton's *Dialogical Model of Explanation* [19–21]. In this type of dialogue an exchange of information takes place between an Explainer and Explainee with the purpose of explaining certain assertions.

5 Dialectical Explanation for Query Failure

In what follows we describe a simple dialectical system of explanation based on the work of [19–21]. Our system is custom-tailored for the problem of *Query Failure Explanation* under ICR-semantics in inconsistent knowledge bases with rule-based language. Our *dialectical explanation* involves two parties: the *User* (the explainee) and the *Reasoner* (the explainer). The *User* wants to understand why the query is not ICR-entailed and the *Reasoner* provides a response aiming at showing the reason why the query is not ICR entailed. We model this explanation through a dialogue composed of moves (speech acts) put forward by both the *User* and the *Reasoner*.

Example 7 (Motivating Example Count. Example 1) Consider the following conjunctive query $Q(x) = \forall x\, works\_in(x, Statistics)$ stands for "Get me all the employees in the department of statistics". The system returns as a result the set of answers $A = \{Carlo, Luca, Jane\}$ under ICR-semantics. The *User* expected that the query $Q(Linda) = Works\_in(Linda, Statistics)$ is also ICR-entailed, that means "Linda" is a part of the answers of $Q(x)$. In addition, the *User* knows also a set of arguments that support his/her expectation denoted by $Arg^+(Q(Linda)) = \{\omega\}$ such that $\omega = (\{works\_in(Linda, Statistics)\}, works\_in(Linda, Statistics))$ which stands for "Linda is an employee in the department of statistics since it is specified as a fact". Now the *User* wants to understand why $Q(Linda)$ is not ICR-entailed. A dialectical explanation is represented as follows:

1. *User*: Why not $Q(Linda)$ given that ω?
2. *Reasoner*: Because $(\{Works\_in(Linda, Accounting)\}, Works\_in(Linda, Accounting))$ (is a counterargument denoted by ω_1)
3. *User*: Why $Works\_in(Linda, Accounting)$?
4. *Reasoner*: Because $uses\_office(Linda, o_1)$ and $locate\_in(o_1, Accounting)$ so $works\_in(Linda, Accounting)$
5. *User*: how's that can be a problem?
6. *Reasoner*: The following negative constraint is violated
   $n_3 = \forall x \forall y \forall z\ (works\_in(x, y) \wedge works\_in(x, z)) \rightarrow \bot$.

We denote by $Arg^+(Q)$ the set of all arguments that support the query Q, namely $a \in Arg^+(Q)$ if and only if $\mathrm{Conc}(a) \models Q$.

This dialogue is governed by rules (pre/post conditions, termination and success rules) that specify what type of moves should follow the other, the conditions under which the dialogue terminates, and when and under which conditions the explanation has been successfully achieved (success rules).

In what follows we define types of moves that can be used in the dialogue.

Definition 8 (*Moves*) A move is a 3-tuple $m = \langle ID, I, \omega \rangle$ such that:

1. m is an **explanation request**, denoted by m^{ERQ} if and only if $ID = User$, I is a query Q and ω is an argument supporting Q.
2. m is an **explanation response**, denoted by m^{ERP} if and only if $ID = Reasoner$, I is an argument supporting Q and ω is an argument such that ω attacks I.
3. m is a **follow-up question**, denoted by m^{FQ} if and only if $ID = User$, I is an argument and ω is either $\mathrm{Conc}(I)$ or an argument ω_1 that supports Q such that $(\omega, \omega_1) \in \mathrm{Att}$.
4. m is a **follow-up answer**, denoted by m^{FA} if and only if $ID = Reasoner$, I is an argument and ω is either a deepening $\mathbb{D}$ or a clarified argument $\mathbb{C}(I)$.

The explanation request $m^{\mathrm{ERQ}} = \langle User, Q, \omega \rangle$ is an explanation request made by the *User* asking "*why the query Q is not ICR-entailed while there is an argument ω asserts the entailment of Q*", an explanation response $m^{\mathrm{ERP}} = \langle Reasoner, \omega, \omega_1 \rangle$ made by the *Reasoner* is an explanation for the previous inquiry by showing that the argument ω (that supports Q) is the subject of an attack made by ω_1. The *User* also can ask a follow-up question if the *Reasoner* provides an explanation. The follow-up question $m^{\text{FQ}} = \langle User, \omega_1, \omega \rangle$ is a compound move, it can represent a need for *deepening* (the *User* wants to know why the argument ω_1 is attacking the argument ω) or the need for clarification (how the argument ω_1 comes to a certain conclusion). To distinguish them, the former has $\omega = \text{Conc}(\omega_1)$ and the latter has ω as an argument. A follow-up answer $m^{\text{FA}} = \langle Reasoner, \omega_1, \omega_1' \rangle$ is also a compound move. Actually, it depends on the follow-up question. It shows the argument ω_1 that needs to be deepened (resp. clarified) and its deepening (resp. clarification) by the deepening mapping $\mathbb{D}(\omega_1, \omega)$ (resp. clarification mapping $\mathbb{C}(\omega)$) in Definition 4 (resp. Definition 6).

In what follows we specify the structure of dialectical explanation and the rules that have to be respected throughout the dialogue.

Definition 9 (*Dialectical Explanation*) Given a QFEP $\mathscr{P}$. A dialectical explanation $\mathscr{D}_{exp}$ for $\mathscr{P}$ is a non-empty sequence of moves $\langle m_1^s, m_2^s, ..., m_n^s \rangle$ where $s \in \{\text{ERQ}, \text{FQ}, \text{ERP}, \text{FA}\}$ and $i \in \{1, ..., n\}$ such that:

1. The first move is always an explanation request m_1^{ERQ}, we call it an opening.
2. $s \in \{\text{ERQ}, \text{FQ}\}$ if and only if i is odd, $s \in \{\text{ERP}, \text{FA}\}$ if and only if i is even.
3. For every explanation request $m_i^{\text{ERQ}} = \langle User, I_i, \omega_i \rangle$, I_i is the query Q to be explained and ω_i is an argument supporting Q and for all m_j^{ERQ} such that $j < i$ $\omega_i \neq \omega_j$.
4. For every explanation response $m_i^{\text{ERP}} = \langle Reasoner, I_i, \omega_i \rangle$ such that $i \geq 1$, $I_i = \omega_{i-1}$ and $\omega_i = \omega'$ such that $(\omega', I_i) \in \text{Att}$.
5. For every follow-up question $m_i^{\text{FQ}} = \langle User, I_i, \omega_i \rangle$, $i \geqslant 1$, $I_i = \omega_{i-1}$ and ω is either I_{i-1} or $\text{Conc}(\omega_{i-1})$.
6. For every follow-up answer $m_i^{\text{FA}} = \langle Reasoner, I_i, \omega_i \rangle$, $i \geqslant 1$, $I_i = I_{i-1}$ and $\omega_i = \mathbb{D}(I_i, \omega_{i-1})$ or $\omega = \mathbb{C}(I_i)$.

We denote by $Arg_{user}(\mathscr{D}_{exp})$ the set of all arguments put by the *User* in the dialogue.

As stated, a dialectical explanation is a sequence of moves put forward by either the *User* or the *Reasoner*. In the following example we model the dialectical explanation of Example 7 and we show the encoding of different moves.

Example 8 (Cont. Example 7) The opening in line (1) (first move) is an explanation request $m_1^{\text{ERQ}} = \langle User, Q(Linda), \omega \rangle$, the *Reasoner* responded with an explanation response $m_2^{\text{ERP}} = \langle Reasoner, \omega, \omega_1 \rangle$ then the *User* asked a follow-up question seeking clarification $m_3^{\text{FQ}} = \langle User, \omega_1, \text{Conc}(\omega_1) \rangle$ to know why $\text{Conc}(\omega_1) = $ "Linda is working in department of Accounting". The *Reasoner* responded with a follow-up answer for clarification $m_4^{\text{FA}} = \langle Reasoner, \omega_1, \mathbb{C}(\omega_1) \rangle$ showing a line of reasoning that led to the conclusion $\text{Conc}(\omega_1)$. Another type of follow-up question is presented in line (5) in which the *User* asked a follow-up question for deepening $m_5^{\text{FA}} = \langle Reasoner, \omega_1, \omega \rangle$ wondering how the fact that "working in department of Accounting contradicts that Linda works in Department of statistics". The *Reasoner* responded with a follow-up answer $m_6^{\text{FA}} = \langle Reasoner, \omega_1, \mathbb{D}(\omega_1, \omega) \rangle$ such that $\mathbb{D}(\omega_1, \omega) =$ "An employee cannot work in two different departments at once". The key point of such responses is that it reflects the intuition described in Sect. 2 about ICR-semantics which dictates that if there is a conflicting information about the query Q then it cannot be deduced.

Every dialogue has to respect certain conditions (protocol). Theses conditions organize the way the *Reasoner* and the *User* should put the moves. For each move we specify the conditions that have to be met for the move to be valid (preconditions). We also specify the conditions that identify the next moves (postconditions).

Definition 10 (*Pre/Post Condition Rules*) Given a QFEP $\mathscr{P}$ and a dialectical explanation $\mathscr{D}_{exp}$ for $\mathscr{P}$. Then, $\mathscr{D}_{exp}$ has to respect the following rules:

Explanation request:

- Preconditions: The beginning of the dialogue or the last move of the *Reasoner* was either an explanation response or a follow-up answer.
- Postconditions: The next move must be an explanation answer.

Explanation response:

- Preconditions: The last move by the *User* was an explanation request.
- Postconditions: The next move must be either another explanation request (it may implicitly means that the *User* had not understood the previous explanation) or a follow-up question.

Follow-up question:

- Preconditions: The last move by the *Reasoner* was an explanation response or this follow-up question is not the second in a row.
- Postconditions: The next move must be a follow-up answer.

Follow-up answer:

- Preconditions: The last move by the *User* was a follow-up question.
- Postconditions: The next move must be an explanation request (it may implicitly means that the *User* had not understood the previous explanation).

Beside the previous rules, there are termination rules that indicate the end of a dialectical explanation.

Definition 11 (*Termination Rules*) Given a QFEP $\mathscr{P}$ and a dialectical explanation $\mathscr{D}_{exp}$ for $\mathscr{P}$. Then $\mathscr{D}_{exp}$ terminates when the *User* puts an empty explanation request $m^{\mathrm{ERQ}} = \langle User, \emptyset, \emptyset \rangle$ or when $Arg_{User}(\mathscr{D}_{exp}) = Arg^+(Q)$.

The rules in Definitions 9, 10 and 11 state that the *Reasoner* is always **committed** to respond with an explanation response, the *User* then may indicate the end of the dialogue by an *empty explanation request* (Definition 11) declaring his/her understanding, otherwise starts another explanation request (this indicates that he/she has not understood the last explanation) or asks a follow-up question, the *User* cannot ask more than two successive follow-up questions. If the *User* asks a follow-up question then the *Reasoner* is **committed** to a follow-up answer. When the *User* asks for another explanation he/she cannot use an argument that has already been used. If the *User* ran out of arguments and he/she has not yet understood, the dialogue ends (Definition 11) and the explanation is judged unsuccessful. It is important to notice that when the *Reasoner* wants to answer the *User* there may be more than one argument to choose. There are many "selection strategies" that can be used in such case (for instance, the shortest argument, the least attacked argument, etc.), but their study is beyond the scope of the paper.

In what follows we elaborate more on the success and the failure rules.

Definition 12 (*Success Rules*) Given a QFEP $\mathscr{P}$ and a dialectical explanation $\mathscr{D}_{exp}$ for $\mathscr{P}$. Then $\mathscr{D}_{exp}$ is successful when it terminates with an empty explanation request $m_i^{\text{ERQ}} = \langle User, \emptyset, \emptyset \rangle$, otherwise it is unsuccessful.

A dialectical explanation is judged to be successful if the *User* terminates the dialogue voluntarily by putting an empty explanation request. If the *User* has used all arguments supporting Q then he/she is forced to stop without indicating his/her understanding, in this case we consider the explanation unsuccessful.

By virtue of the equivalence between ICR-semantics and argumentation presented in Sect. 3, the existence of response is always guaranteed. This property is depicted in the following proposition.

Proposition 1 (Existence of response) *Given a QFEP $\mathscr{P}$ and a dialectical explanation $\mathscr{D}_{exp}$ for $\mathscr{P}$. Then for every $m_i^s \in \mathscr{D}_{exp}$ such that $s \in \{\text{ERQ}, \text{FQ}\}$ and $1 \leq i \leq |\mathscr{D}_{exp}|$, the next move m_{i+1}^s such that $s \in \{\text{ERP}, \text{FA}\}$ always exists.*

Proof For the move $m_i^{\text{ERQ}} = \langle User, Q, \omega \rangle$ the query Q is not ICR-entailed therefore the argument ω that supports Q is not sceptically accepted, hence there is always an argument ω' such that $(\omega', \omega) \in \text{Att}$. Thus we can construct the following explanation response move: $m_{i+1}^{\text{ERP}} = \langle Reasoner, \omega, \omega' \rangle$ such that $s \in \{\text{ERP}\}$.

For the move m_i^{FA} the proof is immediate since the mappings *deepening* $\mathbb{D}$ and *clarification* $\mathbb{C}$ are total functions.

Another issue is the finiteness of the dialectical explanation. It is not hard to conclude that a dialectical explanation is finite (from success and termination rules).

Proposition 2 (Finiteness) *Given a QFEP $\mathscr{P}$ and a dialectical explanation $\mathscr{D}_{exp}$ for $\mathscr{P}$. $\mathscr{D}_{exp}$ is finite.*

6 Conclusion

In this paper we have presented a dialogical approach for explaining boolean conjunctive queries failure, designated by Query Failure Explanation Problem (QFEP), in an inconsistent ontological knowledge base where inconsistency is handled by inconsistency-tolerant semantics (ICR) and issued from the set of facts. The introduced approach relies on both (i) the relation between ontological knowledge base and logical argumentation framework and (ii) the notions of argument deepening and clarifications. So, through a dialogue, the proposed approach explains to the *User* how and why his/her query is not entailed under ICR semantics.

For future work, we aim at generalizing the explanation to general conjunctive queries and studying the proposed explanation in the context of other inconsistency-tolerant semantics.

Acknowledgments The research leading to these results has received funding from the European Community's Seventh Framework Programme (FP7/2007–2013) under the grant agreement n°FP7-265669-EcoBioCAP project.

References

1. Arioua, A., Tamani, N., Croitoru, M.: Query failure explanation in inconsistent knowledge bases an argumentation approach: extended abstract. In: 5th International Conference on Computational Models of Argument 2014, to appear, (2014)
2. Baader, F., Brandt, S., Lutz, C.: Pushing the el envelope. In: Proceedings of IJCAI 2005, (2005)
3. Baget, J.-F., Mugnier, M.-L., Rudolph, S., Thomazo, M.: Walking the complexity lines for generalized guarded existential rules. In: Proceedings of IJCAI'11, pp. 712–717 (2011)
4. Bienvenu, M.: On the complexity of consistent query answering in the presence of simple ontologies. In: Proceedings of AAAI (2012)
5. Bienvenu, M., Rosati, R.: Tractable approximations of consistent query answering for robust ontology-based data access. In: Proceedings of IJCAI'13, pp. 775–781. AAAI Press (2013)
6. Borgida, A., Calvanese, D., Rodriguez-Muro, M.: Explanation in the dl-lite family of description logics. In: Meersman, R., Tari, Z. (eds.) On the Move to Meaningful Internet Systems: OTM 2008. LNCS, vol. 5332, pp. 1440–1457. Springer, Berlin (2008)
7. Calì, A., Gottlob, G., Lukasiewicz, T.: Datalog+/−: a unified approach to ontologies and integrity constraints. In: Proceedings of the 12th International Conference on Database Theory, pp. 14–30. ACM (2009)
8. Calì, A., Gottlob, G., Lukasiewicz, T.: A general datalog-based framework for tractable query answering over ontologies. Web Semant. Sci. Serv. Agents World Wide Web **14**, 57–83 (2012)
9. Calvanese, D., De Giacomo, G., Lembo, D., Lenzerini, M., Rosati, R.: Tractable reasoning and efficient query answering in description logics: the dl-lite family. J. Autom. Reasoning **39**(3), 385–429 (2007)
10. Calvanese, D., Ortiz, M., Šimkus, M., Stefanoni, G.: Reasoning about explanations for negative query answers in dl-lite. J. Artif. Intell. Res. **48**, 635–669 (2013)
11. Caminada, M., Amgoud, L.: On the evaluation of argumentation formalisms. Artif. Intell. **171**(5), 286–310 (2007)
12. Croitoru, M., Vesic, S.: What can argumentation do for inconsistent ontology query answering? In: Liu, W., Subrahmanian, V., Wijsen, J. (eds.) Scalable Uncertainty Management, vol. 8078 of LNCS, pp. 15–29. Springer, Berlin (2013)

13. Dung, P.M.: On the acceptability of arguments and its fundamental role in nonmonotonic reasoning, logic programming and n-person games. Artif. Intell. **77**(2), 321–357 (1995)
14. Garcia, A.J., Chesñevar, C.I., Rotstein, N.D., Simari, G.R.: Formalizing dialectical explanation support for argument-based reasoning in knowledge-based systems. Expert Syst. Appl. **40**(8), 3233–3247 (2013)
15. Lembo, D., Lenzerini, M., Rosati, R., Ruzzi, M., Savo, D.F.: Inconsistency-tolerant semantics for description logics. In: Proceedings of the Fourth International Conference on Web Reasoning and Rule Systems, RR'10, pp. 103–117. Springer, Berlin (2010)
16. Lukasiewicz, T., Martinez, M.V., Simari, G.I.: Complexity of inconsistency-tolerant query answering in datalog+/-. In: Meersman, R., Panetto, H., Dillon, T., Eder, J., Bellahsene, Z., Ritter, N., Leenheer, P.D., Dou, D. (eds.) In: Proceedings of the 12th International Conference on Ontologies, Databases, and Applications of Semantics, 10–11 Sept, 2013, vol. 8185 of LNCS, pp. 488–500. Springer, (2013)
17. McGuinness, D.L., Patel-Schneider, P.F.: Usability issues in knowledge representation systems. In: Proceedings of AAAI-98, pp. 608–614 (1998)
18. Poggi, A., Lembo, D., Calvanese, D., De Giacomo, G., Lenzerini, M., Rosati, R.:. Linking data to ontologies. In: Journal on Data Semantics X, pp. 133–173. Springer, Berlin (2008)
19. Roth-Berghofer, T., Schulz, S., Bahls, D., Leake, D.B. (eds.) Explanation-Aware Computing, Papers from the 2007 AAAI Workshop, Vancouver, British Columbia, Canada, 22–23 July, 2007, vol. WS-07-06 of AAAI Technical Report. AAAI Press (2007)
20. Walton, D.: A new dialectical theory of explanation. Philos. Explor. **7**(1), 71–89 (2004)
21. Walton, D.: A dialogue system specification for explanation. Synthese **182**(3), 349–374 (2011)

Benchmarking Grammar-Based Genetic Programming Algorithms

Christopher J. Headleand, Llyr Ap Cenydd and William J. Teahan

Abstract The publication of Grammatical Evolution (GE) led to the development of numerous variants of this Grammar-Based approach to Genetic Programming (GP). In order for these variants to be compared, the community requires a rigorous means for benchmarking the algorithms. However, GP as a field is not known for the quality of its benchmarking, with many identified problems, making direct comparisons either difficult or impossible in many cases. Aside from there not being a single, agreed-upon, benchmarking test, the tests currently utilised have shown a lack of standardisation. We motivate the following research by identifying some key issues with current benchmarking approaches. We then propose a standardised set of metrics for future benchmarking and demonstrate the use of these metrics by running a comparison of three Grammar-Based Genetic Programming methods. We conclude the work by discussing the results and proposing directions for future benchmarking.

1 Motivation

The benchmarking of Genetic Programming (GP) algorithms is a heavily debated topic. While there are standard problems which are typically used for comparison, the community has often discussed whether these challenges provide useful data.

One of the principle concerns raised is over a lack of standardisation which makes comparison across papers difficult or impossible [16]. Multiple papers describe different set-up conditions, and there has also been a lack of standard measures to evaluate the algorithms against.

With the development of Grammatical Evolution [9], its benchmarking raised additional concerns (discussed in Sect. 2.2) as the use of a grammar adds further

C.J. Headleand (✉) · L.A. Cenydd · W.J. Teahan
Bangor University, Bangor, Wales
e-mail: c.headleand@bangor.ac.uk

L.A. Cenydd
e-mail: eese60d@bangor.ac.uk

W.J. Teahan
e-mail: w.j.teahan@bangor.ac.uk

© Springer International Publishing Switzerland 2014
M. Bramer and M. Petridis (eds.), *Research and Development in Intelligent Systems XXXI*, DOI 10.1007/978-3-319-12069-0_9

complications to the benchmarking process. As there are now several variants of this class of algorithm [3–5, 10, 13], the area is on the verge of being considered a discrete sub-field of GP which motivates us to consider some of the concerns with current benchmarking practice.

This research will explore the following objectives:

1. Discuss concerns identified in the literature along with the current practice of the benchmarking of Genetic Programming, focussing specifically on Grammar-Based variants.
2. Propose a standardised set of measures towards a new benchmarking strategy for future developments.
3. Run a benchmarking exercise of three comparative Grammar-Based algorithms using the proposed measures.

2 GP Benchmarking Concerns

2.1 Better Benchmarks Survey

At the GECCO conference in 2012 a position paper addressed the need to re-evaluate current benchmarking strategies [8]. The authors argued that this could be achieved through community feedback, an argument that led to a community survey later that year. We have selected some of the issues from that survey to highlight some of the GP communities feelings in regards to current benchmarking practices.

79 members of the GP community responded to the survey [16] with over 80 % identifying that they run experiments to measure algorithm performance, and around 70 % of practitioners responding that they performed well known GP benchmarks in their work. Furthermore, 94 % of the GP community who responded to the questionnaire answered yes when asked whether they believed that the current experiment regime had disadvantages. Of that group, 85 % responded that the lack of standardisation made it difficult/impossible to compare results across papers. Finally, 84 % of the community members who answered the survey were in favour of the creation of a standardised GP benchmark suite.

While the authors were discouraged from pursuing the development of a new benchmarking suite, as an outcome of this work the authors propose a blacklist of problems to be avoided, and claim broad support for this blacklist from the community. This is based on the free-text response to question 23, "Can you suggest any existing benchmark problems you think should not be part of a benchmark suite?". However, a close look at this dataset (available online [15]) indicates that there was very little consensus within the responses, with the most agreed upon response only receiving 6 nominations. This is a distinctly small number to claim "broad support" for blacklisting, and calls into question the validity of the proposed 'blacklist'.

2.2 Grammar Search Space

There have been concerns over how various problems have been implemented in benchmarking exercises, preventing a fair comparison between algorithms. This has been proven to be the case in regards to the Santa Fe Trail, where studies have demonstrated key errors in the literature. In [12] the authors note that for the Santa Fe Trail for be used as a benchmarking test "the search space of programs must be semantically equivalent to the set of programs possible within the original Santa Fe Trail definition". The argument is presented that GE has not been fairly benchmarked against GP in the Santa Fe Trail problem as the grammar used is not equivalent to the original search space.

This was evaluated by Georgiou [2] who tested GE against both the grammar described in its original benchmarking paper [9] (BNF-Oneil) and a grammar based on Koza's original search space (BNF-Koza). The study demonstrated significant differences between the two visually similar grammars. Notably, while BNF-Oneil was more successful at finding a solution, the solutions generated were typically less efficient than the solutions generated using BNF-Koza. This supports the claim that the search spaces used were not semantically equivalent, and thus, direct comparisons cannot be made.

In addition [12], it has been noted that the documented limit on the number of steps allowed for an Artificial Ant to complete the trail has also been subjected to some error. Koza's original GP paper fixed the limit to 400 steps, which was assumed to be a mistake in later work [7] where the maximum limit was increased to 600 steps. A single GE publication [9] claims to have used the 600 step limit, which is assumed to be a mistype, as future publications discussed a 800 step limit. These claims of mistaken reporting are confirmed in [2] where, through simulations, it was proven that the best GP solution identified by Koza required 545 steps (exceeding the 400 steps quoted), and the best GE solution [9] required 615 steps (exceeding the 600 steps quoted).

Despite this, in the survey discussed in the previous section, the majority of participants voted that certain settings should be open to change during benchmarking which would include the number of steps. Yet it is an important consideration, as GP variants may not be comparable with identical settings; for example, a swarm based method may better suit a larger population than an evolution based approach.

3 Experimental Design

In the background to this work, we discussed some concerns over Grammar-Based Genetic Programming benchmarking. One of the main areas we identified was the lack of standardisation. In this work, we seek to address this concern by recommending a suite of five comparison metrics. It is worth noting that while we are focused on Grammar Based methods, these concerns have primarily come from the general GP field.

The metrics have been selected due to both there existence in the literature. For this study, three algorithms have been selected for comparison, Grammatical Evolution, Grammatical Herding and Seeded Grammatical Evolution.

3.1 Grammatical Evolution

Grammatical Evolution (GE) is a genetic algorithm based automatic programming method where candidate solutions are evolved over repeated generations. The system was designed in part to avoid some of the complications present in genetic programming (an alternate program synthesis method). By manipulating a bit string before mapping via usually a context free grammar, the candidate solution can be evolved without the need to directly manipulate code (see Fig. 1).

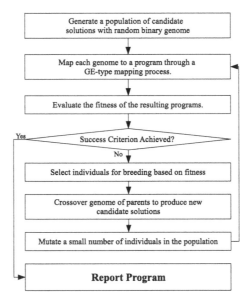

Fig. 1 An algorithmic representation of the Grammatical Evolution algorithm. A population of genomes consisting of a fixed length list of binary codons (typically 8 bits) is randomly generated. Each genome is then mapped via a grammar to produce a program. These programs are then evaluated and provided with a fitness score as with a typical genetic algorithm. Crossover and mutation are applied directly to the binary genomes (the genotype) not to the generated programs (phenotype)

3.2 Grammatical Herding

Grammatical Herding (GH) is a swarm-based automatic programming algorithm, which takes its inspiration from Particle Swarm Optimisation (PSO) and the grazing behaviour of horses; for a detailed explanation, see [4]. Each agent in GH is defined by a bit-string, which represents a location in the search space defined by a fixed number of 8-bit coordinates, equivalent to the genome of codons in Grammatical evolution. The agents move around the search space by selecting a position between themselves and a randomly selected individual with a high fitness. The position is weighted by the respective fitnesses of the agents (Fig. 2).

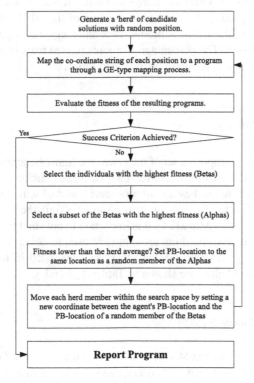

Fig. 2 Visual representation of the Grammatical herding algorithm. A herd of candidate solutions are generated with random initial positions. These co-ordinates within the search space are mapped through a GE-type mapping process to produce a program which is evaluated. The fittest agents are set as Betas, and the fittest members of the Betas as set as Alphas. If the agents fitness is lower than the herd average their position is set to that of one of the Alphas. The agents are then moved within the search space by moving between a position the agent has already visited, and the the best location found by a random Beta

3.3 Seeded Grammatical Evolution

Seeded Grammatical Evolution (sGE) is a hybrid algorithm which uses GH to pre-seed the population of GE. The purpose of the approach is to attempt to capitalise on the qualities of both algorithms, GH's ability to quickly evaluate a search space and the optimisation quality of GE [5]. In sGE, the population is transferred from one algorithm to the other at a point in the solution loop called 'handover' when a specific criterion was achieved. One criterion was a fitness threshold, a user defined fitness level (for problems with a known fitness), which, if achieved by an agent in the population, the handover event would be activated. Additionally an iteration threshold was included to provide a possible escape, if the seeding algorithm (GH) was unable to reach the desired fitness threshold. By including an iteration threshold, it ensured that if GH was unable to find a solution of a high enough fitness, the problem would be transferred to GE which may be more successful. It was also discovered that the best solutions were produced when only a portion of the GH herd was transferred through handover to GE, with additional, randomly generated genomes added to complete the population. The percentage of agents passed from one algorithm to the other during handover is referred to as the 'seed ratio'.

3.4 Selected Problems

In this section, we will propose a set of measures for future benchmarking exercises. For the purpose of discussion, we have selected three benchmark tests which we believe to be comparable, as they are all problems that deal with agents with similar sensory-motor capabilities. These tests are the Santa Fe Trail, the Los Altos Hills Trail [6] and the Hampton Court Maze [14]. It is important to note that these challenges are not proposed as ideal benchmarking tests, but are used to demonstrate the proposed measures. The settings for GE, GH and sGE we have used in the benchmarking experiments described below are shown in Tables 1, 2 and 3.

Table 1 Settings used for the Santa Fe Trail

Algorithm	Settings
GE	Population = 1000; Max Generation = 50; Mutation Probability = 0.01; min Genome Length = 15 codons; max Genome Length = 30 codons; Wrap Limit = 10;
GH	Population = 1000; Max Generation = 50; Mutation = Alpha Wandering; Genome Length = 30 codons; numBetas = 25; numAlphas = 5; Wrap Limit = 10;
sGE	GH Settings = GH; GE Settings = GE; Fitness Threshold = 30; Iteration Threshold = 10; Seed Ratio = 25 %

Table 2 Settings used for the Los Altos Hills Trail

Algorithm	Settings
GE	Population = 2000; Max Generation = 50; Mutation Probability = 0.01; min Genome Length = 50 codons; max Genome Length = 100 codons; Wrap Limit = 10;
GH	Population = 2000; Max Generation = 50; Mutation = Alpha Wandering; Genome Length = 100 codons; numBetas = 35; numAlphas = 10; Wrap Limit = 10;
sGE	GH Settings = GH; GE Settings = GE; Fitness Threshold = 30; Iteration Threshold = 10; Seed Ratio = 25 %

Table 3 Settings used for the Hampton Court Maze

Algorithm	Settings
GE	Population = 100; Max Generation = 25; Mutation Probability = 0.01; min Genome Length = 15 codons; max Genome Length = 25 codons; Wrap Limit = 10;
GH	Population = 100; Max Generation = 25; Mutation = Alpha Wandering; Genome Length = 25 codons; numBetas = 10; numAlphas = 2; Wrap Limit = 10;
sGE	GH Settings = GH; GE Settings = GE; Fitness Threshold = 30; Iteration Threshold = 10; Seed Ratio = 25 %

3.4.1 Santa Fe Trail

The Santa Fe Trail (SFT), also called the artificial ant problem, is a common benchmark. The problem is to generate instructions for an artificial ant so that it can successfully traverse along a non-uniform path through a 32 by 32 toroidal grid (see Fig. 3). The generated program can use three movement operations, move, which moves the ant forward a single square, turn-right and turn-left which turn the ant 90 degrees to face a new direction, with each of these operations taking one time step. A sensing function ifelse food-ahead looks to the square directly ahead of the ant and executes one of two arguments based on whether the square contains food or not. These functions and operations are represented in the grammar detailed in Listings 1. The fitness is calculated by counting the amount of food eaten by the agent before reaching the maximum number of time steps.

Listing. 1 Santa Fe Trail grammar

```
<expr>::=<line>|<expr><line>
<line>::=ifelse food-ahead[<expr>][<expr>]|<op>
<op>::=turn-left|turn-right|move
```

While there is debate on the quality and usefulness of the SFT benchmark [12, 16], there are also arguments in its favour. Firstly, it is one of the most commonly used benchmarks, providing historical results which span the entire history of Genetic

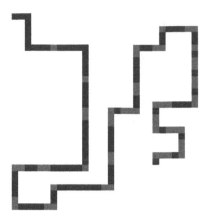

Fig. 3 The Santa Fe Trail benchmark environment, a 32 by 32 toroidal grid (a grid that wraps both *vertically* and *horizontally*). The *dark grey* patches represent the 89 pieces of food along the trail, the *light grey patches* represent the gaps in the trail. The fitness is calculated by counting the amount of food eaten along the trail. Importantly food does not regenerate, so the food at each *grey square* can only be eaten once

Programming. Secondly, whilst the trail could be considered a "toy problem", new algorithms have continued to produce improved solutions to the problem, demonstrating that there is still room for continued research. A final argument is that despite the perceived simplicity of the problem, GP algorithms have failed to solve the benchmark reliably, for example, GE is only capable of an average success rate of 10 % [2].

3.4.2 Los Altos Hills Trail

The Los Altos Hills (LAH) benchmark is a more difficult variation of the Artificial ant problem introduced by Koza [6]. The objective is similar to the SFT, to generate a computer program capable of navigating an artificial ant to find all 157 pieces of food along a trail located in a 100×100 toroidal grid. The ant uses the same operations as with the Santa Fe Trail problem (see Listing 1). The LAH trail begins with the same irregularities, in the same order as the SFT. Then it introduces two new kinds of irregularity that appear toward the end of the trail. As with the SFT, the fitness is calculated based on the amount of food eaten before the maximum step limit is reached (Fig. 4).

3.4.3 Hampton Court Maze

The Hampton Court Maze (HCM) problem, (first described in [14] and later proposed as a benchmark in [2]) is a simple, connected maze, designed on a 39 by 23 grid (see Fig. 5). The objective of the problem is to generate a program capable of guiding

Fig. 4 The Los Altos Hills Trail benchmark environment, a 100 by 100 toroidal grid. The *dark grey patches* represent the 157 pieces of food along the trail, the *light grey* represent the gaps in the trail

Fig. 5 The Hampton Court Maze benchmark environment, a 39 by 23 grid with walls. The *dark grey patches* represent the walls of the maze which are impassable by the agent. The *light grey squares* represent the entrance (*bottom* of the maze) and the goal (*centre* of the maze)

an agent from the entry square (light grey square at the bottom of the maze) to the centre (second light grey square identifies the entrance).

The grammar used is based on the SFT grammar (see Listing 2), with a key variation, the single sensing condition food-ahead is replaced with three possible sensing conditions. These conditions are wall-ahead?, wall-left? and wall-right?, which return true if a wall is detected in the sensed direction.

The fitness is calculated as a value between 0 and 1 using the following function.

$$F = 1/(1 + (D/S))$$

where F is the fitness of the agent, D is the distance of the agent to the centre of the maze, and V is the number steps in the path traversed.

Listing. 2 Hampton Court grammar

```
<expr>::=<line>|<expr><line>
<line>::=ifelse<condition>[<expr>][<expr>]|<op>
<condition>::=wall-ahead?|wall-left?|wall-right?
<op>::=turn-left|turn-right|move
```

4 Proposed Standardised Measures

In [1], four main categories of problem are discussed, notably, Black Art Problems, Programming the Unprogrammable (PTU), Commercially Useful New Inventions (CUNI) and Optimal Control. Different problems may also suit algorithms with different qualities which may not be considered "state of the art" by current measures.

We propose a standardised set of five metrics discussed over the following sections, and demonstrate the insights they provide by running a benchmark comparison of the three algorithms discussed in Sect. 3. An experiment of 10,000 runs were completed for each of the three algorithms against each of the three benchmark tests discussed in the previous section.

4.1 Metric 1: Success Rate

This measure is a percentage, which represents how often the algorithm is capable of producing a full fitness result (on problems with a known full fitness). This metric is one of the most commonly used in the literature as it provides a good measure of the reliability of the algorithm. For problems with an unknown 'full fitness', we propose that a threshold level should be used, with all experiments capable of achieving or exceeding that threshold being considered successful.

The data in Table 4 shows us that, using the settings defined in the previous section, GE achieves the strongest results on the SFT, though it under preforms on the HCM. Similarly, while sGE produced the weakest results on the SFT, it out performs GE and GH on the HCM. This result shows that multiple problems should be considered when benchmarking as certain algorithms may be better fitted for different domains. Also, none of the algorithms were capable of producing a full-fitness solution to the LAT.

4.2 Metric 2: Average Fitness

As demonstrated with our first metric, typically a full fitness solution (if full fitness is known) may only be found a relatively small amount of the time, even on problems that are perceived to be simple such as the Santa Fe Trail. However, there are examples

Table 4 Metric 1: success rate for GE, GH and sGE

	GH (%)	GE (%)	sGE (%)
Santa Fe Trail	6.21	32.95	3.25
Los Altos Hills Trail	0	0	0
Hampton Court Maze	4.87	2.41	8.94

Table 5 Metric 2: average fitness results for GE, GH and sGE

	GH	GE	sGE
Santa Fe Trail	54.50	74.98	59.51
Los Altos Hills Trail	42.16	46.75	62.44
Hampton Court Maze	99.23	98.18	99.25

where a full fitness solution may not be required, and in these cases an understanding of the typical "average fitness" may provide a better insight.

The average fitness results (Table 5) demonstrate that on the SFT, sGE is capable of producing higher fitness results than GH, though GE is the most reliable overall. In addition, the results identify one possible flaw with the fitness function currently defined for the HCM, notably that it biases towards large values, with little fidelity in the upper fitness values, possibly explaining the poor performance in Metric 1.

4.3 Metric 3: Solution Development

The fourth proposed metric examines how quickly an algorithm develops its solutions (as opposed to Metrics 1 and 2 which gauge overall success) by exploring how many generations/iterations it takes, on average, to achieve 25, 50, 75 and 100 % fitness. However, this again relies upon knowing a full fitness value for the problem, so, as with Metric 1, we propose that a threshold system is used for problems where the full fitness is unknown. By examining this data, we can compare the number of evaluations required to produce high fitness solutions in order to gauge the relative "responsiveness" of the algorithm to the problem.

Comparing the algorithms using the metric (see Table 6 and Fig. 6), we see that GH produces high fitness solutions in less generations than sGE or GE . This trend can be seen across all three problems, although the small number of generations in the HCM makes this hard to observe. GE typically takes longer to develop solutions than GH (more than double the number of generations to reach 25 and 50 % fitness) though it discovers full fitness solutions in a similar number of generations. In comparison, sGE is quicker than GE but less so than GH on the SFT and HCM, though it is the best performing on the LAH.

Table 6 Metric 3: solution development results for GE, GH and sGE

	Santa Fe Trail			Los Altos Hills Trail			Hampton Court Maze		
	GH	GE	sGE	GH	GE	sGE	GH	GE	sGE
25 % of full fitness	4	13	7	7	12	7	1	2	1
50 % of full fitness	12	31	21	–	–	47	1	3	2
75 % of full fitness	26	35	31	–	–	–	2	3	4
100 % of full fitness	45	46	42	–	–	–	3	5	5

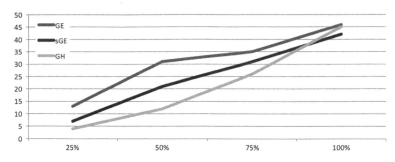

Fig. 6 The average number of generations each of the three algorithms require to achieve increasing levels of fitness for the Santa Fe Trail

4.4 Metric 4: Best Solution (Highest Quality)

The final proposed metric is one of the most often used in the literature, where the best solution from all simulation runs is reported. This metric demonstrates the quality of solutions that an algorithm is capable of producing against a specific benchmark. For the problems we have evaluated, the highest quality solution is the one capable of achieving full fitness with the least number of steps.

From the results generated by this metric (Table 7), we can observe that whilst GE produced the best solution for the SFT, GH and sGE produced better results in the HCM (the LAH not being solved by any of the algorithms). However, better results have been reported in the literature, for example, our sGE paper describes a solution capable of completing the SFT in 303 steps. Clearly, focusing too closely just on the best solution leads to the danger that other qualities of an algorithm are overlooked.

4.5 Metric 5: Average Quality of Successful Runs

This metric concerns the average 'quality' of successful solutions produced by the benchmarked algorithm, which may not be applicable for all benchmarking problems. For artificial ant problems, this value is simply the average number of steps taken for the ant to complete the problem.

Table 7 Metric 4: best solution results for GE, GH and sGE

	GH	GE	sGE
Santa Fe Trail	405	347	386
Los Altos Hills Trail	–	–	–
Hampton Court Maze	602	690	602

Table 8 Metric 5: average quality of successful runs results for GE, GH and sGE

	GH	GE	sGE
Santa Fe Trail	607.16	553.72	642.93
Los Altos Hills Trail	–	–	–
Hampton Court Maze	682.87	883.42	672.46

The analysis of Metric 5 (see Table 8) further supports the conclusions gained from Metric 1. Notably, that whilst GE achieves superior results on the SFT, it underperforms on the HCM in comparison to GH and sGE.

5 Conclusions and Next Steps

Three principle objectives have been addressed in this research: firstly, to identify some of the concerns which have been raised in regards to the benchmarking of three Grammar-Based Genetic Programming algorithms (Grammatical Evolution, Grammatical Herding and Seeded Grammatical Evolution); secondly, to propose a standardised set of metrics for future benchmarking to address one of the current concerns; and finally, to run a benchmarking exercise to demonstrate the insights that can be gained through these measures. An additional contribution has been the provision of further benchmarking data in order to compare the three algorithms.

The work has demonstrated that a standardised set of comparison metrics can aid in comparison between algorithms. Also, by including a more diverse set of metrics than those currently used, algorithms can be better matched to specific problems by understanding their unique qualities. For example, while sGE typically produced the poorest results in success rates and average quality, it was the only algorithm to achieve over 50 % fitness in the Los Altos Hills.

Throughout this work, we have referred to the 'better benchmarks' survey [16] and the subsequent analysis. In this work, 84 % of GP community members who answered the survey were in favour of the creation of a standardised GP benchmark suite, mainly because it would simplify comparisons between techniques. Additionally, devising high quality benchmarks has been identified as an important issue for the foreseeable future of the GP development [11]. However, the authors of the survey were discouraged from this course of action, instead, the research proposed a blacklist, which included the Artificial Ant Problem. However, this research has demonstrated even the simple version of this challenge (the SFT) is not solved reliably, and the more complex version (LAH) was significantly more difficult, as none of the evaluated algorithms were capable of producing a full-fitness solution.

During the benchmarking process, we did a large number of runs (a total of 90,000 experiments). This is significantly larger than the typical number of experiments reported in benchmarking studies (typically between 100 and 500 are usually run).

We propose that a much larger sample should be run in future studies as by observing the averages we can avoid inconsistent results. For example, in the study describing sGE, we reported much stronger results based on this work we now believe that these were statistical anomalies, a risk of any stochastic search algorithms.

The next phase of this work will be to engage with the genetic programming community in the development of a standardised benchmarking suite. We propose that this should take the form of an open-source software package to be maintained by the GP community.

Acknowledgments The authors would like to thank HPC Wales for providing their facilities and technical support during the running of the experiments described in this research. Chris Headleand would also like to thank Fujitsu for their ongoing financial support.

References

1. Behera, R., Pati, B.B., Panigrahi, B.P., Misra, S.: An application of genetic programming for power system planning and operation. Int. J. Control Syst. Instrum. **3**(2) (2012)
2. Georgiou, L: Constituent grammatical evolution. PhD thesis, University of Wales, Bangor (2004)
3. Georgiou, L., Teahan, W.J.: Constituent grammatical evolution. In: Proceedings of the Twenty-Second International Joint Conference on Artificial Intelligence-Volume, vol. 2, pp. 1261–1268. AAAI Press (2011)
4. Headleand, C., Teahan, W.J.: Grammatical herding. J. Comput. Sci. Syst. Biol. **6**, 043–047 (2013)
5. Headleand, C., Teahan, W.J.: Swarm based population seeding of grammatical evolution. J. Comput. Sci. Syst. Biol. **6**, 132–135 (2013)
6. Koza, J.R.: Genetic Programming: On the Programming of Computers by Means of Natural Selection. MIT press, Cambridge (1992)
7. Langdon, W.B, Poli, R.: Better trained ants for genetic programming (1998)
8. McDermott, J., White, D.R., Luke, S., Manzoni, L., Castelli, M., Vanneschi, L., Jaskowski, W., Krawiec, K., Harper, R., De Jong, K., et al.: Genetic programming needs better benchmarks. In: Proceedings of the Fourteenth International Conference on Genetic and Evolutionary Computation Conference, pp. 791–798. ACM (2012)
9. O'Neil, M., Ryan, C.: Grammatical evolution. In: Grammatical Evolution, pp. 33–47. Springer, Berlin (2003)
10. O'Neill, M., Brabazon, A.: Grammatical swarm: the generation of programs by social programming. Nat. Comput. **5**(4), 443–462 (2006)
11. O'Neill, M., Vanneschi, L., Gustafson, S., Banzhaf, W.: Open issues in genetic programming. Genet. Program Evolvable Mach. **11**(3–4), 339–363 (2010)
12. Robilliard, D., Mahler, S., Verhaghe, D., Fonlupt, C.: Santa fe trail hazards. In: Artificial Evolution, pp. 1–12. Springer, Berlin (2006)
13. Si, T., De, A., Bhattacharjee A.K.: Grammatical bee colony. In: Swarm, Evolutionary, and Memetic Computing, pp. 436–445. Springer, Berlin (2013)
14. Teahan, W.J.: Artificial Intelligence-Agent Behaviour. BookBoon, London (2010)
15. White, D.R., McDermott, J., Castelli, M., Manzoni, L., Goldman, B.W., Kronberger, G., Jaśkowski, W., OReilly, U., Luke, S.: Free text responses. Online, Sept 2012. http://gpbenchmarks.org/wp-content/uploads/2012/09/SurveyFreeTextResponses.txt
16. White, D.R., McDermott, J., Castelli, M., Manzoni, L., Goldman, B.W., Kronberger, G., Jaśkowski, W., OReilly, U., Luke, S.: Better gp benchmarks: community survey results and proposals. Genet. Program. Evolvable Mach. **14**(1), 3–29 (2013)

The Effects of Bounding Rationality on the Performance and Learning of CHREST Agents in Tileworld

Martyn Lloyd-Kelly, Peter C.R. Lane and Fernand Gobet

Abstract Learning in complex and complicated domains is fundamental to performing suitable and timely actions within them. The ability of chess masters to learn and recall huge numbers of board configurations to produce near-optimal actions provides evidence that *chunking* mechanisms are likely to underpin human learning. Cognitive theories based on chunking argue in favour for the notion of bounded rationality since relatively small chunks of information are learnt in comparison to the total information present in the environment. CHREST, a computational architecture that implements chunking theory, has previously been used to investigate learning in deterministic environments such as chess, where future states are solely dependent upon the actions of agents. In this paper, the CHREST architecture is implemented in agents situated in "Tileworld", a stochastic environment whose future state depends on both the actions of agents and factors intrinsic to the environment which agents have no control over. The effects of bounding agents' visual input on learning and performance in various scenarios where the complexity of Tileworld is altered is analysed using computer simulations. Our results show that interactions between independent variables are complex and have important implications for agents situated in stochastic environments where a balance must be struck between learning and performance.

M. Lloyd-Kelly (✉) · F. Gobet
Department of Psychological Sciences, University of Liverpool, Liverpool L69 3BX, UK
e-mail: m.lloyd-kelly@liverpool.ac.uk

F. Gobet
e-mail: fernand.gobet@liverpool.ac.uk

P.C.R. Lane
School of Computer Science, University of Hertfordshire, College Lane,
Hatfield L69 3BX, UK
e-mail: peter.lane@bcs.org.uk

© Springer International Publishing Switzerland 2014   149
M. Bramer and M. Petridis (eds.), *Research and Development
in Intelligent Systems XXXI*, DOI 10.1007/978-3-319-12069-0_10

1 Introduction

Improving performance in a particular domain is potentially a problem of leviathan proportions depending upon the complexity of the domain in question and the cognitive limitations of an agent. Expert behaviour in complex domains such as chess has been studied extensively and of particular relevance to this discussion is that the difference in performance between chess masters and amateurs does not hinge upon mental ability (the depth of search space when considering an action does not differ significantly between masters and amateurs, for example) but rather on the breadth and quality of knowledge learnt and possessed by masters [5].

Chess' complex and complicated nature[1] results in an enormous state space. Jongman calculates that there are potentially 143.09 bits of information if all possible positions are considered [11], which buttresses Shannon's earlier calculations of a potential space of 10^{43} positions ($2^{143.09}$) [15]. However, Shannon's and Jongman's space includes redundant and implausible positions; de Groot and Gobet attempt to rectify this and calculate that the total space of possible chess positions contains 50 bits of information, giving 10^{15} positions [5]. Given known limitations on human cognition [7], the pertinent question is then: how do chess masters learn and retain such large[2] databases of chess positions?

One solution is *chunking* [3], whereby an agent aggregates many pieces of information from the environment into small units, and uses these chunks as input to its learning or production-rule system. This theory explains how substantial knowledge can lead to an improved ability to extract information from the environment, despite the cognitive limitations mentioned earlier. With respect to chess, Simon and Gilmartin propose that chess masters learn and retain between 10,000 and 100,000 chunks in memory [19], whilst Gobet and Simon propose a figure of 300,000 chunks [9].

The CHREST architecture [8] implements chunking computationally and has provided strong evidence, via simulation, that chunking mechanisms underpin human cognition in domains including board games [2, 5], implicit learning [13], and language acquisition [6, 10]. CHREST itself is a symbolic cognitive architecture, like ACT-R [1] and Soar [12]. However, unlike these other architectures, CHREST must acquire its internal knowledge from experience using *perceptual* chunking. The domains in which CHREST has been applied in the past are notably *deterministic*, i.e. future states and environmental complexity are determined by the actions of the agents situated within it; intrinsic environmental variables, relating to the environment (time allowed to make a decision, the conditions for removal of pieces in a game etc), contribute nothing.

---

[1] Complex in that there are many pieces capable of being moved, complicated in that there may be many possible solutions to a given position.

[2] With respect to both the number of positions and the amount of information in each position.

Simon's proposal of *bounded rationality* [17] complements chunking theory since its assumptions regarding limitations of human cognition can serve to reduce the complexity of an environment for an agent. Note that an agent's rationality may be *voluntary* rather than *obligatory*; de Groot and Gobet show that a chess master's perceptual coverage of a chess board is larger than that of non-masters [5]. Non-masters voluntarily bind their rationality by decreasing their visual-field since this serves to reduce the total complexity of the chess board, which in turn facilitates learning since there is less information to process. However, this has a negative impact on performance, since an optimal solution in a smaller area of the board may no longer be optimal when a larger portion of the board is considered.

In this paper, we implement a set of multi-agent simulations to examine how the performance and learning of agents based on the CHREST architecture and situated in a stochastic environment of varying levels of complexity, are affected when voluntary rationality is bound to different degrees. The stochastic environment in which these agents are situated is the *Tileworld* environment [14], where agents must fill holes with tiles in a two-dimensional world. To enable such behaviour, visual information is used as input to a domain-specific production-rule system to generate actions; visual information and actions are also encoded as "patterns" and are used as input to each agent's CHREST architecture to enable learning during run-time. We place bounds upon voluntary rationality by manipulating an agent's "sight-radius" parameter; this dictates the size of an agent's visual input and thus the amount of information that can be passed as input to the agent's CHREST architecture and production-rule system.

To the best of our knowledge, a pattern-orientated model of learning such as CHREST has never been implemented in Tileworld before. Furthermore, we are not aware of any research that analyses the effects of bounding voluntary rationality to different degrees on an agent's learning and performance in the context of a stochastic environment whose intrinsic and extrinsic complexity can be precisely manipulated.

The paper is structured as follows: Sect. 2 provides additional background to Tileworld and justifies its use in this research; Sect. 3 discusses the CHREST architecture; Sect. 4 describes how CHREST is incorporated into the agent architecture used; Sect. 5 gives an overview of the simulations conducted; Sect. 6 presents and discusses the results from these simulations; Sect. 7 concludes the paper by summarising key points.

2 Tileworld

The Tileworld testbed is intended to provide a multi-agent simulation environment where the meta-level reasoning of agents can be analysed [14]. As mentioned in Sect. 1, the Tileworld environment is stochastic; it is highly parameterized and a substantial degree of control can be exerted over intrinsic environment parameters that affect the environment's complexity. These properties allow experimenters to be exceptionally specific with regard to what factors affect the environment's future state and the actions of agents within the environment itself. Since our

Table 1 Agent, CHREST and environment parameters and descriptions (see Sect. 3 for details of CHREST parameters and Sect. 4 for details of agent parameters)

Parameter	Type	Description
Add link time	CHREST	Time taken to create a visual-action link in LTM
Deliberation time	CHREST	Time taken for the production-rule system to generate an action
Discrimination time	CHREST	Time taken to discriminate a node in LTM
Familiarisation time	CHREST	Time taken to familiarise a node in LTM
Sight radius	Agent	Number of squares the agent can see to the north, south, east and west of itself
Number of agents	Environment	Number of agents situated in the Tileworld
Time limit	Environment	Length of time agents have in the Tileworld before environment is cleared
Tile/hole birth interval	Environment	How much time must pass before a tile/hole has a chance of being "born"
Tile/hole birth prob.	Environment	Probability of a tile/hole being "born" after the value specified in the "Tile/hole birth interval" parameter has elapsed
Tile/hole lifespan	Environment	Length of time a tile/hole is present in the Tileworld for before it is removed

investigation requires a stochastic environment and a high degree of control over parameters which directly influence its stochastic nature, we assert that Tileworld satisfies such requirements and is fit for purpose.

In a typical Tileworld environment a number of tiles, holes and agents exist. Some variations of Tileworld include explicit obstacles, but these have not been included in our implementation since tiles, holes and agents naturally act as obstacles.[3] The goal of agents in Tileworld is to push tiles to fill holes; when this occurs, the "pusher" earns a point and both the tile and hole disappear. One of the crucial intrinsic factors which contributes to the stochastic nature of Tileworld is that the number of tiles and holes is not finite: new tiles and holes can be created at a defined time interval with a defined probability and can disappear after a defined period of time. A comprehensive list of all CHREST, agent and environmental parameters that can be set in our version of Tileworld, can be found in Table 1.

Depending upon the size of the Tileworld used, this intrinsic and extrinsic dynamism resulting from the actions of agents upon artifacts makes the environment immensely complicated and complex (assuming that the number of squares constituting the environment is sufficiently large). Simari and Parsons posit that the total number of states possible in a simplified version of Tileworld with a width of n squares, one agent and only holes (no tiles or obstacles) is $n^2 2^{n^2}$ [16].[4] They note

---

[3] Any Tileworld square can only be occupied by one artifact at most.

[4] The base 2 in the 2^{n^2} term of the expression is derived from the fact that Tileworld squares may be empty or may be occupied by one instance of an artifact class. In Simari and Parsons' version of

that the limit for the tractability of direct calculations by a computer with reasonable resources using a Markov Decision-Theory process to calculate optimal solutions in this simplified Tileworld is around $n = 4$, or $n = 5$. Therefore, Simon's concept of bounded rationality is practically a *necessity* in order for agents to perform any action, optimal or otherwise, when $n > 5$.

The complexity of the Tileworld environment used in these simulations is orders of magnitude greater than that used by Simari and Parsons. Our environment is a two-dimensional (35×35) grid that "wraps" (grid edges are not strict boundaries) and consists of multiple agents, tiles and holes. Following Simari and Parsons calculations, this would mean that there are 4 artifact classes that an agent may encounter (excluding itself): another agent, a tile, a hole and an empty square, resulting in $35^2 4^{35^2}$ or 4×10^{740} possible states. We assert that this degree of complexity, in addition to the ability to be able to precisely control parameters intrinsic to the environment that further alter Tileworld's complexity, provides a suitable test-bed for analysing the interplay between extrinsic and intrinsic environmental complexity on the learning rates and performance of agents situated within such an environment.

Aside from the standard Tileworld rules outlined in [14], in our version of Tileworld only one tile may be pushed at any time by an agent. For example, if an agent has two tiles to its east on consecutive squares, it is not able to push the tile closest to itself east since it is blocked by the tile two squares to the east of the agent.

3 CHREST

CHREST is an example of a symbolic cognitive architecture, with an emphasis on perception and learning. In this section, we present only those details of CHREST needed to understand the behaviour of agents in the simulations outlined; more detailed descriptions are available in [5, 8] and an implementation of CHREST is available at http://chrest.info.

The version of CHREST used is composed of two main components: short-term memory (STM) and long-term memory (LTM). STM and LTM hold chunks of patterns for differing modalities of which there are three: action, visual and verbal (verbal STM/LTM is not utilised in this investigation). To clarify, there is only one LTM with different modality sections whereas there are 3 independent STMs, one for each modality. The size of STM is limited to around 4 discrete chunks whereas LTM size is unlimited. Chunks are retrieved by sorting input patterns generated by the agent through a *discrimination network*. The role of the discrimination network is to sort an incoming pattern to the most relevant part of LTM. The discrimination network acts as a retrieval device and a similarity function, analogous to the hidden layers of a connectionist network, or the RETE network of Soar [12]. Unlike other

---

(Footnote 4 continued)
Tileworld, there is only one artifact class: a hole. Therefore, in their version of Tileworld, a square will only ever be empty or occupied by a hole, which gives the base 2.

cognitive architectures such as Soar and ACT-R [1], CHREST does not discriminate between types of LTM memory such as procedural, declarative or semantic.

Since Tileworld consists of squares which can contain a tile, hole, agent or nothing (empty squares are ignored), visual information perceived by an agent is encoded as an item-on-square triple by the agent's input/output component. This triple consists of: the artifact's type, its x-offset and its y-offset from the agent's current location. Thus, a visual-pattern indicating that a tile is located three squares to the east of an agent is represented as [T 3 0], a hole one square to the north of the agent is represented as [H 0 1] and another agent one square north and one square west of the agent is represented as [A 1 -1]. An agent which can see one tile and one hole in its field of vision represents the visual scene as a pattern that contains a list of triples terminated by a "$" sign: ⟨[T 3 0] [H 0 1] $⟩. Visual patterns descend from the visual root node in LTM and are stored in visual STM.

Action-patterns are again encoded by the agent's input/output system from actions prescribed by the agent's production-rule system. Action-patterns are represented in a similar way to visual-patterns: triples consisting of an action code, direction and number of squares to move along. For example, the action "move north by one square to a tile" would be represented as a triple [MTT 0 1].[5] Action patterns can be learnt as chunks in the same way as visual patterns except that these chunks descend from the action root node in LTM and are stored in the action STM.

Patterns are learnt as chunks using the processes of *familiarisation* and *discrimination*. When a pattern is input to CHREST, it is sorted through the discrimination network in LTM by checking for the presence of that pattern's information on the network's test links. If sorting retrieves a chunk from LTM, the pattern within this chunk is compared with the input pattern to determine if further learning should occur.[6] If the input pattern differs from the stored pattern (the input pattern may indicate a tile rather than a hole, for example), *discrimination* occurs and a new test link and child chunk is added to the network. If a part of the sorted pattern matches the pattern in the retrieved chunk but is more elaborate with respect to the information it contains, then *familiarisation* occurs to add this new information to the retrieved chunk. The two mechanisms of discrimination and familiarisation work together to increase the number of distinct chunks that the model can recognise and to supplement the details of already known chunks with further information.

If a visual-pattern corresponds to a completely familiarised chunk in visual LTM, and the same is true for an action-pattern in action LTM, then it is possible to associate the visual and action chunks in question together to create a visual-action link. In this investigation, these links are produced but not used in the agent's deliberation procedure since we did not want to contaminate an agent's performance with input from pattern-recognition systems. The only independent variables intrinsic to the

---

[5] MTT = "Move To Tile".

[6] Note that the chunk retrieved from LTM is also placed into STM but this version of CHREST does not make use of STM chunks.

agent that should affect its performance should be its sight radius. We mention this feature since it affects an agent's rate of learning: if an agent is creating a visual-action link, it can not discriminate or familiarise other patterns.

4 CHREST in Tileworld

As mentioned in Sect. 1, agents in our simulations use visual-patterns as input to both their production-rule system and LTM discrimination network. These symbolic visual-patterns are created by the agent's input/output component that encodes information which agents "perceive" on squares that fall within their current sight radius (translation and structure of these patterns is discussed in Sect. 3). The visual-patterns generated are then used as input to a goal-driven[7] production-rule system that produces actions which will be performed by the agent after being translated into action patterns and used as input to LTM (again, translation and structure of these patterns is discussed in Sect. 3). As such, CHREST agents in Tileworld undertake an explicit deliberation and means-end reasoning procedure. The execution cycle used by CHREST agents is outlined in Sect. 4.1 and the production-rule system is described in Sect. 4.2. Note that learnt visual or action-patterns play no role in the deliberation procedure of agents, they are simply added to the CHREST agent's LTM discrimination network. Figure 1 illustrates the combination of the agent and CHREST architecture for clarification.

The production-rule system used by agents in the Tileworld environment implemented in this investigation can create up to 17 actions: the "move-randomly", "move-to-tile", "move-around-tile", "push-tile" actions (of which there are four variations each: north, south, east and west) and the "remain-stationary" action of which

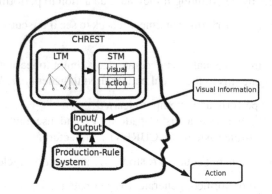

Fig. 1 Illustration of agent and CHREST architecture combination

---

[7] The agent's goals are implicit, i.e. goals are not explicitly represented in any data structure available to CHREST agents.

there is only one variation. Actions that are generated by agents are loaded for execution and performed after a certain amount of time has elapsed (this value is controlled by the "deliberation-time" parameter, δ, that is set by the experimenter). So, if an agent loads an action, α, for execution at simulation time, t, the agent will perform α at time $t + \delta$, simulating time passing whilst an agent decides upon what action to perform given its current visual information. Note that an agent will not generate additional visual-patterns between t and $t + \delta$ so learning is directly affected by deliberation.

As mentioned by Simari and Parsons in [16], an agent's *intention reconsideration* strategy weighs heavily upon the performance of an agent in the Tileworld environment. If the strategy implemented causes excessive intention reconsideration, then the agent will waste effort by constantly deliberating about what to do next rather than performing an action [20]. Informally, the intention reconsideration strategy implemented states: if the visual-pattern representing the current state of the agent's observable environment is equal to the visual-pattern which represented the state of the agent's observable environment used to generate the currently scheduled action, then execute the scheduled action, otherwise, generate a new action. Thus, the agent's production-rule system uses the most up-to-date visual information obtainable from the environment to inform action generation, but does not react to changes in the environment that occur between deciding upon what action to perform and performing the action. A more formal description of this intention reconsideration strategy is given in Sect. 4.1.

4.1 CHREST Agent Execution Cycle

The agent execution cycle runs for every time increment in the Tileworld environment. Agents begin by determining if they have an action to perform.

1. An action, α, is to be performed at time, t: check to see if the current time is equal to t.

    a. The current time is equal to t: generate and attempt to learn a visual pattern, V' and compare this to the visual pattern used to generate α, V.

        i. $V = V'$: perform and attempt to learn α.
        ii. $V \neq V'$: generate a visual-pattern, V, and use this as input to the production-rule system and CHREST architecture.

    b. The current time is not equal to t: stop current execution cycle.

2. No action is to be performed: generate a visual-pattern, V, and use this as input to the production-rule system and CHREST architecture.

4.2 Production-Rule System

Given that the ultimate goal of agents in Tileworld is to push tiles into holes thus earning the agent a point, the production-rule system implemented takes a visual-pattern, V, as input and follows the procedure outlined below. Points at which actions are generated are highlighted in bold for clarity:

1. V indicates that the agent is surrounded, i.e. squares immediately north, east, south and west of the agent are occupied by non-movable tiles, holes or other agents. **Agent generates a "remain-stationary" action.**
2. V indicates that the agent is not surrounded and that tiles and holes are nearby: determine which hole is closest to the agent, H, then determine the tile that is closest to to H, T.

   a. Agent is 1 square away from T and can push it closer to H from the agent's current position: **Agent generates a "push-tile" action.**
   b. Agent is 1 square away from T but the agent's current position means that it can't push T closer to H: **Agent generates a "move-around-tile" action.**
   c. Agent is more than 1 square away from T **Agent generates a "move-to-tile" action.**

3. V indicates that the agent is not surrounded and that tiles are nearby but holes are not: determine distance of T from the agent.

   a. Agent is 1 square away from T: Agent faces T and attempts to push T along the agent's current heading (if the agent had to turn east to face T then the agent will attempt to push T east). **Agent generates a "push-tile" action.**
   b. Agent is more than 1 square away from T: **Agent generates a "move-to-tile" action.**

4. V indicates that the agent is not surrounded but can't see any tiles: **Agent generates "move-randomly" action.**

5 Simulation Overview

We study the effect of bounding voluntary rationality on learning and performance using scenarios where the intrinsic and extrinsic complexity of the environment is directly manipulated. Our experimental hypothesis states that, as environment complexity increases, bounding an agent's rationality to a greater degree will impinge on learning rates but improve performance. Two dependent variables are measured: to measure performance, we use the score of agents (how many holes the agent filled with tiles), and to measure the amount of information the agent has learned, we use the number of nodes in an agent's visual LTM. We do not consider the size of action LTM in our measure of learning since the maximum size of this LTM modality is

Table 2 Agent, CHREST and Tileworld parameter values

Parameter	Varied/constant	Value(s)
Add link time	Constant	1 s
Deliberation time	Constant	1 s
Discrimination time	Constant	10 s
Familiarisation time	Constant	2 s
Sight radius	Varied	2 (below avg.), 3 (avg.), 4 (above avg.)
Number of agents	Varied	2 (below avg.), 4 (avg.), 8 (above avg.)
Time limit	Constant	14,400 s
Tile/hole birth interval	Constant	1 s
Tile/hole birth prob	Varied	0.1 (below avg.), 0.5 (avg.), 0.9 (above avg.)
Tile/hole lifespan	Varied	80 s (below avg.), 40s (avg.), 20 s (above avg.)

17 and given the length of time agents are situated in the Tileworld for (a simulated time of 4 h), it is always the case that every action-pattern is learnt.

Since Tileworld's stochastic nature is controlled by a number of parameters (see Sect. 2 and Table 1), some parameter values were altered and others kept constant. Table 2 delineates whether each parameter was varied or kept constant and what the parameter's value(s) was(were) set to.[8]

To alter intrinsic environment complexity, the values for "tile/hole birth prob." and "tile/hole lifespan" parameters were altered: higher tile/hole birth prob. values and lower tile/hole lifespan values equate to greater environment complexity, since more tiles/holes will appear but for shorter periods of time. One may expect the value for the "tile/hole birth interval" to also be varied in this case; however, the complexity of the environment can be significantly modified by varying the values for the "tile/hole birth prob." and "tile/hole lifespan" parameters. To alter extrinsic environment complexity, we varied the "number of agents" parameter: the higher the number of agents, the greater the complexity of the environment, since introducing more agents should result in more interactions with artifacts in the environment, thus increasing environment dynamism.

Parameters which are varied in Table 2 have values that are grouped into three levels of complexity: below average, average and above average; for ease of reference, these complexity levels are given numerical values and used in Figs. 2 and 3 in Sect. 6: below average = 1, average = 2 and above average = 3. The values for the "number of agents" and "tile/hole lifespan" were determined by taking the average value (4 and 40, respectively) and either halving or doubling to produce the below average or above average counterpart. Values for the "sight-radius" parameter were derived by taking the minimum sight radius possible as the below average value[9]

---

[8] All parameters concerning time are specified in seconds.

[9] Minimum "sight-radius" parameter value is 2 since agents must be able to see at least 1 square in front of a tile, so that its ability to be pushed can be determined.

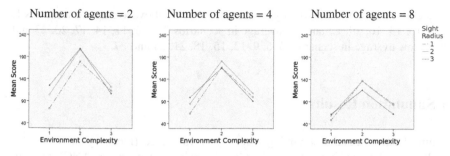

Fig. 2 Average score as a function of number of agents, complexity of the environment and sight radius

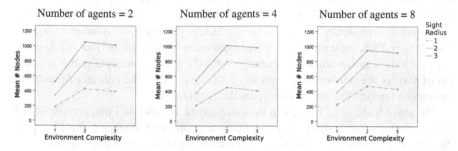

Fig. 3 Average number of visual LTM nodes as a function of number of agents, complexity of the environment and sight radius

and then adding 1 for each increase in complexity. Values for the "tile/hole birth prob." parameter were derived by simply taking the median probability, 0.5, as the average complexity value and then taking the lowest/highest values possible without guaranteeing tile/hole birth since this would significantly skew the results.

Values for parameters which are kept constant in Table 2 are justified thus: "add link time", "discrimination time" and "familiarisation time" parameter values were taken from [18]. We chose 1 s for the "deliberation-time" parameter value since this coincides with the value chosen for the "tile/hole birth interval" parameter so that an agent may need to reconsider its planned action due to the appearance of a new tile or hole. After preliminary testing of the simulations we deemed 4 h suitable enough for the "time-limit" parameter value since this gives agents ample time to learn and act within the environment.

In total, 27 scenarios were simulated and each repeated 10 times, permitting the collection of sufficient data so that a rigorous statistical analysis can be performed; values for dependent variables were averaged across agents for each scenario. Scenarios can be segregated into blocks of 9 where intrinsic complexity is altered: below average in scenarios 1–9; average for scenarios 10–18; above average for scenarios 19–27. Scenarios can then be segregated into blocks of 3 where extrinsic complexity is altered: above average in scenarios 1–3, 10–12 and 19–21; average in scenarios 4–6, 13–15 and 22–24; above average in scenarios 7–9, 16–18 and 25–27.

Alterations to bounded rationality do not form blocks: above average in scenarios 1, 4, 7, 10, 13, 16, 19, 22 and 25; average in scenarios 2, 5, 8, 11, 14, 17, 20, 23 and 26; below average in scenarios 3, 6, 9, 12, 15, 18, 21, 24 and 27.

6 Simulation Results

Figure 2 shows results pertaining to agent performance (mean agent score). To analyse the data, we carried out a $3 \times 3 \times 3$ analysis of variance (ANOVA), with environment complexity, number of agents and sight radius as between-subject variables. All one, two and three-way interactions were statistically significant ($p < 0.001$): sight radius $F(2, 243) = 128.3$, number of players $F(2,243) = 1,832.0$, environment complexity $F(2,243) = 6,109.1$, sight radius $\times$ number of agents $F(4,243) = 75.5$, sight radius $\times$ environment complexity $F(4,243) = 142.5$, number of agents $\times$ environment complexity $F(4,243) = 66.2$ and sight radius $\times$ number of players $\times$ environment complexity $F(8,243) = 7.0$. The presence of multiple interactions means that the pattern of results is complex.

The main features of Fig. 2 can be summarised as follows. First, complexity of the environment has a non-linear effect on mean score, with average complexity producing the highest scores. Second, as the number of players—and thus complexity—increases, mean score decreases. Third, sight radius, number of players and complexity of the environment interact in an intricate but critical way. With a below average level of environment complexity, the largest sight radius obtains the highest scores irrespective of the number of players, although minimally so with 8 players. With an above average level of environment complexity, the largest sight radius always obtains the lowest mean score. The pattern of results is less stable with an average level of complexity of the environment. As the number of players increases from 2 to 8, sight radius incurs a premium: the largest sight radius moves from obtaining the highest mean scores (although minimally so) to obtaining the lowest mean scores. Intuitively, it is as if reducing sight radius as complexity increases allows agents to filter out information and to be more selective in what they give their attention to. This is exactly what is predicted by Simon's theory of bounded rationality [17].

Figure 3 shows results pertaining to learning (mean number of nodes). As with the analysis of performance, the three main effects, the three two-way interactions and the three-way interaction were all statistically significant ($p < 0.001$): sight radius $F(2,243) = 42,187.5$, number of players $F(2,243) = 12.0$, environment complexity $F(2,243) = 31,474.6$, sight radius $\times$ number of agents $F(4,243) = 176.1$, sight radius $\times$ environment complexity $F(4,243) = 1,151.7$, number of agents $\times$ environment complexity $F(4,243) = 49.3$ and sight radius $\times$ number of players $\times$ environment complexity $F(8,243) = 19.3$.

Compared to performance, results concerning learning are simpler to interpret; two main patterns are apparent. First, the mean number of nodes learnt sharply increases from a below average level of environment complexity to an average level (an increase of about 100 % in each case), and then slightly decreases from an average level of environment complexity to an above average level. Second, a large sight radius always produces a higher number of nodes, followed by a medium sight radius. Although number of agents has a statistically significant effect, this effect was very small compared to the effect of the other two independent variables and are picked up by the ANOVA due to the high statistical power of our simulations.

7 Conclusion

The relationship between complexity and bounded rationality is of considerable interest for artificial intelligence and cognitive science. One approach to this question considers how selective attention and learning, and more specifically chunking, helps humans to cope with the complexity of an environment. Having simulated key phenomena on chess expertise, the cognitive architecture CHREST provides plausible mechanisms to explain the relation between complexity and bounded rationality in a deterministic environment. In this paper, we extended CHREST's theoretical coverage to a stochastic environment, Tileworld, with the presence of multiple agents and the continuous appearance and disappearance of artifacts (tiles and holes)—Tileworld offers a challenging environment to study this question.

The simulations described in this paper looked at the effects of bounding an agent's rationality (sight radius) upon the performance of agents and their learning in context of the Tileworld environment whose complexity is altered by varying one extrinsic environment variable (number of agents situated in the environment) and two intrinsic environment variables (tile/hole birth probability and tile/hole lifespan). A systematic manipulation of these variables provided results of considerable theoretical importance.

In this paper, agents were not able to enjoy the benefits of learning with respect to improving their performance through the use of modified deliberation procedures; the complexity of the results suggests that this strategy was wise. The results not only demonstrate that too much complexity leads to weaker performance, but also that the ability to filter out complexity in taxing environments by reducing sight radius results in an improvement of performance in these environments but at the cost of learning new information. In future research, we plan to establish whether this "smaller-is-better" effect persists when agents can make use of the knowledge they have acquired to improve their deliberation procedures.

The research presented here is in the context of agent-modeling using a symbolic cognitive architecture and, as we have noted previously, there are no previous studies known to us of pattern-learning agents within the Tileworld environment. However, in the wider field of machine learning, there are some parallels to our main finding that bounding rationality can improve performance. The effect of bounding rationality in our domain is to limit the potential set of perceived perceptual patterns; the bounded rationality provides an heuristic bias towards a particular set of patterns which, the bias assumes, will be most useful for learning at that time. A related strategy is found in the area of active learning [4], in which selective sampling is used to identify the next set of unlearnt examples to include in the training pool. Future work will also look at the contrast between the theoretically-driven approaches derived from machine learning, and the heuristically-driven approaches derived from cognitive science.

References

1. Anderson, J.R., Bothell, D., Byrne, M.D., Douglass, S., Lebière, C., Qin, Y.L.: An integrated theory of the mind. Psychol. Rev. **111**(4), 1036–1060 (2004)
2. Bossomaier, T., Traish, J., Gobet, F., Lane, P.C.R.: Neuro-cognitive model of move location in the game of Go. In: Proceedings of the 2012 international joint conference on neural networks (2012)
3. Chase, W.G., Simon, H.A.: Perception in chess. Cogn. Psychol. **4**, 55–81 (1973)
4. Cohn, D., Atlas, L., Ladner, R.: Improving generalization with active learning. Mach. Learn. **15**, 201–221 (1994)
5. de Groot, A.D., Gobet, F.: Perception and Memory in Chess: Heuristics of the Professional Eye. Van Gorcum, Assen (1996)
6. Freudenthal, D., Pine, J.M., Gobet, F.: Simulating the referential properties of Dutch, German and English root infinitives in MOSAIC. Lang. Learn. Dev. **15**, 1–29 (2009)
7. Gobet, F., Lane, P.: Encyclopedia of the Science of Learning, Chapter Bounded Rationality and Learning. Springer, NY (2012)
8. Gobet, F., Lane, P.C.R., Croker, S.J., Cheng, P.C.-H., Jones, G., Oliver, I., Pine, J.M.: Chunking mechanisms in human learning. Trends Cogn. Sci. **5**, 236–243 (2001)
9. Gobet, F., Simon, H.A.: Five seconds or sixty? Presentation time in expert memory. Cogn. Sci. **24**, 651–682 (2000)
10. Jones, G.A., Gobet, F., Pine, J.M.: Linking working memory and long-term memory: a computational model of the learning of new words. Dev. Sci. **10**, 853–873 (2007)
11. Jongman, R.W.: Het Oog Van De Meester. Van Gorcum, Assen (1968)
12. Laird, J.E.: The Soar Cognitive Architecture. MIT Press, Cambridge (2012)
13. Lane, P.C.R., Gobet, F.: Chrest models of implicit learning and board game interpetation. In: Bach, J., Goertzel, B., Ikle, M. (eds.) Proceedings of the Fifth Conference on Artificial General Intelligence, pp. 148–157. Springer, Berlin, Heidelberg (2012)
14. Pollack, M., Ringuette, M.: Introducing the Tileworld: experimentally evaluating agent architectures. In: Eighth National Conference on Artificial Intelligence, pp. 183–189. AAAI Press, Menlo Park (1990)
15. Shannon, C.E.: A chess-playing machine. Philos. Mag. **182**, 41–51 (1950)
16. Simari, G.I., Parsons, S.D.: On approximating the best decision for an autonomous agent. In: Sixth Workshop on Game Theoretic and Decision Theoretic Agents, Third Conference on Autonomous Agents and Multi-Agent Systems, pp. 91–100 (2004)
17. Simon, H.A.: A behavioral model of rational choice. Q. J. Econ. **69**, 99–118 (1955)
18. Simon, H.A.: The Sciences of the Artificial. MIT Press, Cambridge, MA (1969)
19. Simon, H.A., Gilmartin, K.J.: A simulation of memory for chess positions. Cogn. Psychol. **5**, 29–46 (1973)
20. Wooldridge, M., Parsons, S.: Intention reconsideration reconsidered. In: Intelligent Agents V, pp. 63–80. Springer, New York (1998)

Short Papers

An Autopoietic Repertoire

M.J. Wheatman

Abstract This paper presents a strategy for natural language processing in natural language. Using a *concept* as a unit of conversation, defined by a repertoire of phrases, it describes the concept of *autopoiesis*: a repertoire for the construction of repertoires. A minimal repertoire, representing a shopping list app, is described. This is followed by a specification of the autopoietic repertoire, followed by the full repertoire of the shopping list. The advantages of this approach is basically two-fold: a natural language specification is self-evident; moreover, it results in a rich, tiered interface of repertoires supporting repertoires. This paper is validated by an aural information system, publicly available on various mobile platforms.

1 Introduction

Conversation definitions are typically stored in structured data files, often tagged in an XML format, such as AIML [1]. This approach is suitable for current vocal interfaces which are directed at a given task, such as a restaurant or appointment booking system [2, 3], because the conversation can be constructed prior to the operation of the application. The strategy followed in this paper is to store the repertoire definitions themselves as natural language (NL), and have them interpreted by the same language engine which interprets users' utterances. A closed set of phrases, supporting a *concept* by interacting with persistent data, has already been introduced as a *repertoire* in [4]. Further, an online demonstration of this language engine, *enguage*™, implements an aural shopping list [5] which exploits the text-to-speech facilities of mobile platforms. This results in a rich NL interface which can, if necessary, be manipulated at runtime, as demonstrated by the provision of device controls [5], such as the utterance *bigger text*. The one requirement to achieve this is an initial, or bootstrapping, repertoire: autopoiesis.

The notion of *autopoiesis* [6] is the ability of biological systems to self-reproduce, as opposed to the process of *allopoiesis* which is the production of one thing by

M.J. Wheatman (✉)
Yagadi Ltd, Withinreap Barn, Moss Side Lane Thornley, Preston PR3 2ND, UK
e-mail: martin@wheatman.net

© Springer International Publishing Switzerland 2014   165
M. Bramer and M. Petridis (eds.), *Research and Development in Intelligent Systems XXXI*, DOI 10.1007/978-3-319-12069-0_11

another. The typical example is that biological cells are created through cell division, whereas a car factory is required for the creation of motor cars. Thus software systems are akin to genetic systems: both are concerned with the processing of information [7]. The analogy here is of the compiler which is written in the language it compiles. Autopoiesis was only alluded to in [4]; this paper presents a detailed description. The autopoietic repertoire takes inspiration from the sentences *X means Y* and *X implies Y*. A repertoire containing such phrases will be one which informs the interpretation of language, and is one which implements the construction of the data structure used in interpreting utterances. C.S. Peirce noted these cognitive products of interpretation as *Interpretant*, being created during the connection of sign vehicles, *Representamen*, and their referent *Objects* [8]. This project models interpretant as a list, an ordered bag, of named values, or *intentions* [4]. The autopoietic repertoire manages that list.

This paper proposes a definition of natural language given in natural language: it starts by describing a simple example repertoire, *needs*, and how it is used in practice; it describes the thirteen utterances of the repertoire *autopoiesis* which are used to construct repertoires; it then gives the full *needs* repertoire which is used to configure the *enguage* language engine.

2 An Example Repertoire

A simple repertoire presented here which neatly illustrates the vocal/aural interaction achieved by Ref. [5]. It is also used as the example repertoire constructed by the autopoietic repertoire below. It consists of the following three utterances to record, replay and consume an items-of-requirement list:

1. *I need X.*

This phrase may be issued throughout the day, providing the mechanism to add an item to a list. Unless the item is already on the list, the response to this is simply *Ok, you need X*. The token *X* is defined as a being phrase, so it allows the use of qualifiers, such as *an egg* or *some cheese*. The separator *and* can be used to add multiple items—in practice speech-to-text mechanisms do not detect a pause and insert a comma, and we do not naturally express commas. However, this is a simple repertoire: the app does not need to know of the items on the list. It does not conform to a shop's stock list: it is up to the user to define whatever they wish to include.

2. *What do I need?*

This phrase is issued in the shop, presenting the user with the items added to the list.

So far, this aggregation and recall could simply be provided by vocal input to any notepad-type software. The utility of this approach becomes apparent with the ability to remove things from this list, without resorting to the manipulation of a cursor within some text and the delete key. This consists of the following utterance.

3. *I have got X.*

This provides the ability to remove items from the list, and so completes a (minimal) repertoire for the aural manipulation of persistent data.

Finally, a fourth utterance, not entirely necessary for list maintenance, is useful by removing the list altogether, rather than to have to empty all items individually:

4. *I have everything.*

While there are many possible improvements, we now have a useful repertoire—the complete management of persistent data by the issuing of four utterances in a conversation [9]. While this may seem straightforward, the link between the utterances and the meaning remains arbitrary through the implementation of a repertoire.

3 The Autopoiesis Repertoire

The problem addressed by this paper is that while the *iNeed* repertoire, above, could be hardcoded in a relatively small program, such as found in [10]; as a cultural artefact language is autopoietic, suggesting its representation might also generated. The mechanism for representing such a repertoire is an implementation of the sign model [8]. A sign is modeled in [4] as an utterance pattern and the *interpretant*—a list of named values, or *intentions*, that represent action resulting from a match: *think* to recast the utterance in a more specific manner; *perform* to act upon on persistent data; *reply* to formulate an appropriate response. Each of these intentions has an alternative: *elseThink, elsePerform* and *elseReply*, which are enacted if the previous *think* is *infelicitous* [11] i.e. does not have a successful outcome.

The autopoiesis repertoire is directed at the management of the list of intentions used in the interpreting of given utterances. A repertoire of thirteen utterances are required to construct all interpretant. It builds on two examples of raw meaning definition: *"X" means "Y"*; and, *"X" implies "Y"*. By *raw* here it means they may be heard as part of a conversation over the meaning of another conversation. The utterances are as follows:

1. *X means Y.*

This utterance appends onto the list of intentions the intention to think Y when X is matched with a given utterance. Implied by this is that Y is understood but X is a new pattern. It is included as it is a natural way of expressing this form of decomposition. Currently, this works on complete sentences; however, it is the intention in this project to change the semantics of this intention to work on sub-sentence units of meaning, which will circumvent some of the options in the configuration file.

2. *X implies Y.*

This utterance adds the think intention Y onto the beginning of the list of intentions when X is matched with a given utterance, to have it mediate first. X may also be an already understood pattern. Again, it is included as it is a natural way of expressing this form of decomposition.

3. *X implies Y, if not, say Z.*

Similar to 2, above, but this adds two intentions to the front of the list of intentions: firstly the *elseReply* intention Z; then, the *implies* intention, as in 2.

4. *On X, think Y.*
5. *On X, perform Y.*
6. *On X, reply Y.*

These utterances append the *think, perform* and *reply* intentions Y onto the list of intentions for a given pattern X. Currently, 4 is entirely identical to utterance 1 in this repertoire, simply with a different pattern; however as described above, there may be a need to modify the semantics of 1.

7. *Then on X, think Y.*
8. *Then on X, perform Y.*
9. *Then on X, reply Y.*

These utterances append a subsequent *think* intention onto the list of intentions.

10. *Then on X, if not, think Y.*
11. *Then on X, if not, perform Y.*
12. *Then on X, if not, reply Y.*

These utterances append subsequent *think, perform* and *reply* intentions to the list of intentions, to be interpreted if the outcome of the previous *think* intention is infelicitous. The notion of *infelicitous* outcome is taken from [11].

13. *Finally on X, perform Y.*

This is useful for housekeeping at the end of interpretation.

4 Example System

The simple repertoire given at the beginning of this paper can be encoded in the autopoietic repertoire thus. Firstly, there are four main utterances which deal with third-person statements.

```
On "X needs PHRASE-Y.", perform "list exists X needs Y".
Then on "X needs PHRASE-Y.", reply "I know".
Then on "X needs PHRASE-Y.", if not, perform "list add X needs
Y".
Then on "X needs PHRASE-Y.", reply "ok, X needs Y".

On "X does not need PHRASE-Y.", perform "list removeAll X needs
Y".
Then on "X does not need PHRASE-Y.", if not, reply "No, X does
not need Y".
Then on "X does not need PHRASE-Y.", reply "ok".

On "what does X need?", perform "list get X needs".
Then on "what does X need?", if not, reply "X does not need
anything".
Then on "what does X need?", reply "X needs ...".

On "X has everything.", perform "list delete X needs".
Then on "X has everything.", reply "ok".
```

Secondly, there are six first-person utterances which transform into these third-person statements.

```
On "I         need  PHRASE-X.",  think "_user needs X".
On "I do not   need  PHRASE-X.",  think "_user does not need X".
On "I         have  PHRASE-X.",  think "_user does not need X".
On                       "?",  think "what does _user need?".
On        "I have everything.",  think "_user has everything".
"What do i need." means "what does _user need?".
```

Finally, there are two miscellaneous phrases required to translate some idiosyncrasies into third person statements.

```
"X has PHRASE-Y."   means "X does not need Y"
"what does X need." means "what does X need?". # dropping of "?"
```

There are other phrases in the production iNeed app, which are superfluous to the discussion here.

5 Conclusion

This paper shows that not only can we interact with persistent data vocally using NL utterances, we can also represent these repertoires in an autopoietic repertoire. Not all of the autopoietic repertoire is used in this paper; however, it uses enough to demonstrate construction of a simple app. For a more complex example, that given in [4] shows this approach is capable of subtle, conceptual examples, in particular *if we are holding hands, whose hand am I holding?* Further, [4] shows how contextual information can be used to deal with ambiguity—a demonstration of which is beyond the scope of this paper. Processing is distributed through client-side interpretation. The term *meta-repertoire* is avoided since it implies a different syntax to the one it describes: autopoietic utterances are interpreted by the same machine. Indeed the notions of syntax and semantics are also avoided because they imply dualism approach—that meaning is found in structure: *the plane has landed* has the same syntax, but an entirely different meaning, to *the Eagle has landed*. Finally, the use of utterances in the configuration of a program also leads to succinct applications—the iNeed app is essentially an interpreter and a 40-line script.

References

1. Wallace, R.S.: Artificial Intelligence Markup Language (AIML) Version 1.0.1, Oct 2011. http://www.alicebot.org/TR/2011 (2011). Retrieved Apr 2014
2. VocalIQ Ltd.: Restaurant booking demo. http://youtu.be/tgYpiHjjo8Q. Retrieved Apr 2014, Uploaded 13 June (2013)
3. VocalIQ Ltd.: Appointment booking demo. http://youtu.be/jzaDyAAWOvI. Retrieved Apr 2014, Uploaded 13 June (2013)
4. Wheatman, M.J.: If we are holding hands, whose hand am I holding: an autopoietic conceptual analysis system. Proceedings of the 3rd International Conference on Logistics, Informatics and Service Science (LISS 2013), Reading (2013)
5. Yagadi Ltd.: I Need: A Vocal Shopping List. http://bit.ly/1fIrEdZ. Retrieved Apr 2014
6. Maturana, H.R., Varela, F.: Autopoiesis and Cognition: The Realization of the Living. Springer, Dordrecht (1980)
7. Winograd, T., Flores, F.: Understanding Computers and Cognition: A New Foundation for Design. Addison Wesley, Boston (1986)
8. Peirce, C.S.: In: Hartshorne, C., Weiss, P. (eds.) Collected Papers of C.S. Peirce, vol. 2, pp. 2.227–2.306. Harvard University Press, Cambridge (1935–1957).
9. Wheatman, M.J.: A semiotic model of information systems. Proceedings of the 13th International Conference on Informatics and Semiotics in Organizations: Problems and Possibilities of Computational Humanities, Leeuwarden, The Netherlands, 4–6 Jul (2011).
10. Weizenbaum, J.: ELIZA—A computer program for the study of natural language communication between man and machine. Commun. ACM **9**, 36–45 (1966)
11. Austin, J.L.: How To Do Things With Words. Oxford University Press, Oxford (1962)

De-risking Fleet Replacement Decisions

Anne Liret, Amir H. Ansaripoor and Fernando S. Oliveira

Abstract This paper outlines the different modelling approaches for realizing sustainable operations of asset replacement. We study the fleet portfolio management problem that could be faced by a firm deciding ahead which vehicles to choose for its fleet. In particular it suggests a model that enables generating a plan of vehicle replacement actions with cost minimization and risk exposure simulation. It proposes to use conditional value at risk (CVaR) to account for uncertainty in the decision process, and to use clusters modelling to align the generated plan with vehicle utilization.

1 Introduction

As assets age, they generally deteriorate, resulting in rising operating and maintenance (O&M) costs and decreasing salvage values. Moreover, newer assets that have a better performance may be available in the marketplace. Therefore, service organizations that maintain fleets of vehicles or specialized equipment, need to decide when to replace vehicles composing their fleet. These equipment replacement decisions are usually based on a desire to minimize fleet costs and are often motivated by the state of deterioration of the asset and by technological advances [3]. The recent volatility of fossil fuel prices and the increasing concerns regarding the global warming have drawn attention to the need to reduce fossil fuel energy consumption. Positive aspect is that Electric Vehicle (EV) is not affected by uncertainties arising from supply side of fossil fuels; however there are barriers for their adoption due to the lack of planning systems able to address firstly the other uncertainty

A. Liret (✉)
British Telecom, Paris, France
e-mail: anne.liret@bt.com

A.H. Ansaripoor · F.S. Oliveira
ESSEC Business School, Singapore, Singapore
e-mail: amirhossein.ansaripoor@essec.edu

F.S. Oliveira
e-mail: b00183496@essec.edu

© Springer International Publishing Switzerland 2014
M. Bramer and M. Petridis (eds.), *Research and Development in Intelligent Systems XXXI*, DOI 10.1007/978-3-319-12069-0_12

factors- namely market prices, carbon emission prices, regulations, technology's life cycle- and, secondly, the limited mileage these vehicles may be driven on a single charge, the cost of battery (leasing cost), and finally the need for management systems to assess a good trade-off between low running cost, and shorter life cycle.

Section 2 studies models which underlie the decision support system that automatically generates a plan for replacing vehicles in parallel. Section 3 concludes the paper.

2 Classifying Fleet (Asset) Replacement Models

The models can be categorized into two main groups based on different fleet characteristics: homogenous and heterogeneous, on one hand, parallel versus serial on the other hand. Parallel replacement deals with a multitude of economically interdependent assets that operate in parallel. The paper focuses on heterogeneous parallel models. We categorize 2 replacement policies: (1) Static Policy where all decisions are made at first period (2) Dynamic Policy where decisions are updated at every period. In fleet use case, both policies would be evaluated, when facing the requirement to replace a fixed stock of vehicles with some combinations of Petrol, Diesel and Electric Vehicle (EV), and while modelling the constraint for the battery's technology life cycle, the impact of uncertainties in the market due to fuel and carbon prices, and the impact on carbon emissions compensation cost.

1. **From the correlation matrix for actual fuel prices** from Jan. 2000 to Dec. 2012, we derive that Diesel and Petrol prices have a very high correlation (0.876), but negative correlation with Electricity (−0.126). The price of electricity is the price of each charge for 22 kwh battery. This suggests that we can use electric vehicles to hedge risk of fuel price rises, but that petrol and diesel vehicles, due to the high correlation, are exposed to the same sources of risk and are not very useful to reduce each other's risk exposure. Then, we can generate the simulated scenarios for fuel prices with the above correlation matrix using expected forecasted prices from 2013 to the end of 2020.
2. **As a first approach, we have used Portfolio Theory** to compute the risk associated to a portfolio of vehicles, while meeting a fixed mileage total driven each year. However, in a situation where two technologies are highly correlated, the Portfolio theory won't generate a good diversification.
3. **Our second approach consists on modelling a Parallel Replacement problem**. The limitation of the latter is the combinatorial nature [2] forcing to make hypothesis to simplify the problem (such as fixed replacement costs or Static Policy). Thus we design a variant, the **Parallel Heterogeneous Asset Leasing Replacement Model**, where the assets are bounded by common budget and demand constraints, and a fixed cost is charged in any period in which there exists a replacement. The model is general however it does not handle the uncertainty factors. When adapting it for fleet replacement, future economic or technical factors, and costs such as lease prices, fuel prices, fuel and electricity consumption rates, will have

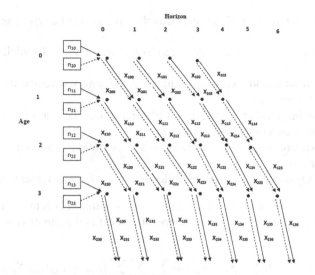

Fig. 1 Parallel heterogeneous asset leasing replacement model. Challengers are denoted by different arcs and different source (initial fleet) nodes

to be deterministic functions of time and vehicle type [2, 4]. The notation and formulation is more easily described by the network in Fig. 1. For the sake of simplicity this figure represents the case of two asset types that are available to meet the demand ($I = 2$). Figure 1 considers two types of technologies: the fossil fuel technology (Defender) and the new engine technology (Challenger). Nodes are labelled (a, t) with a being the age of the asset in years and t the planning time period in years. Flow X_{iat} represents the number of assets type i leased ($a = 0$) or assets type i in use ($a > 0$) from the end of year t to the end of year $t + 1$, in which the asset is of age $a + 1$.

4. **In our fleet case study**, the leasing period is assumed to be 4 years. So, the model is represented with $A = 3$ and $T = 6$. At the end of time horizon $T = 6$ all the assets are retired thus in the last 4 years of the planning horizon, no leasing of new cars should be planned. Assets are either provided from the initial fleet, represented as flow from supply nodes n_{ia}, or must be leased or in use (X_{i0t}). An asset when reaches age A must be retired. For meeting the associated demand in each period, the retired assets should be replaced by leasing new assets. In Fig. 1, each type of assets is represented a different arrow type. Let's now adapt the model for fleet replacement. There would be as many types as technologies (Petrol, Diesel, Electric). The indices in the model are the types of vehicle, $i \in \{1, 2\}$, the maximum age of vehicles in years, $a \in A$; $A = \{0,1, 2,..., A\}$, and the time periods (year), $t \in T$; $T = \{0, 1,..., T\}$. The decision variables include X_{iat} the number of type i, age a vehicles which are currently leased in year t, and P_{it} the number of type i vehicles which are leased at the beginning of year t, X_{iat} and P_{it} cannot have negative values. Actually, the number of new leased cars at the beginning of each year t, P_{it} equals X_{i0t}. The parameters are:

- **The expected utilization** (miles travelled per year) of a type i, age a vehicle in year t (miles/year), u_{iat},
- **The expected demand** (miles need to be travelled by all vehicles) in year t (miles), d_t,
- **The available budget** (money available for leasing new vehicles) in the beginning of year t, b_t,
- **The initial number of vehicles** of type i, age a at the start of first year, h_{ia},
- **The lease cost** of a type i vehicle, l_i,
- **The expected operating (running) cost**—per mile—of a type i, age a vehicle in year t, o_{iat},
- **The CO$_2$ emissions cost**—per mile—of a type i, age a vehicle, e_{ia}.

The objective function which we want to minimize (1) is the sum of leasing costs for the period $(T - 3)$ and the operating cost to the end of year T.

$$\text{Min} \sum_{i=0}^{I} \sum_{t=0}^{T-3} (l_i P_{it}) + \sum_{i=0}^{I} \sum_{a=0}^{A} \sum_{t=0}^{T} [o_{iat} + e_{ia}] u_{iat} X_{iat} \tag{1}$$

Equation (2) shows that the leasing costs cannot exceed the annual budget. Equation (3) requires that the total miles travelled by all used vehicles meet the annual demand. Equation (4) describes that the total number of vehicles in the first year should be equal to the initial condition of the system.

$$\sum_{i=i}^{I} l_i \cdot P_{it} \le b_t \quad \forall t \in \{0, 1, 2, ..., T - 3\} \tag{2}$$

$$\sum_{a=0}^{A} \sum_{i=}^{I} X_{iat} u_{iat} \ge d_t \quad \forall t \in \{0, 1, 2, .., T - 3\} \tag{3}$$

$$X_{ia0} = h_{ia} \quad \forall i \in I, \forall a \in A \tag{4}$$

Equation (5) represents the flow equation in which the number of the cars at each year equals to the number of new leased cars plus the number of cars belonged to the previous year.

$$X_{iat} = P_{it} + X_{i(a-1)(t-1)} \quad \forall i \in I, \forall a \in A, \forall t \in T \tag{5}$$

5. **Clustering the fleet portfolio**: We are using clustering approach (based on K-means algorithm) to identify homogenous subgroups in the population of vehicles. The system identifies the clusters of vehicles driving the same range of mileage, taking comparable capacity or type of technology. This allows identifying homogeneous groups of underutilized vehicles in a population (cluster 1 in Table 1), and high mileage cluster (cluster 4). Moreover, clustering for the

Table 1 Cluster analysis for mileage and fuel consumption data for small vans

Types of vehicles	Cluster	Monthly mileage		Fuel consumption per 100 km (l)		No. of vehicles	% of total
		Mean	S.D.	Mean	S.D.		
Small vans	1	581.42	267.77	6.34	0.44	135	12
	2	789.18	254.85	5.31	0.25	381	34
	3	845.63	255.72	4.46	0.37	297	26
	4	1,605.01	313.58	4.96	0.43	324	28
	Combined	1,011.74	470.53	5.11	0.66	1,137	100

capacity of vehicle type allows the selection of the specific technology better fit for a given type of vehicle working under the cluster conditions. Using cluster analysis enables us to identify patterns of utilisation and of fuel consumption compared to average mileage driven (cluster "combined").

6. **Multi-stage stochastic programming model** proposed and proved by Ref. [1] uses Conditional Value at Risk (CVaR) to measure the risk exposure and a constrained optimisation problem to model the carbon emission costs and leasing cost. The solution is the optimal number of vehicles to be leased, considering expected cost and a risk tolerance parameter during the planning horizon. Reference [1] includes a proof of time-consistency for its dynamic risk measure. By time-consistency [5], decisions will be a function of the history of the data process available, at the time t, when decisions are made. Time-consistency brings the advantage of computing optimal policies that do not contradict each other when you move along the time axis. The model of [1] is fast to compute (a matter of minutes) and can be used to plan the optimal dynamic policy. As for the Parallel Replacement model, the out coming plan is a tree of nodes, where each node represents a point in time when an action on fleet has to be performed.

7. **Application**: Vehicle capacity is an important characteristic because the technicians have to drive a vehicle having enough capacity. We do cluster analysis for each type of vehicle. We've got a tree of scenarios generated from the prices forecast. At each node we have generated a random vector of normal distribution for fuel prices. We use a normal distribution for fuel consumption and mileage driven in each utilization rate-based cluster (as in Table 1). For CO_2 prices we have assumed a uniform distribution between [5, 20]. The number of vehicles at each node is a stochastic parameter, which we map to a normal distribution function of the cluster size. Table 2 shows an example of resulting plan for Small Vans. Symbol ω stands for the risk tolerance parameter (0 meaning maximum risk exposure allowance, 1 being maximum risk reduction). Depending on ω, the system would lease different number of vehicles per year.

Results of experiments suggest that (1) by clustering, the cost and risk per vehicle can be reduced compared to global decision; (2) the expected cost per vehicle is not reduced for high mileage clusters; (3) EVs contribute to risk minimization goal for

Table 2 Optimal number of vehicles and technologies plan for all clusters of small vans

		Cluster 1			Cluster 2			Cluster 3			Cluster 4		
ω		0	0.5	1	0	0.5	1	0	0.5	1	0	0.5	1
Year	2014	52	54	76	171	185	22	141	150	18	144	165	205
	2015	41	44	39	99	104	10	74	75(1)	76	88(1	87	92
	2016	30	25(4)	6	75	59(5)	9	54(1)	47(2)	6	54(9	50(5)	3
	2017	12	1(9)	15	35	2(25)	38	27	7(16)	29	20(9	4(1)(13)	24
Techno		D	D(P)	D	D	D(P)	D	D(P)	D(P)	D	D(E)	D(E)(P)	D

D stands for "Diesel", P for "Petrol", and E for "Electric"

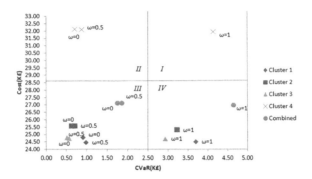

Fig. 2 Fleet portfolio clustering effect on risk and expected cost for small vans, in quadrants

high mileage usage only (more than 1,000 miles/month). Figure 2 outlines that risk value for "combined cluster"—that is the global fleet planning—is always higher than each cluster-based risk value (for each ω).

3 Conclusion

We have presented a model that helps addressing asset replacement problem and its particular use case in field service fleet. The combinatorial aspect of the replacement operation is complicated by the fact that the matching of vehicle types and running technology depends on the variation in usage over days. The solution of this multi-objective combinatorial optimization problem is in the form of a ranking of the technology and brands based on the most economical, ecological and risk-affordable choice.

References

1. Ansaripoor, A.H., Oliveira, F., Liret, A.: An improved decision support system for sustainable fleet replacement. J. Oper. Res. Soc. **237**(2), 701–712 (2014)
2. Hartman, J.C.: The parallel replacement problem with demand and capital budgeting constraints. Naval Res. Logistics (NRL). **47**(1), 40–56 (2000)
3. Hartman, J.C.: A note on "a strategy for optimal equipment replacement". Prod. Planning Control **16**(7), 733–739 (2005)
4. Karabakal, N., Lohmann, J.R., Bean, J.C.: Parallel replacement under capital rationing constraints. Manag. Sci. **40**(3), 305–319 (1994)
5. Shapiro, A.: On a time consistency concept in risk averse multistage stochastic programming. Oper. Res. Lett. **37**(3), 143–147 (2009)

Reliability and Effectiveness of Cross-Validation in Feature Selection

Ghadah Aldehim and Wenjia Wang

Abstract Feature selection is increasingly important in data analysis and machine learning in the big data era. However, how to use the data in feature selection has become a serious issue as the conventional practice of using ALL the data in feature selection may lead to selection bias and some suggest to use PART of the data instead. This paper investigates the reliability and effectiveness of a PART approach implemented by cross validation mechanism in feature selection filters and compares it with the ALL approach. The reliability is measured by an Inter-system Average Tanimoto Index and the effectiveness of the selected features is measured by the mean generalisation accuracy of classification. The experiments are carried out by using synthetic datasets generated with a fixed number of relevant features and varied numbers of irrelevant features and instances, and different level of noise, to mimic some possible real world environments. The results indicate that the PART approach is more effective in reducing the bias when the dataset is small but starts to lose its advantage as the dataset size increases.

1 Introduction

Feature selection (FS) is generally applied as a pre-processing step in machine learning and data mining and has become increasingly important, especially when dealing with high dimensionality in big data. Numerous techniques have been developed and applied in various problems.

There is, however, a long on-going argument in the field about how the data should be used when carrying out FS [1, 4, 7, 10]. The central issue is whether all the data should be used in FS only once before a modelling approach, or just some parts of the data should be used in FS. The ALL approach became almost a de-facto practice in FS because FS is viewed as a mere pre-processing step before analysis and the ALL

G. Aldehim (✉) · W. Wang
The School of Computing Sciences, University of East Anglia, Norwich NR4 7TJ, UK
e-mail: g.aldehim@uea.ac.uk

W. Wang
e-mail: wenjia.wang@uea.ac.uk

© Springer International Publishing Switzerland 2014
M. Bramer and M. Petridis (eds.), *Research and Development
in Intelligent Systems XXXI*, DOI 10.1007/978-3-319-12069-0_13

179

approach increases the chance of selecting all the relevant features and then helps to build better models [7, 8]. However, the ALL approach may produce overoptimistic results, as it may peek at the test set before the evaluation. This is called feature subset selection bias. Some studies [1, 4, 7, 10] have discussed this issue and attempted to address it by using the PART approach. Nevertheless, the PART approach may lead to underestimate the relevant features under some conditions. Also, these studies were mostly done on rather specific problem domains such as on wrapper-based feature selection algorithms or genome wide analysis and thus their findings are limited. Therefore, it is important to evaluate these two approaches systematically and determine their reliability and effectiveness under various circumstances.

2 The ALL and PART Approaches

The ALL approach uses all the instances in the step of feature selection while the PART approach only uses training instances partitioned from the data in feature selection.

With the ALL approach, all the data of a given dataset is used once in FS stage and then the selected features are used as the input variables to generate classifiers with a common procedure. While, with the PART approach, the dataset is partitioned and some parts are used for FS and also used as the training set for inducing classifiers. This study employs the K-fold cross validation mechanism in the PART approach. It works as follows: $(K - 1)$ folds are used as the training data for each filter; the selected features are used as the inputs for the classification base learner to build the classifier with the same $(K - 1)$ folds of the data; then remaining fold is used as a validation set to test the classifier. This procedure is repeated round-robin for K times.

3 Reliability and Effectiveness Measures

The reliability of a feature selection method in this context is measured by computing the degree of the agreement, or similarity, between a set of the selected features and a given set of the desired features. A few similarity measures available are reviewed by [3], but these measures are used for evaluating the internal stability of one FS process, as in the PART approach. They cannot be used directly in the ALL approach because the entire dataset is used only once and only one subset of the features is selected. Moreover, with these measures we cannot compare the subset produced from each FS process with the optimal features.

In this study, we introduce a measure that is modified from the Inter-system Average Tanimoto Index [11] by removing the influence of the total number of the features.

$$IATI(X_s^1, X_s^2) = \frac{1}{k_i \cdot k_j} \sum_{i=1}^{k_i} \sum_{j=1}^{k_j} \frac{|X_i^1 \cap X_j^2|}{|X_i^1 \cup X_j^2|} \qquad (1)$$

where, X_s^1 and X_s^2 are the two sets of the feature subsets selected by two methods respectively; $X = \{x_1, ..., x_N\}$; X_i, X_j are the subsets of the selected features; X_i, $X_j \subset X$; k_i and k_j are the number of folds used to generate X_s^1 and X_s^2.

The reliability is calculated once for the ALL approach and K times for the PART approach with K-fold cross validation. The effectiveness of the selected features is measured by the average classification accuracy of the classifiers generated with the selected features by the ALL or the PART approaches.

4 Experimental Design

In order to carry out the empirical investigations systematically, we take advantages of using artificial data because we can pre-set the desired features and other parameters systematically. Table 1 lists the synthetic datasets generated for a classification problem with a fixed number (10) of the relevant features, varied numbers of irrelevant features, the number of instances, and levels (5 %, 10 %) of noise.

Four commonly used filters: ReliefF [9], Gain Ratio [6], Correlation-based Feature Selection (CFS) [2] and Fast Correlation Based Filter (FCBF) [12] are chosen in this experiment. Naive Bayes (NB) classifier is chosen as a base learner for evaluating the effectiveness.

A 10-fold cross validation strategy is used in the FS stage with the PART approach as well as in the classification stage of both approaches. Moreover, each experiment

Table 1 The synthetic data sets with 10 relevant features at different strengths. S the number of instances, N the total number of features, N_I the number of irrelevant features

Dataset	S	N	N_I	Dataset	S	N	N_I
S1	100	100	90	S2Noise5	1,000	100	90
S2	1,000	100	90	S2Noise10	1,000	100	90
S3	10,000	100	90	S5Noise5	1,000	1,000	990
S4	100	1,000	990	S5Noise10	1,000	1,000	990
S5	1,000	1,000	990	S8Noise5	1,000	10,000	9,990
S6	10,000	1,000	990	S8Noise10	1,000	10,000	9,990
S7	100	10,000	9,990	S8	1,000	10,000	9,990

is repeated ten times with different shuffling random seeds to assess the consistency and reliability of the results. In total, 11,200 models were built in our experiments. The average accuracy as well as the average similarity will be presented in the final results.

5 Experimental Results and Dissection

Figure 1 gives the summary of the results of the reliability measured by the IATI for all the synthetic datasets. They show that the PART and ALL approaches are mostly similar with all the filters on almost all the datasets, except for few cases. One case is S4-PART versus S4-ALL, where the number of instances is small (100) and the number of the irrelevant features is relatively very large, (990) 99 times higher than the number of relevant features. Another case is that when the noise level is increased to 10 %, the PART approach appeared to be slightly worse than the ALL approach.

In addition, it is worth noting how the similarity in S2Noise5 and S2Noise10 datasets with both the PART and ALL approaches decreases quite significantly from 0.81 to 0.64 when the noise level increased from 5 to 10 %, while there is almost no difference between S5 and S8 with 10 % noise than with 5 % noise. Therefore, we can say that datasets with small numbers of samples (as S2) can be easily affected by noise more than the data with high numbers of samples.

Accordingly, we can note that the number of samples plays the most important role. As we can observe, if the number of samples is low, it will be hard for all the filters to select a high number of relevant features; also, we notice an increasing tendency to select more irrelevant features. Additionally, the results indicated that increasing the number of the irrelevant features in the dataset can have a quite strong adverse effect on the performance of filters as it also increases the chance of choosing irrelevant features.

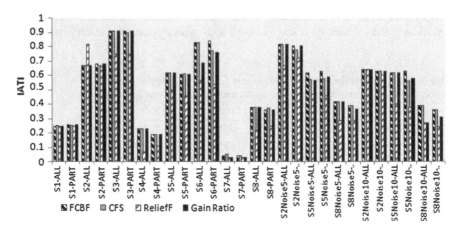

Fig. 1 IATI comparison between subsets of features selected by the four filters and the relevant features

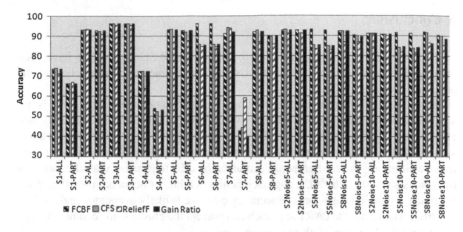

Fig. 2 The average test accuracy of NB classifiers trained with the features selected by the Filters using the PART and ALL approaches on the synthetic datasets

Figure 2 shows the average test accuracy of the Naive Bayes (NB) classifiers trained with the features selected by both approaches. The best classification accuracy was obtained by S3-PART as well as S3-ALL, which has $S = 10,000$ and $N_I = 90$. While, the worst classification accuracy as well as the lowest similarity was obtained by S7-PART, as this dataset has $S = 100$ and $N_I = 9,990$. Among these two datasets, we can see various classification accuracy results, varying based on two factors in general: the number of samples and the number of irrelevant features.

We can clearly observe that the ALL approach has a higher accuracy than the PART approach on the datasets with small samples (as S1, S4 and S7). The NB classifiers trained with S7-ALL in particular greatly outperform those trained with S7-PART by about 47.2 % on average in terms of accuracy, while both approaches give very low similarity, as we can see in Fig. 1; this case simulates the problem of micro-array datasets, which have high dimensionality with small numbers of samples. On the other hand, the PART and ALL approaches obtained similar accuracy on the remaining datasets, which have medium or high numbers of samples.

Further observations reveal that NB classifiers trained with S2, S5 and S8 (without adding any noise) have achieved a higher accuracy than that of the classifiers with the datasets added 5 and 10 % noise. Moreover, The above figure shows a little decrease in the accuracy by increasing the noise rate. For example, S2Noise5-ALL has (93.04) with most of the filters, while S2Noise10-ALL has (91.19) with all the filters.

6 Conclusion

In this paper, the differences between the PART and ALL approaches have been investigated in terms of similarity and classification accuracy on 14 synthetic datasets generated.

In conclusion, when the dataset contains a large number of samples there is not a noticeable difference between these two approaches in terms of reliability and effectiveness. When dataset is small, the ALL and PART approaches have almost similar reliability compared with the desired features, but the actual features selected may be different. Their difference is evaluated by effectiveness of classification and the results show that, the ALL approach achieves a higher accuracy than the PART approach, which indicates that the accuracy estimate is likely overstated and the bias occurred. Therefore, the PART approach can prevent the bias to some extent although its superiority decreased with increasing samples.

References

1. Ambroise, C.: Selection bias in gene extraction on the basis of microarray gene-expression data. Proc. Nat. Acad. Sci. **99**(10), 6562–6566 (2002)
2. Hall, M.: Correlation-based feature selection for machine learning, The University of Waikato, Hamilton (1999)
3. He, Z., Yu, W.: Stable Feature Selection for Biomarker Discovery. (2010). doi:10.1016/j.compbiolchem.2010.07.002
4. Lecocke, M., Hess, K.: An empirical study of univariate and genetic algorithm-based feature selection in binary classification with microarray data. Cancer Inform. **2**(3), 313–327 (2006)
5. Liu, H., Yu, L.: Toward integrating feature selection algorithms for classification and clustering. Knowl. Data Eng. IEEE Trans. **17**(4), 491–502 (2005)
6. Quinlan, J.: C4.5: Programs for Machine Learning, Morgan kaufmann, Burlington (1993)
7. Refaeilzadeh, P., Tang, L., et al.: On comparison of feature selection algorithms. Proceedings of AAAI Workshop on Evaluation Methods for Machine Learning II (2007)
8. Refaeilzadeh, P., Tang, L., et al.: Cross-validation. Encyclopedia of Database Systems, pp. 532–538. Springer, Berlin (2009)
9. Robnik-Sikonja, M., Kononenko, I.: Theoretical and empirical analysis of ReliefF and RReliefF. Mach. Learn. **53**(1), 23–69 (2003)
10. Singhi, S., Liu, H.: Feature subset selection bias for classification learning. Proceedings of the 23rd International Conference on Machine learning, ACM (2006)
11. Somol, P., Novovicova, J.: Evaluating stability and comparing output of feature selectors that optimize feature subset cardinality. Pattern Anal. Mach. Intell. IEEE Trans. **32**(11), 1921–1939 (2010)
12. Yu, L., Liu, H.: Efficient feature selection via analysis of relevance and redundancy. J. Mach. Learn. Res. **5**, 1205–1224 (2004)

Self-reinforced Meta Learning for Belief Generation

Alexandros Gkiokas, Alexandra I. Cristea and Matthew Thorpe

Abstract Contrary to common perception, learning does not stop once knowledge has been transferred to an agent. Intelligent behaviour observed in humans and animals strongly suggests that after learning, we self-organise our experiences and knowledge, so that they can be more efficiently reused; a process that is unsupervised and employs reasoning based on the acquired knowledge. Our proposed algorithm emulates meta-learning in-silico: creating beliefs from previously acquired knowledge representations, which in turn become subject to learning, and are further self-reinforced. The proposition of meta-learning, in the form of an algorithm that can learn how to create beliefs on its own accord, raises an interesting question: can artificial intelligence arrive to similar beliefs, rules or ideas, as the ones we humans come to? The described work briefly analyses existing theories and research, and formalises a practical implementation of a meta-learning algorithm.

1 Introduction

A question often asked, is what could be done with the knowledge of a domain, after it has been acquired. Raymond Cattell hypothesised that *Crystallised Intelligence* [1] is the key to acquiring knowledge in a supervised or unsupervised manner, and that there exists a connection between learning and what takes place after: reusing the acquired knowledge or skill. The connection of learning and the processes taking place after are the focus of our paper, as it describes the transition from learning to meta-learning, the ability to manipulate and reshape existing knowledge in order to optimise its reuse. Meta learning in psychology was first described by Donald

A. Gkiokas (✉) · A.I. Cristea
Computer Science Department, University of Warwick, Coventry CV4 7AL, UK
e-mail: a.gkiokas@warwick.ac.uk

A.I. Cristea
e-mail: a.i.cristea@warwick.ac.uk

M. Thorpe
Mathematics Institute, University of Warwick, Coventry CV4 7AL, UK
e-mail: M.Thorpe@warwick.ac.uk

© Springer International Publishing Switzerland 2014
M. Bramer and M. Petridis (eds.), *Research and Development in Intelligent Systems XXXI*, DOI 10.1007/978-3-319-12069-0_14

Maudsley [2] in 1979. In Computer Science, meta-learning has been considered a sub-field of machine learning used automatically on meta-data [3, 4]. Psychology and Computer Science display different approaches to the same idea: *some process* takes place after learning, probably *consciously* [2]. Existing research on meta-learning in artificial intelligence is scarce; Vanschoren and Blockeel [3] examine the computational aspects of meta-learning in data mining, yet they emphasise the usage of synthetic knowledge representation. In an analogue manner, we propose the employment of conceptual graphs [5, 6], as the knowledge representation structure. We do so because the knowledge acquisition agent from our previous work [7], already implements conceptual graphs, but also because they are based on *common logic*, and parsimoniously capture the structural and relational information of the generated knowledge, in the form of a belief without any excess of other meta-data.

State of the art research done for IBM by Vilalta and Drissi [4], best describes the key components of a meta-learning agent or system. They emphasise classification as being an important part, but most importantly describe meta-learning as a fusion of inductive learning, classification and knowledge representation. Their work is of a more theoretical nature, as they focus on structured and unstructured data, as well as cross-domain performance. Yet, their most important insight is that meta-learning is a self-adaptive processes relying on experience and knowledge.

2 Objectives

Our foremost objective is the generalisation of knowledge, or in this case, conceptual graphs. The knowledge generated is not *discovered* in the traditional sense; it is not a search or exploration, as is often the case with artificial intelligence architectures, reasoners or problem solvers, such as UIMA [8]. The first *goal is to compress the accumulated knowledge* for a specific cluster of highly correlated knowledge, in more generalised or abstract forms, that can represent a group, or parts of it. Our expectation of doing so, is achieving better learning performance while reducing knowledge instances. The algorithm must self-assess and evaluate its own beliefs, and self-reward its learning progress, without the intervention of an external party. The positively reinforced beliefs, should be apt to changes and non-monotonic; if one contradicts another they should not cancel each-other out, whereas belief generation should become better as learning converges.

The input is the set of conceptual graphs G, stored in the agent's memory; where g_n is any conceptual graph, positively rewarded by a user; that set is first classified into a set of clusters K, by using vector space model [9] and K-Means [10, 11]. Each *cluster* in set K contains a sub-set of G that is attributionally similar (e.g., a high frequency of similar words). Following classification, every graph g_n in cluster k_n, is examined semantically and compared to other graphs in its cluster k_n, so as to induce a new set of generalised graphs (called beliefs) g_n'.

$$Size(g_n') < Size(k_n) \quad \text{and} \quad g_n' \not\subset k_n. \tag{1}$$

The conditions of the output (1) are that the result must be smaller than the initial number of conceptual graphs, and the result must **not** contain any identical conceptual graph (but only *similar*). Any new derived conceptual graph g', *may or may not* depend, on a single graph or more, from k_n. The most important objective of the system, is to learn what semantic distance is acceptable for the generation of a new belief through induction. We employ reinforcement learning for this purpose [12], in an episodic manner, with the aim to converge on reasoning: *what semantic similarity is acceptable for the generation of a belief.* Rewarding of the generated beliefs is done posteriorly, as the belief needs be tested against new incoming conceptual graphs; in that sense the agent is self-reinforcing its decision to create that belief, express by a series of policies Q as shown in (2), for the episode that constructed the belief g'_n. Constants α and γ are the learning and discount rates, whilst R is the reward of the next state.

$$Q(s_t, a_t) \leftarrow Q(s_t, a_t) + \alpha[R_{(t+1)} + \gamma Q(s_{t+1}, a_{t+1}) - Q(s_t, a_t)]. \qquad (2)$$

To obtain meta-learning, we follow the phases as defined by Vivalta et al. [4]. The first phase of meta-learning they propose is characterised by a clustering operation. We fuse K-Means [10], and vector space model [9] (VSM) and use the methodology described in [11] for estimating the number of clusters K. The VSM constructs a sparse matrix M of token-pattern frequencies, using the frequency of appearance of a token within a graph, and each graph is associated with a positively rewarded conceptual graph. The vector space model, given as input the proportionally vectorised $V(g_n)$ of the queried graph, will return as output a vector of degrees of similarity $S(g_n)$, with respect to the (graph index) columns of matrix M, as shown in (3).

$$S(g_n) = \frac{M * V(g_n)}{\|M\| \|V(g_n)\|}. \qquad (3)$$

The output vector $S(g_n)$ represents the similarity of the input graph g_n to all other graphs in memory. For each VSM output $S(g)$, we *min-max* normalise it between *0* and *1*, and then append it as a row into a matrix A. We use K-Means, by minimizing (4) with respect to 'cluster centers' μ_i, on data set $\{x_j\}_{j=1}^n$ where x_i are the similarity values of graph g_n to all other graphs.

$$\sum_{j=1}^{n} \min_{i=1,...,K} \|x_j - \mu_i\|^2. \qquad (4)$$

The partition is then defined by associating each x_j with the cluster center closest to it, i.e., $k_i = \{x_j : \|x_j - \mu_i\| \leq \|x_j - \mu_m\|$ for all $m = 1, 2, \ldots, K\}$. The number of clusters are estimated by the number of dominant eigenvalues of the matrix A where $A(i, j) = \|x_i - x_j\|^2$. Semantic approximation takes the form of induction or generalisation, by iterating or traversing a semantic graph, starting from a query node, and moving towards super-ordinates. WordNet [13] implements its "senses" in a *Hasse* graph, a mono-directional and ordered semantic graph. In inductive logic, a traversal from n to n' (often denoted as $[n, n']$), where n is in a different layer from n', the originating node n will *most likely* inherit the properties and attributes of n'. Induction is essential to meta-learning, when producing generalisations, as expressed by Vivalta et al. [4]. Using WordNet, we acquire the super-ordinate graph of node n and n' (known as hypernyms). Establishing a semantic path from one another, is done by iterating super-ordinates and their layers. Finding a direct path from $[n, n']$ might not always be possible, in which case, the algorithm uses a breadth-first search, by finding if nodes n and n' share a common super-ordinate. For the first common super-ordinate obtained, the semantic distance, quantifiable by value $\delta_{[n,n']}$ is calculated, as shown in (5).

$$\delta_{[n,n']} = w_{s_i}\left(\sum d_{[n,n']}\right). \quad \text{and} \quad w_{s_i} = \frac{s_i - s_{min}}{s_{max} - s_{min}}. \tag{5}$$

The WordNet sense s_i is min-max normalised to w_{s_i}, as shown in (5), and $\sum d_{[n,n']}$ (5) is the sum of traversed semantic graph layers. The normalised sense weight w_{s_i}, biases towards most frequently appearing WordNet senses, since the topmost frequent is presented first, and least frequent is returned last [13]. Semantic distance is employed directly by the meta-learning algorithm, as a means to estimate the substitutability of a node n, by its super-ordinate n'.

3 Meta-Learning

Generating beliefs in the form of conceptual graphs, is the second step. After having acquired clusters of *similar* conceptual graphs, the algorithm iterates conceptual graphs in a cluster k_n. Each conceptual graph g_i in cluster k_n, is compared to another graph g_k also from k_n. Calculating the similarity of g_i and g_k, involves summing all their vertex similarities expressed by the value $\sum \delta[g_i^n, g_k^n]$. A second condition is that g_i and g_k are *isomorphically equivalent*, in addition to being *semantically related*. Isomorphism, a key component in graphs [5], requires that g_i and g_k are *structurally identical* with regards to their edges (graph similarity). Yet, graph g_i may have an edge to a vertex to which g_k doesn't. If, however, there exists a semantic path between the vertices, meaning that the connected vertex n by edge e in g_i, is *similar* to the connected vertex n' by edge e' in graph g_k, induction is possible. Therefore, in addition to $\delta_{[n,n']}$, we examine *isomorphic similarity* by iterating all edges e of n and all edges e' of n'. Edge similarity is accumulative and expressed by $\varepsilon_{[n,n']}$, as shown in (6).

$$\epsilon_{[n,n']} = \sum \delta_{[e_i(n),e_i(n')]}. \tag{6}$$

If both semantic and isomorphic similarity are satisfied for g_i and g_k, then a new graph g'_{ik} will be created, representing a *belief*. That *belief* is created, by replacing the *common super-ordinate* found for semantic path $[n, n']$ when calculating $\delta_{[n,n']}$. Those criteria (node delta value and isomorphism) are two of the Markov properties used by reinforcement learning, in the *decision making* (third) phase. The Markov decision process (MDP), controlled by reinforcement learning, creates an *episode*: a sequence of *states* s_t, chained together by *actions* a_t. A state s_t is described and *partially observable* by three components: conceptual graph distributional similarity (with respect to another graph) normalised from 0 to 1, node semantic distance expressed by $\delta_{[n,n']}$, and isomorphic variance expressed by $\varepsilon_{[n,n']}$. An action a_t may be one of following decisions: abort the process, or substitute nodes n and n' with their common super-ordinate k. The reward $R_{(t+1)}$ is the reward of the next state, and is given to the algorithm at a later time, by cross-evaluating the belief g'_{ik}. As per the reinforcement learning literature [12], the reward is given only to terminal states. When creating a belief g'_{ik}, the policy $Q(s_t, a_t)$ updates its value, whilst trying to converge on the optimal decision. Evaluating actions is done by rewarding positively the episode that created the belief g'_{ik} when a new conceptual graph g_z (newly acquired knowledge) verifies it. Verification is performed by finding that graph g_z could substitute any or all of its nodes with the ones in belief g'_{ik}, and by establishing that they are isomorphically equivalent. The graph g_z must belong to the same cluster k_n, from which g'_{ik} was generated, in order to be eligible for evaluation. Self-reinforcement essentially strengthens the algorithm's policy and decision making process, making it learn how to best create beliefs.

4 Conclusions

In this paper we have presented a novel meta-learning algorithm, created via a fusion of existing artificial intelligence models and technologies. We've formulated the theoretical and algorithmic foundations of the meta-learning model, and systematically explained how it would work under training and testing. The proposed algorithm constructs meta-knowledge representations, by generalising and abstracting existing knowledge into beliefs, based on induction. The learning mechanism is based on a robust reinforcement learning algorithm, whereas the knowledge representation scheme used is Sowa's conceptual graphs. Finally, classification is done via a fusion of K-means and vector space model, whereas Semantics are obtained via Word-Net. By designing and describing this algorithm, we've taken every precaution to adhere to previous theoretical and practical work: adaptive behaviour, learning via induction, semantics, classification, knowledge representation, all emphasised key characteristics from previous researchers involved in state-of-the-art frameworks.

References

1. Cattell, R.B.: Abilities: Their Structure, Growth, and Action. Houghton Mifflin, New York (1971). ISBN-10: 0395042755
2. Maudsley, D.B.: A Theory of Meta-Learning and Principles of Facilitation: An Organismic Perspective, University of Toronto, Toronto (1979)

3. Vanschoren, J., Blockeel, H.: Towards understanding learning behavior. In: Proceedings of the Annual Machine Learning Conference of Belgium and The Netherlands, Benelearn, pp. 89–96 (2006)
4. Vilalta, R., Drissi, Y.: A perspective view and survey of meta-learning. Artif. Intell. Rev. **18**, 77–95 (2002)
5. Sowa, J.F.: Knowledge Representation: Logical. Brooks Press, Philosophical and Computational Foundations (1999). ISBN:10:0534949657
6. Chein, M., Mugnier, M.: Graph-based Knowledge Representation. Springer, Berlin (2009). ISBN:978-1-84800-285-2
7. Gkiokas, A., Cristea, A.I.: Training a cognitive agent to acquire and represent knowledge from RSS feeds onto conceptual graphs, IARIA COGNITIVE 2014, pp. 184–194. Venice, Italy, 25 (May 2014)
8. Ferrucci, D., Lally, A.: UIMA: an architectural approach to unstructured information processing in the corporate research environment. J. Nat. Lang. **10**(3–4), 327–348 (2004)
9. Turney, P., Pantel, P.: From frequency to meaning: Vector space models of semantics. J. Artif. Intell. Res. **37**, 141–188 (2010)
10. MacQueen, J.B.: Some methods for classification and analysis of multivariate observations. In: Proceedings of 5th Berkeley Symposium on Mathematical Statistics and Probability, Berkeley, vol. 1, pp. 281–297. University of California Press, California (1967)
11. Girolami, M.: Mercer kernel-based clustering in feature space. Neural Netw. IEEE Trans. **13**(3), 780–784 (2002)
12. Sutton, R.S., Barto, A.G.: Reinforcement Learning: An Introduction. MIT press, Cambridge (1998). ISBN-10:9780262193986
13. Fellbaum, C.: WordNet: An Electronic Lexical Database (Language, Speech, and Communication). MIT Press, Cambridge (1998). ISBN-10:026206197X

Applications and Innovations
in Intelligent Systems XXII

Best Application Paper

Combining Semantic Web Technologies with Evolving Fuzzy Classifier eClass for EHR-Based Phenotyping: A Feasibility Study

M. Arguello, S. Lekkas, J. Des, M.J. Fernandez-Prieto and L. Mikhailov

Abstract In parallel to nation-wide efforts for setting up shared electronic health records (EHRs) across healthcare settings, several large-scale national and international projects are developing, validating, and deploying electronic EHR-oriented phenotype algorithms that aim at large-scale use of EHRs data for genomic studies. A current bottleneck in using EHRs data for obtaining computable phenotypes is to transform the raw EHR data into clinically relevant features. The research study presented here proposes a novel combination of Semantic Web technologies with the on-line evolving fuzzy classifier eClass to obtain and validate EHR-driven computable phenotypes derived from 1,956 clinical statements from EHRs. The evaluation performed with clinicians demonstrates the feasibility and practical acceptability of the approach proposed.

1 Introduction

The adoption of *electronic health records* (EHRs) systems is at the heart of many international efforts to improve the safety and quality of healthcare [1]. A major benefit of EHRs as defined by the International Organisation for Standardisation (ISO) [2] is the ability to integrate patient's clinical data across different healthcare institutions (e.g. primary and secondary care). Economically developed countries like Australia, New Zealand, United Kingdom, France, and United States have launched projects that aim at setting up shared EHRs among various healthcare providers to improve care coordination [3]. Nowadays, the most prominent standards for the exchange of EHRs are: CEN/ISO EN13606 [4] and Health Level Seven (HL7) Clinical Document Architecture (CDA) [5].

M. Arguello (✉) · S. Lekkas · L. Mikhailov
University of Manchester, Manchester M13 9PL, UK
e-mail: m.arguello@computer.org; arguellm@cs.man.ac.uk

J. Des
SERGAS, 15703 Lugo, Spain

M.J. Fernandez-Prieto
University of Salford, Salford M5 4WT, UK

© Springer International Publishing Switzerland 2014
M. Bramer and M. Petridis (eds.), *Research and Development in Intelligent Systems XXXI*, DOI 10.1007/978-3-319-12069-0_15

In parallel to the above-mentioned efforts to achieve interoperability of EHRs across healthcare settings, several large-scale national and international projects, like eMERGE [6], CICTR [7], and SHARP [8] are developing tools and technologies for identifying patient cohorts using EHRs data. A key component in this process is to develop, validate, and deploy electronic EHR-oriented phenotype algorithms that aim at large-scale use of EHRs data.

EHR phenotyping uses data from EHRs with the aim of identifying individuals or populations with a condition or clinical profile, the so-called computable phenotype. EHR-driven phenotype definitions may include data from different sources, for example, clinical narratives. Shivade et al. [9] performed a literature review to pin down approaches that aim at automatically identifying patients with a common phenotype. According to their review, after the extraction of relevant terms and concepts from free text reports, either a rule—or machine learning-based model is used to classify patients into cohorts. Shivade et al. [9] emphasise that few studies use Semantic Web technologies for phenotyping. Likewise, Shivade et al. [9] stress that very few studies explore automated rule mining for phenotyping.

The research study presented here proposes a novel combination of Semantic Web technologies (OWL [10], SPARQL [11], and SWRL [12]) with the on-line evolving fuzzy classifier eClass [13] to obtain EHR-driven computable phenotypes from 125 HL7 CDA consultation notes that contain 1,956 clinical statements. The paper also reports the results of an evaluation performed, which demonstrates the feasibility and practical acceptability of the approach proposed that automatically generates Takagi-Sugeno fuzzy rules of first order [13].

2 Research Background and Related Work

A phenotype is defined as: "*the observable expression of an individual's genotype*" [14]. In turn, the genotype is defined as: "*the specific DNA sequence at a given location*" [15]. As Schulze and McMahon [16] highlight, while genotypes affect proteins, cells, and biological pathways; phenotypes are more easily observed as manifestations (symptoms and signs) of the presence of a disease. Hripcsak and Albers [17] notice that the emerging studies employing large-scale EHR data use a two-step approach: (1) a phenotyping or feature extraction step, which transforms the raw EHR data into clinically relevant features; and (2) more traditional analysis step, which uses the features extracted for discovering/measuring associations (e.g. genotype-phenotype associations if EHR are linked to DNA biorepositories or biobanks) or identifying individuals that match research criteria for clinical trials (i.e. assessing patient eligibility).

Ontologies are the backbone of the Semantic Web, for they provide a shared understanding of a domain of interest and are particularly suitable for knowledge exchange and integration. The key components of the Semantic Web include RDF [18] as the basic data model, OWL [10] for expressive ontologies, and SPARQL query language [11]. Until now, few recent studies adopt Semantic Web technologies for phenotyping. Cui et al. [19] developed an ontology-based epilepsy cohort identification system. Pathak et al. [20] present a system architecture that leverages Semantic Web technologies for phenotyping and illustrate its viability with a case study for diabetes type 2 [21]. Pathak et al. [22] investigate how Semantic Web technologies (RDF and SPARQL) can aid the discovery of genotype-phenotype associations with EHR-linked to DNA biobanks. More specifically, the Pathak et al. [22] represent EHR data (diagnosis and procedures) in RDF and use federated queries (multiple SPARQL endpoints) to enable the discovery of gene-disease associations in individuals genotyped for Diabetes Type 2 and Hypothyroidism.

Shivade et al. [9] recognise that extracting meaningful pieces of information from EHRs and consolidating them into a coherent structure is paramount for automatically identifying patient cohorts satisfying complex criteria. However, defining even a small number of phenotypes can take a group of institutions years [17]. Hripcsak and Albers [17] emphasise that despite advances in ontologies and language processing for phenotyping, the process of feature extraction remains largely unchanged since the earliest days [23].

Besides popular machine learning and statistical analysis methods for determining phenotypes, other approaches have also been explored [9]. Tatari et el. [24] used multi-agent fuzzy systems to identify patients with a high risk of breast cancer. It should be noted that fuzzy classification approaches, such as [25, 26], generally require the data to be processed in off-line mode, as a batch.

eClass and FLEXFIS-Class are on-line evolving fuzzy classifiers [27]. Both of them follow a well-known option for achieving data-driven approximation models that lies in the usage of Takagi-Sugeno fuzzy models [28]. As Angelov et al. [27] remark: (a) the term "evolving" means that new rules, structures, and so on are evolved during on-line processing based on 'new' data samples; and (b) they should not be confused with "evolutionary" proposals (sometimes also called "evolving" [29]), which are usually processed in off-line mode, as a batch.

Both eClass and FLEXFIS-Class methods are designed to work on a per-sample basis and are thus one-pass and incremental [27]. They both are evolving in the sense that their structure (fuzzy rule-base) is not fixed, and can grow and shrink [13, 30]. Classifiers from the eClass family [13] can start learning "from scratch", while classifiers from the FLEXFIS family [30] require a certain amount of data for initialisation prior to the on-line incremental operation. As having a clinician interpreting EHR data for a certain number of patients can be labour-intense or unfeasible, eClass seems better suited for EHR-based phenotyping. eClass has been already successfully used with medical data [31–33]. To our best knowledge, eClass has not been used with EHR data.

3 Fuzzy Pattern Recognition for EHR-Based Phenotyping

The starting point of this research is a set of 125 consultation notes that are formatted according to the EHR standard HL7 CDA. HL7 CDA documents derive their machine processable meaning from the HL7 Reference Information Model (RIM) [34] and use the HL7 V3 [35] data types. The RIM and the V3 data types provide a powerful mechanism for enabling CDA's incorporation of concepts from standard coding systems such as SNOMED CT [36] and LOINC [37]. According to CDA Release 2 [5], a CDA document section can contain a single narrative block (free text) and any number of CDA entries (coded information, e.g. HL7 RIM Observations). We built on our previous work, and thus, XML-based CDA sections and entries are mapped to ontology instances in OWL 2 [10].

Pattern recognition can be seen as a sequence of some steps [38], namely (1) data acquisition; (2) feature selection; and (3) classification procedure. Figure 1 shows an architecture overview of our approach with the key components involved. Each of the three major steps can be summarised as follows:

1. *Data acquisition*—this step relies on the OWL converter (see Fig. 1) to obtain a formal semantic representation of clinical statements as stated by the XML-based HL7 CDA standard specification. This step builds on our previous work [39], and takes advantage of the OWL's import capability to deal with ontological information from multiple documents.
2. *Feature selection*—this step firstly uses the query engine ARQ for Jena [40] to execute SPARQL queries that retrieve individuals (ontological instances) with specific characteristics from the OWL model, and it builds on our previous work [39]. Secondly, the feature selection filter F-score [41] is applied.
3. *Classification procedure*—this step employs the evolving fuzzy classifier eClass, which exploits the first order Takagi-Sugeno (TS) fuzzy model [28], where the consequents of the fuzzy rule are linear classifiers. Although eClass makes sole use of its knowledge in its fuzzy form, a defuzzification process is also employed to achieve transparency. Despite this defuzzification process, the straightforward linguistic interpretability of the TS rules generated makes unavoidable for clinicians to possess some understanding of TS fuzzy models. And thus, the TS rule translator component is introduced to transform TS rules into SWRL [12]

Fig. 1 Approach overview: fuzzy pattern recognition for EHR-based phenotyping

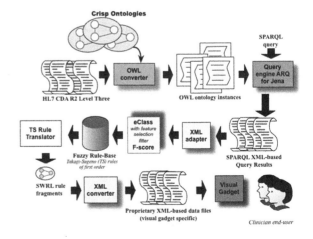

rule fragments according with some *disease-specific indicators* defined by an epidemiologist to aid the clinical interpretation of the TS rules. An XML converter is then added to convert the SWRL rule fragments into proprietary XML-based fragments that are visual gadget specific. The clinicians interact with visual gadgets from amCharts [42] and remain oblivious of the underlying transformations, i.e. from TS fuzzy rules of first order to SWRL rule fragments.

3.1 Fuzzy Pattern Recognition Using eClass

This section provides an overview of the *evolving fuzzy classifier* eClass introduced that tackles *knowledge generation* from EHR *data streams in real-time* and is used for EHR-based phenotyping.

Let Z be a set of 125 HL7 CDA consultation notes with established diagnosis belonging to the *Assessment and Plan sections*. These established diagnosis (118 CDA entries) are considered as known classification labels to classes from the set $\Omega = \{w_1,...,w_C\}$. Clinical statements from the *History of Present Illness section* (405 CDA entries) or the *Physical Findings section* (1,433 CDA entries) for each CDA consultation note can be mapped to numeric values of the features from a set $X = \{x_1,...,x_n\}$. Let S^n denote the feature space generated by the set X. The fuzzy pattern recognition task relies on fuzzy decision (FD) rules, where each input feature vector $x \in S^n$ may be attached to a set of classes with different degrees of membership.

The first order Takagi-Sugeno (TS) fuzzy models [28] are a special group of rule-based models with fuzzy antecedents and functional consequents that follow from the Takagi-Sugeno-Kang reasoning method [43]. The fuzzy rule base that describes the non-linear evolving classifier eClass can be described as a set of TS fuzzy rules of first order, where the consequents of the fuzzy rule are linear classifiers. These TS fuzzy rules are the FD rules for the fuzzy pattern recognition task and follow the form:

$$R^i: \text{IF } (x_1 \text{ is } A_1^i) \text{ AND } ... \text{ AND } (x_n \text{ is } A_n^i) \text{ THEN } y_c^i = f^i \qquad (1)$$

In Formula 1, R^i is the *i*th TS fuzzy rule (FD rule). In the antecedent part (if-part) of the TS fuzzy rule $x = [x_1, x_2, ..., x_n]^T$ is the n-dimensional (input) feature vector (also known as data sample). A feature vector contains discrete numerical values that are mapped to clinical statements from the *History of Present Illness section* or the *Physical Findings section* of a patient's CDA consultation note. A_j^i denotes the antecedent fuzzy sets, $j \in [1, n]$. In the consequent part (then-part) of the TS fuzzy rule, $y^i = [y_1^i, y_2^i, ..., y_C^i]$ is the C-dimensional fuzzy output, $i \in [1, N_C]$ and $c \in [1, C]$, where N_C is the number of rules per class c and C is the number of classes from the set Ω.

The structure of the eClass is thus formed by sets of TS fuzzy rules in such a way that there is at least one TS fuzzy rule per class. As suggested in [13, 44] and further employed in [32], the so called "winner-takes-all" defuzzification is applied to determine the correct class, which is the usual choice in classification problems. Hence, in Eq. 2, Class^i is the class label of the ith TS fuzzy rule, which corresponds to the class c with the highest possibility $y_c{}^i$.

$$\text{Class}^i = \text{argmax}_{1 \leq c \leq C}(y_c{}^i) \qquad (2)$$

For Gaussian-like antecedent fuzzy sets, a fuzzy set $A_j{}^i$ of the ith TS fuzzy rule, $j \in [1, n]$, is defined by the $\mu^i{}_j$ membership function (MF). The MF defines the spread of the antecedent and the zone of influence of the ith rule; where x^{i*} is the prototype (centroid or focal point) of the ith TS fuzzy rule antecedent. Hence, $\mu^i{}_j$ measures the degree of proximity of a data sample x^t to the prototype of a TS fuzzy rule.

Generally, the problem of identification of a first order TS model is divided into two sub-tasks [28, 43]: (1) learning the antecedent part of the model (see Eq. 1), which consists of determination of the focal points of the rules and the spreads of the MFs; and (2) learning the parameters of the linear subsystems of the consequents.

3.2 Introducing a Novel Online Feature Selection Filter

How to successfully achieve dimensionality reduction in conjunction with *eClass* is another novelty of the current research study. In general, dimensionality reduction methods are said to follow two basic baselines [45]: *filter* methods and *wrapper* methods. However both of them present serious drawbacks when dealing with online data streams. A third alternative to tackle dimensionality reduction, which also appears in literature, is based on F-Score [41]. This study adopts the extended version of the F-Score method as proposed by Lekkas and Mikhailov [32], which can deal with more than two classes. What remains to be asked from the study presented in [32] is whether the temporal F-Scores can be further used as a criterion to reduce the original dimensionality of the problem in online mode. Given the set of scores for every tth input sample, how can the resultant filtered input be more meaningful for an eClass model to learn from? To answer this question, we propose a *threshold based condition*. Let $F^t = \{F_1{}^t, \ldots, F_n{}^t\}$ be the set of F-Scores of the tth input sample x^t, and f^t the tth filtered input sample. The latter can be obtained by using the condition states in Formula 3.

$$\forall j, j \in [1, n]: \text{IF}(F_j^t > h * \max(F^t))\text{THEN} (f_j^t \leftarrow x_j^t) \text{ ELSE } (f_j^t \leftarrow 0) \qquad (3)$$

Formula 3 expresses the following condition: *every feature that has a score greater than a fraction of the maximum of the tth scores must be retained, otherwise it can be disregarded from the tth filtered input f.* In Formula 3, h is a threshold constant with reliable values in the range [0.001, 0.05]. The purpose of h is to restrain a portion of the maximum temporal score, leading to a very flexible online feature selection scheme. It is flexible because also considers when none of the features should be removed, for example when they scored alike. Hence, it is unlikely to cause loss of information as opposed to suddenly selecting the k top most-ranking features (according to what is suggested in [45]).

3.3 TS Fuzzy Rules: Clinically Aided Interpretation and Visualisation

The linguistic interpretability of the generated TS fuzzy rules by clinicians has been proved difficult as they are not computer scientists familiarised with fuzzy sets that involve cluster prototypes (focal points) or fuzzy membership function that defines the degree of closeness (or proximity) of a sample to a cluster prototype. The following excerpt exemplifies the straightforward linguistic interpretability of a TS fuzzy rule:

IF (x1 isAround[+/− 0.468] 0.000) AND (x2 isAround[+/− 0.468] 0.000) AND … AND (x17 isAround[+/− 0.468] 0.000) THEN Class = 2

In [46], different categories of *clinical indicators* can be found. According to [46], *disease-specific indicators* are diagnosis-specific. On the one hand, the consequents of the TS fuzzy rules used take into account the diagnosis, where the process of learning the consequent part of a fuzzy rule is supervised, and thus, the adoption of *disease-specific indicators* seems appropriate. On the other hand, the generated TS fuzzy rules can be plotted into 2D graphs to represent the *cluster centroid* and the *cluster radius*, and thereby, the introduction of the qualifiers *distal* and *proximal* to capture the graphical implications of the TS fuzzy rules can be seen as properly justified. In our proposal, we introduce "proximal *disease-specific indicators*" and "distal *disease-specific indicators*". However, these two terms are well known to epidemiologists. To illustrate this: 'smoking' is a well known *proximal disease-specific indicator* for lung cancer, while 'working in a coal mine' is a well known distal *disease-specific indicator* for lung cancer. Thus, the *proximal* and *distal disease-specific clinical indicators* introduced facilitate a quick understanding and easy access to key information capture in the *evolving fuzzy rule-base* that is dynamically obtained.

With the aim of providing a clinical interpretation for the TS fuzzy rules that is easy to understand by clinicians, firstly, the set of TS fuzzy rules is transformed into a set of rules in SWRL [12]. This transformation is feasible due to common characteristics among the two rule languages: the rules are written as antecedent-consequent pairs and the rules have conjunctive antecedents. For example, part of an antecedent of a TS fuzzy rule can be the following: *IF (x1 isAround[+/− 0.269] 0.000) AND (x2 isAround[+/− 0.269] 0.000) AND (x3 isAround[+/− 0.269] 0.000) AND ...* To transform the above-mentioned TS fuzzy rule antecedent to a rule antecedent in SWRL, the features (variables) $\{x_1, ..., x_n\}$ are reinterpreted as SNOMED CT clinical findings and two OWL ObjectProperties are being introduced (isAround and hasFocalPoint). To further illustrate this: (x1 isAround[+/− 0.269] 0.000) from a TS fuzzy rule is mapped to the following in SWRL: Lid_adhesions(?x1) ^ isAround(?x1, 0.269) ^ hasFocalPoint(?x1, 0.000).

The success of Ajax [47], shorthand for Asynchronous JavaScript and XML, has brought an increasing amount of visual gadgets, most of them generic enough to be used in different contexts, where typically JavaScript provides the means to allow end-user interaction. An attractive feature that is increasingly commonly available among these visual gadgets, such as *amCharts* [42], is to distinguish between *settings* and *data*. This means that the settings could be configured just once for a particular use, and then, different data sets can be used without reconfiguration. Furthermore, *settings* and *data* are usually stored in two separated XML-based files, and therefore, it is possible to build on-fly XML-based *data files* upon user request. The current approach (see Fig. 1) incorporates XML converters to "translate" fragments of SWRL rules into an XML-based file (XML-based data files of visual gadgets).

4 Experimental Results and Clinicians' Evaluation

The experiments conducted exploit the architecture described in Sect. 3 (see Fig. 1 for details), and adopts the buffering technique proposed by Lekkas and Mikhailov [33], where data samples are processed in time windows and not in batch. The buffering technique applied considers buffers that can store 5, 10, or 20 data samples.

The experiments performed consider two datasets:

- Dataset I—the data samples contain discrete numerical values that are mapped to clinical statements from the *History of Present Illness sections* of patients' CDA consultation notes; and
- Dataset II—the data samples contain discrete numerical values that are mapped to clinical statements from the *Physical Findings sections* of patients' CDA consultation notes.

The clinical statements from both the *History of Present Illness sections* and the *Physical Findings sections* of patients' CDA consultation notes mostly state the presence or absence of SNOMED CT clinical findings. And thus, the XML adapter (see Fig. 1) translate the XML-based results of SPARQL queries into: (a) variables that correspond to SNOMED CT clinical findings; and (b) discrete numerical values associated to the variables that mostly correspond to the presence or absence of SNOMED CT clinical findings.

The two datasets (dataset I and dataset II) come from 125 anonymised CDA consultation notes that have established diagnosis (CDA entries) related to 'Red Eye' [48], which involves 8 diseases coded in SNOMED CT. For both datasets, eClass starts 'from scratch' with an empty fuzzy rule base and no pre-training; and subsequently, evolves its structure from the data stream. The execution time of the dataset I is 203 ms; while the dataset II is executed in 515 ms. This difference is due to differences in dimensionality. Dataset I has 17 features, while Dataset II has

37 features. Hence, for the same amount of data samples, it is expected that as the number of features increases, so does the execution time. Eight TS fuzzy rules are automatically generated for dataset I and nine TS fuzzy rules for dataset II.

Figure 2 shows for dataset I the learning rate of eClass with the online feature selection filter F-score from [32] (left hand-side); the expansion rate of the fuzzy rule base (RB) for increasing time step (middle); and the area of dimensional reduction achieved using eClass with the filter F-Score (right hand-side), where a number of features may be discarded for every time-step.

Figure 3 shows for dataset II the learning rate of eClass (left hand-side) and the expansion rate of fuzzy RB for increasing time step (right hand-side).

To validate the proposal, visual representations of the *distal* and *proximal disease-specific indicators* introduced are enabled by means of amCharts [42] visual gadgets. It was soon observed the utility of visualising more than one diagnosis simultaneously for ease of comparison. Figure 4 shows the visual representation of the *distal disease-specific indicators* obtained for three diagnoses related to 'Red Eye'. The clinical indicators introduced aid the interpretation of the TS rules obtained for these three diagnoses from symptoms (CDA entries) of the *History of Present Illness sections* of CDA consultation notes, i.e. EHR data. In order to measure the suitability of the EHR-based phenotypes obtained, Newton et al. [6] use precision and recall. In the same vein, we conducted an evaluation with a physician with more than fifteen years of experience in clinical practice and we calculated accuracy. Table 1 reports the accuracy of the TS rules obtained for the eight diseases (SNOMED CT established diagnoses) related to 'Red Eye' with EHR data, i.e. dataset I and dataset II. It should

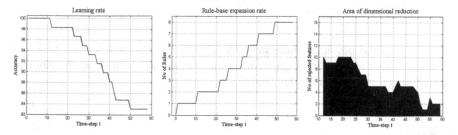

Fig. 2 Dataset I—learning rate of eClass with F-Score; expansion rate of fuzzy RB; and area of dimensionality reduction

Fig. 3 Dataset II—learning rate of eClass and expansion rate of fuzzy RB

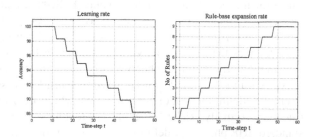

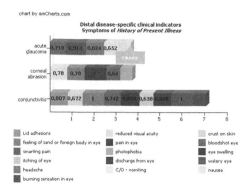

Fig. 4 Cross-comparing distal disease-specific indicators for three diseases for 'Red Eye'

Table 1 Physician evaluation—accuracy of the TS rules automatically generated

Established diagnosis (SNOMED CT)	Accuracy for dataset I (TP + TN)/(P + N) (%)	Accuracy for dataset II (TP + TN)/(P + N) (%)
Conjunctivitis	82.35	83.78
Conjunctival hemorrhage	76.47	97.29
Corneal ulcer	94.12	81.08
Herpes simplex dendritic keratitis	94.12	89.19
Photokeratitis	88.23	94.59
Scleritis and episcleritis	88.23	94.59
Acute angle-closure glaucoma	94.12	78.38
Anterior uveitis	82.35	78.38

be noted that the physician understood the *disease-specific indicators* introduced with the help of an epidemiologist without any further explanation. At the moment of writing, more evaluations are planned.

5 Discussion

The growing adoption of EHRs favours EHR-based genetic studies that aggregate/collect phenotype information as a by-product of routine healthcare. Dealing with EHRs data implies dealing with subjective information and missing data. Besides the heterogeneity in EHRs, one of the current challenges towards a systematic study of clinical phenotypes is achieving standards-based executable phenotype definitions [49]. To further illustrate this: Richesson et al. [50] report seven phenotype definitions for diabetes mellitus, which are used for diabetes cohort identification. Actually, Richesson et al. [50] recognise that currently, there is a lack of standard EHR phenotype definitions for most chronic conditions, including diabetes.

This research study adopts eClass, which can cope with real-time classification of streaming data from EHRs. Three main benefits arise from the experiments performed to prove the suitability of eClass for EHR-based phenotyping: (1) when using eClass the number of fuzzy rules as well as the number of classes (main diagnoses from EHRs) can change and do not need to be fixed; (2) eClass can start either "from scratch" (with an empty fuzzy rule base) or with some pre-specified set of fuzzy rules; and (3) if the class label is not provided, the existing fuzzy rule base can generate the predicted class, and thereby, a patient can be assigned to a certain cohort. These three benefits make of eClass a powerful tool for identifying patient phenotype cohorts using EHRs data. The main drawback of eClass, despite the defuzzification process, is the lack of straightforward clinical interpretability of the TS fuzzy rules generated, which makes unavoidable for clinicians to possess some understanding of TS fuzzy models.

Our proposal incorporates Semantic Web technologies for: (1) extracting clinical statements (symptoms, signs, and established diagnoses) from EHRs, by means of SPARQL queries that retrieve OWL individuals (ontological instances) as part of the feature extraction step; and (2) aiding the clinical interpretation and visualisation of the evolving fuzzy rules by means of *disease-specific indicators* that are incorporated into the OWL ontology and exploited by the SWRL rules that are mapped to the TS fuzzy rules generated from the EHRs data.

It should be noted that our approach advocates for managing the fuzziness directly within Semantic Web components, as Ciamarella et al. [51], instead of extending OWL or SWRL to deal with fuzziness. Examples of the latter can be found in Stoilos et al. [52] and Pan et al. [53].

In our study, the *disease-specific indicators* introduced are incorporated into crisp OWL ontologies and SWRL rules. These clinical indicators emerge from close collaboration with epidemiologists and favour a straightforward validation by healthcare professionals of the EHR-driven computable phenotypes obtained. Nowadays, validation of EHR-based phenotypes remains an important aspect of their use.

6 Conclusion

The growing adoption of EHRs favours EHR-based genetic studies that aggregate/collect phenotype information as a by-product of routine healthcare. EHR phenotyping uses data from EHRs with the aim of identifying individuals or populations with a condition or clinical profile, the so-called computable phenotype. The emerging studies for identifying computable phenotypes employ large-scale EHR data and use a two-step approach. Even with advances and new approaches, the first step that is feature extraction remains labour-intense and time consuming. Furthermore, the heterogeneity in EHRs means that phenotype validation is an important aspect of phenotype use, particularly when there is a lack of agreement about phenotype definitions for the same condition.

The research study presented here proposes a novel combination of Semantic Web technologies (OWL, SPARQL, and SWRL) with the on-line evolving fuzzy classifier eClass to obtain and validate EHR-driven computable phenotypes derived from 125 HL7 CDA consultation notes containing 1,956 clinical statements. The evaluation performed demonstrates the feasibility and practical acceptability of the approach proposed to automatically generate and validate EHR-based phenotypes. In our approach, the healthcare professionals remain unaware of the underlying technologies that support fuzzy pattern recognition to enable EHR-based phenotyping. Hence, consistent phenotype assessment may be feasible, fostering the possibility of reducing variation in phenotype definitions.

References

1. Cresswell, K.M., Worth, A., Sheikh, A.: Comparative case study investigating sociotechnical processes of change in the context of a national electronic health record implementation. Health Inform. J. **18**, 251–270 (2012)
2. ISO/TR 20514 Technical Report: EHR Definition, Scope, and Context (2005). Available via ISO http://www.iso.org/iso/catalogue_detail.htm?csnumber=39525. Accessed May 2014
3. Metzger, M.H., Durand, T., Lallich, S., Salamon, R., Castets, P.: The use of regional platforms for managing electronic health records for the production of regional public health indicators in France. BMC Medical Informatics and Decision Making. Springer, Berlin (2012)
4. CEN/ISO EN13606, http://www.en13606.org/the-ceniso-en13606-standard. Accessed May 2014
5. Dolin, R.H., Alshuler, L., Boyer, S., Beebe, C., Behlen, F.M., Biron, P.V., Shabo, A.: HL7 clinical document architecture, release 2. Am. Med. Inform. Assoc. **13**, 30–39 (2006)
6. Newton, K.M., Peissig, P.L., Kho, A.N., Bielinski, S.J., Berg, R.L., Choudhary, V., Basford, M., Chute, C.G., Kullo, I.J., Li, R., Pacheco, J.A., Rasmussen, L.V., Spangler, L., Denny, J.C.: Validation of electronic medical record-based phenotyping algorithms: results and lesson learned from the eMERGE network. Am. Med. Inform. Assoc. **20**, 147–154 (2013)
7. Anderson, N., Abend, A., Mandel, A., Geraghty, E., Gabriel, D., Wynden, R., Kamerick, M., Anderson, K., Rainwater, J., Tarczy-Hornoch, P.: Implementation of a deidentified federated data network for population-based cohort discovery. J. Am. Med. Inform. Assoc. **19**, 60–67 (2011)
8. Strategic Health IT Advanced Research Projects (SHARP), http://www.healthit.gov/policy-researchers-implementers/strategic-health-it-advanced-research-projects-sharp. Accessed May 2014
9. Shivade, C., Raghavan, P., Fosler-Lussier, E., Embi, P.J., Elhadad, N., Johnson, S.B., Lai, A.M.: A review of approaches to identifying patient phenotype cohorts using electronic health records. Am. Med. Inform. Assoc. **21**, 221–230 (2014)
10. OWL 2 Web Ontology Language, http://www.w3.org/TR/owl2-primer/. Accessed May 2014
11. SPARQL, http://www.w3.org/TR/rdf-sparql-query/. Accessed May 2014
12. SWRL, http://www.w3.org/Submission/2004/SUBM-SWRL-20040521/. Accessed May 2014
13. Angelov, P., Zhou, X., Klawonn, F.: Evolving fuzzy rule-based classifiers. In: Proceedings of IEEE Symposium on Computational Intelligence in Image and Signal Processing, pp. 220–225 (2007)
14. Wojczynski, M.K., Tiwari, H.K.: Definition of phenotype. Adv. Genet. **60**, 75–105 (2008)
15. Denny, J.C.: Mining electronic health records in the genomics era. PLoS Comput. Biol. **8**, 395–405 (2012)

16. Schulze, T.G., McMahon, F.J.: Defining the phenotype in human genetic studies: forward genetics and reverse phenotyping. Hum. Hered. **58**, 131–138 (2004)
17. Hripcsak, G., Albers, D.J.: Next-generation phenotyping of electronic health records. J. Am. Med. Inform. Assoc. **20**, 117–121 (2013)
18. RDF, http://www.3.org/RDF. Accessed May 2014
19. Cui, L., Bozorgi, A., Lhatoo, S.D., Zhang, G.Q., Sahoo, S.S.: EpiDEA: Extracting structured epilepsy and seizure information from patient discharge summaries for cohort identification. In: Proceedings of AMIA Annual Symposium, pp. 1191–1200 (2012)
20. Pathak, J., Kiefer, R.C., Chute, C.G.: Using semantic web technologies for cohort identification from electronic health records for clinical research. AMIA Summits Transl. Sci. Proc. **2012**, 10–19 (2012)
21. Pathak, J., Kiefer, R.C., Bielinski, S.J., Chute, C.G.: Mining the human phenome using semantic web technologies: a case study for type 2 diabetes. In: Proceedings of AMIA Annual Symposium, pp. 699–708 (2012)
22. Pathak, J., Kiefer, R.C., Bielinski, S.J., Chute, C.G.: Applying semantic web technologies for phenome-wide scan using an electronic health record linked Biobank. J. Biomed. Semant. **3**, 1–17 (2012)
23. Warner, H.R.: Knowledge sectors for logical processing of patient data in the HELP system. In: Proceedings of International Conference on Interactive Techniques in Computer-Aided Design, pp. 401–404 (1978)
24. Tatari, F., Akbarzadeh-T, M.-R., Sabahi, A.: Fuzzy-probabilistic multi agent system for breast cancer risk assessment and insurance premium assignment. J. Biomed. Inform. **45**, 1021–1034 (2012)
25. Goncalves, L.B., Velasco, M.M.B.R., Pacheco, M.A.C., De Souza, F.J.: Inverted hierarchical neuro-fuzzy BSP system: a novel neuro-fuzzy model for pattern classification and rule extraction in databases. IEEE Trans. Syst. Man Cybern. (Part C) **16**, 236–248 (2006)
26. Loo, C.K., Rao, M.V.C.: Accurate and reliable diagnosis and classification using probabilistic ensemble simplified fuzzy ARTMAP. IEEE Trans. Knowl. Data Eng. **17**, 1589–1593 (2005)
27. Angelov, P., Lughofer, E., Zhou, X.: Evolving fuzzy classifiers using different model architectures. Fuzzy Sets Syst. **159**, 3160–3182 (2008)
28. Takagi, T., Sugeno, M.: Fuzzy identification of systems and its applications to modelling and control. IEEE Trans. Syst. Man Cybern. **15**, 116–132 (1985)
29. Gomez, J., Gonzalez, F., Dasgupta, D., Nasaroui, O.: Complete expression tree for evolving fuzzy classifier systems with generic algorithms. In: Proceedings of North American Fuzzy Information Processing Society Conference on Fuzzy Logic and the Internet, pp. 469–474 (2002)
30. Lughofer, E., Klement, E.: FLEXFIS: A variant for incremental learning of tagaki-sugeno fuzzy systems. In: Proceedings of 14th IEEE International Conference on Fuzzy Systems, pp. 915–920 (2005)
31. Xydeas, C., Angelov, P., Chiao, S., Reoullas, M.: Advances in classification of EEG signals via evolving fuzzy classifiers and dependant multiple HMMs. Comput. Biol. Med. **36**, 1064–1083 (2005)
32. Lekkas, S., Mikhailov, L.: Breast cancer diagnosis based on evolvable fuzzy classifiers and feature selection. In: Proceedings of 28th International Conference on Innovation Techniques and Applications of Artificial Intelligence, pp. 185–195 (2008)
33. Lekkas, S., Mikhailov, L.: Evolving fuzzy medical diagnosis of Pima Indians diabetes and of dermatological diseases. Artif. Intell. Med. **50**, 117–126 (2010)
34. HL7 RIM, http://www.hl7.org/implement/standards/rim.cfm. Accessed May 2014
35. HL7 V3 data types, http://www.hl7.org/implement/standards/product_brief.cfm?product_id=264. Accessed May 2014
36. IHTSDO, SNOMED CT Editorial Guide (January 2014 International Release). Available via IHTSDO http://www.ihtsdo.org/eg.pdf. Accessed May 2014
37. LOINC, http://www.loinc.org/. Accessed May 2014

38. Pedrycz, W.: Fuzzy sets in fuzzy recognition: methodology and methods. Pattern Recogn. **23**, 121–146 (1990)
39. Arguello, M., Fernandez-Prieto, M.J., Des, J.: Extracting and visualising clinical statements from electronic health records. In: Research and Development in Intelligent Systems XXX, pp. 307–320. Springer, Berlin (2013)
40. Query engine ARQ for Jena, http://jena.apache.org/documentation/query/. Accessed May 2014
41. Chen, Y.W., Lin, C.J.: Combining SVMs with various feature selection strategies. Stud. Fuzziness SoftComput. **207**, 315–324 (2006)
42. Am Charts, http://www.amcharts.com/. Accessed May 2014
43. Angelov, P., Filev, D.: An approach to online identification of Tagaki-Sugeno fuzzy models. IEEE Trans. Syst. Man Cybern. **34**, 484–498 (2004)
44. Angelov, P., Zhou, X.: Evolving fuzzy rule-based classifiers from data streams. IEEE Trans. Fuzzy Syst. Spec. Issue Evol. Fuzzy Syst. **16**, 1462–1475 (2008)
45. Perkins, S., Theiler, J.: Online feature selection using grafting. In: Proceedings of the Twenty-First International Conference on Machine Learning, pp. 592–599 (2003)
46. Mainz, J.: Defining classifying clinical indicators for quality improvement. Int. J. Qual. Health Care **15**, 523–530 (2003)
47. Ajax, http://adaptivepath.com/publications/essays/archives/000385.php. Accessed May 2014
48. Fraser, S.R., Asaria, R., Kon, C.: Eye Know How. BMJ Books, London (2001)
49. Li, D., Endle, C.M., Murthy, S., Stancl, C., Suesse, D., Sottara, D., Huff, S.M., Chute, C.G., Pathak, J.: Modeling and executing electronic health records driven phenotyping algorithms using the NQF quality data and JBoss drools engine. AMIA Summits Transl. Sci. Proc. **2012**, 532–541 (2012)
50. Richesson, R.L., Rusincovitch, S.A., Wixted, D., Batch, B.C., Feinglos, M.N., Miranda, M.L., Hammond, W.E., Califf, R.M., Spratt, S.E.: A comparison of phenotype definitions for diabetes mellitus. J. Am. Med. Inform. Assoc. **20**, 319–326 (2013)
51. Ciaramella, A., Cimino, M., Marcelloni, F., Straccia, U.: Combining fuzzy logic and semantic web to enable situation-awareness in service recommendation. In: Database and Expert Systems Applications, pp. 31–45. Springer, Berlin (2010)
52. Stoilos, G., Stamou, G., Tzouvaras, V., Pan, J.Z., Horrocks, I.: Fuzzy OWL: Uncertainty and the semantic web. In: Proceedings of 1st International Workshop on OWL Experiences and Directions (2005)
53. Pan, J.Z., Stoilos, G., Stamou, G., Tzouvaras, V., Horrocks, I.: f-SWRL: A fuzzy extension of SWRL. J. Data Seman. **6**, 28–46 (2006)

Evolutionary Algorithms/Dynamic Modelling

Rail-Freight Crew Scheduling with a Genetic Algorithm

E. Khmeleva, A.A. Hopgood, L. Tipi and M. Shahidan

Abstract This article presents a novel genetic algorithm designed for the solution of the Crew Scheduling Problem (CSP) in the rail-freight industry. CSP is the task of assigning drivers to a sequence of train trips while ensuring that no driver's schedule exceeds the permitted working hours, that each driver starts and finishes their day's work at the same location, and that no train routes are left without a driver. Real-life CSPs are extremely complex due to the large number of trips, opportunities to use other means of transportation, and numerous government regulations and trade union agreements. CSP is usually modelled as a set-covering problem and solved with linear programming methods. However, the sheer volume of data makes the application of conventional techniques computationally expensive, while existing genetic algorithms often struggle to handle the large number of constraints. A genetic algorithm is presented that overcomes these challenges by using an indirect chromosome representation and decoding procedure. Experiments using real schedules on the UK national rail network show that the algorithm provides an effective solution within a faster timeframe than alternative approaches.

1 Introduction

While international trade continues to expand, businesses are striving to increase reliability and reduce their environmental impact. As a result, demand for rail freight increases every year and rail-freight carriers attempt to maximize their efficiency.

E. Khmeleva (✉) · A.A. Hopgood · L. Tipi · M. Shahidan
Sheffield Business School, Sheffield Hallam University, Howard Street,
Sheffield S1 1WB, UK
e-mail: e.khmeleva@shu.ac.uk

A.A. Hopgood
e-mail: a.hopgood@shu.ac.uk

L. Tipi
e-mail: l.tipi@shu.ac.uk

M. Shahidan
e-mail: m.shahidan@shu.ac.uk

© Springer International Publishing Switzerland 2014   211
M. Bramer and M. Petridis (eds.), *Research and Development
in Intelligent Systems XXXI*, DOI 10.1007/978-3-319-12069-0_16

The crew cost constitutes 20–25 % of the total rail-freight operating cost and is second only to cost of fuel. Therefore even a small improvement in the scheduling processes can save a company millions of pounds a year.

The CSP in the rail-freight industry is the problem of constructing a schedule for a train driver. Each schedule contains instructions for the driver of what he or she should do on a particular day. Within the industry, the driver's schedule is called a *diagram*. Each diagram should cover all the trains driven by a driver in a given day. It must start and end at the same station and obey all labour laws and trade union agreements. These rules regulate the maximum diagram duration, maximum continuous and aggregate driving time in a diagram, and minimum break time.

All drivers are located in *depots* where they start and finish their work. Depots are distributed fairly evenly across the UK. Sometimes in order to connect two trips that finish and start at different locations, a driver has to travel on a passenger train, taxi or a freight train driven by another driver. The situation of a driver travelling as a passenger while on duty is called *deadheading*. The cost of deadheading varies and depends on the means of transportation and business agreements between operating companies. Despite the potential cost, deadheading is sometimes inevitable and it can benefit the overall schedule [1].

Due to employment contract terms, the drivers are paid the same hourly rate for any time spent on duty regardless of the number of hours they have actually been driving the train. Moreover, in accordance with collectively bargained contracts, each driver has a fixed number of working hours per year, so the company is obliged to pay for all the stated hours in full even if some of the hours are not utilized. Paid additional overtime hours can be worked at the driver's discretion. Thus it is in the best interests of the company to use the agreed driving hours in the most efficient and economical way.

Taking all of this into consideration, the operational objectives for the diagrams are:

1. Minimize a number of unused and excess contract hours at the end of the year with a minimum spread of durations of the diagrams. All diagrams will therefore be of duration close to the average 8.5 h, i.e. the annual contract hours divided by the number of the working days.

$$T_{diagram} = T_{driving} + T_{deadheading} + T_{break} + T_{idle}$$

$$T_{diagram} \rightarrow T_{average}$$

2. Maximize the throttle time, i.e. the proportion of the work shift that is actually spent driving a train. It excludes time for deadheading and waiting between trips.

$$Throttle\ Time = \frac{T_{driving}}{T_{diagram}}$$

2 Approaches to Crew Scheduling

The CSP is usually solved in two stages. At the first stage, all possible diagrams satisfying the industrial constraints are enumerated. At the second stage, only the set of diagrams that covers the entire schedule in the most cost-effective way is identified. Diagrams are usually modelled as binary vectors (Fig. 1) where '1' denotes that the trip i is included in the diagram j, otherwise '0' is inserted. Each diagram has its own cost. The deadhead journeys are displayed by including the same trip in more than one diagram. In the rest of the article the terms diagram and column will be used interchangeably.

Although the generation of the diagrams can be performed in a simple and relatively straightforward manner using various graph search and label-setting techniques [2], finding an optimal set of diagrams may be highly time-consuming. The problem boils down to the solution of the 0–1 integer combinatorial optimization set covering problem (SCP):

$$Minimize \sum_{j=1}^{m} c_j x_j$$

$$Subject\ to : \sum_{i=1}^{n} a_{ij} x_j \geq 1$$

$$x_j \in \{0, 1\}$$
$$i = 1, 2 \ldots n\ trips$$
$$j = 1, 2 \ldots m\ diagrams$$

where a_{ij} is a decision variable indicating whether a trip i is included in the diagram j; x_j shows if the diagram is included in the schedule; c_j is the cost of the diagram.

2.1 Branch-and-Price

The complete enumeration of all possible diagrams is likely to be impractical due to the large geographical scope of operations, the number of train services, and industry regulations. Linear programming methods such as branch-and-price [3, 4]

	Diagram1	Diagram2	Diagram3	Diagram4
Trip1	1	0	0	1
Trip2	0	1	1	0
Trip3	0	1	0	1
Trip4	0	1	0	1
Trip5	1	1	0	0

Fig. 1 Diagrams

have been popular for the solution of medium-sized CSPs in the passenger train and airline industries [5]. These methods usually rely on a column-generation approach, where the main principle is to generate diagrams in the course of the algorithm, rather than having them all constructed a priori. Despite the ability of the algorithm to work with an incomplete set of columns, the column generation method alone does not guarantee an integer solution of SCP. It is usually used in conjunction with various branching techniques that are able to find the nearest integer optimal solution. However, this approach is not quite suitable for the CSP in rail freight, where the possible number of diagrams tend to be considerably higher.

2.2 Genetic Algorithms

Linear programming (LP) has been used for CSP since the 1960s [6] but genetic algorithms (GAs) were introduced more recently [7]. GAs have been applied either for the production of additional columns as a part of column generation [6] or for the solution of SCP from the set of columns generated prior to the application of a GA [9–12], but there are not yet any reports of them solving both stages of the problem. Since the diagrams are generated outside the GA in advance, the GA cannot change or add new columns. The GA is therefore confined to finding only good combinations from a pre-determined pool of columns.

For the solution of CSP with a GA, chromosomes are normally represented by integer or binary vectors. Integer vector chromosomes contain only the numbers of the diagrams that constitute the schedule. This approach requires knowledge of the minimum number of diagrams in the schedule and this information is usually obtained from the lower bounds. Lower bounds are usually acquired through the solution of LP relaxation for SCP [13]. Since the number of diagrams tends to be higher than the lower bound, Costa et al. [14] have suggested the following approach. In the first population the chromosomes have a length equal to the lower bound. Then, if a solution has not been found within a certain number of iterations, the length of the chromosome increases by one. This process repeats until the termination criteria are met.

In the binary vector representation, each gene stands for one diagram. The figure '1' denotes that the diagram is included in the schedule, otherwise it is '0'. Such chromosomes usually consist of several hundred thousand genes, and only around a hundred of them appear in the final solution. The number of diagrams can be unknown and the algorithm is likely to need a large number of iterations in order to solve the problem.

The application of genetic operators often violates the feasibility of the chromosomes, resulting in certain trips being highly over-covered (i.e. more than one driver assigned to the train) or under-covered (i.e. no drivers assigned to the train). One way of resolving this difficulty is to penalize the chromosome through the fitness function in accordance with the number of constraints that have been violated. However, the development of the penalty parameters can be problematic as in some cases it is impossible to verify them analytically and they are usually designed experimentally [15]. The penalty parameters are therefore data-dependent and likely to be inapplicable to other industries and companies. Moreover, the feasibility of the entire population is not guaranteed and might be achieved only after a large number of iterations.

Another more straightforward approach to maintaining the feasibility is to design heuristic "repair" operators. These operators are based on the principles "REMOVE" and "INSERT". They scan the schedule and remove certain drivers from the over-covered trips and assign those drivers to under-covered journeys [13, 15]. This procedure might have to be repeated several times, leading to high memory consumption and increased computation time.

3 GA-Generated Crew Schedules

3.1 Initial Data

The process starts with a user uploading the freight train and driver data (Fig. 2). Each train has the following attributes: place of origin, destination, departure and arrival time, type of train and route code. The last two attributes indicate the knowledge that a driver must have in order to operate a particular train. The system also stores information about the drivers, i.e. where each driver is located and his or her

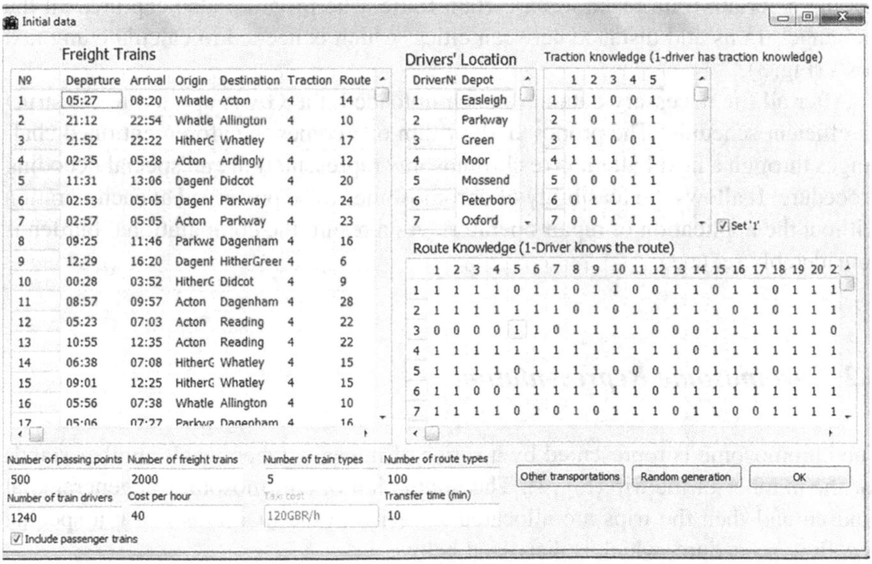

Initial data

Freight Trains

Nº	Departure	Arrival	Origin	Destination	Traction	Route
1	05:27	08:20	Whatle	Acton	4	14
2	21:12	22:54	Whatle	Allington	4	10
3	21:52	22:22	HitherG	Whatley	4	17
4	02:35	05:28	Acton	Ardingly	4	12
5	11:31	13:06	Dagent	Didcot	4	20
6	02:53	05:05	Dagent	Parkway	4	24
7	02:57	05:05	Acton	Parkway	4	23
8	09:25	11:46	Parkwɛ	Dagenham	4	16
9	12:29	16:20	Dagent	HitherGreei	4	6
10	00:28	03:52	HitherG	Didcot	4	9
11	08:57	09:57	Acton	Dagenham	4	28
12	05:23	07:03	Acton	Reading	4	22
13	10:55	12:35	Acton	Reading	4	22
14	06:38	07:08	HitherG	Whatley	4	15
15	09:01	12:25	HitherG	Whatley	4	15
16	05:56	07:38	Whatle	Allington	4	10
17	05:06	07:27	Parkwɛ	Dagenham	4	16

Drivers' Location

Traction knowledge (1-driver has traction knowledge)

DriverNº	Depot		1	2	3	4	5
1	Easleigh		1	1	1	0	0	1
2	Parkway		2	1	0	1	0	1
3	Green		3	1	1	0	0	1
4	Moor		4	1	1	1	1	1
5	Mossend		5	0	0	0	1	1
6	Peterboro		6	0	1	1	1	1
7	Oxford		7	0	0	1	1	1

☑ Set '1'

Route Knowledge (1-Driver knows the route)

	1	2	3	4	5	6	7	8	9	10	11	12	13	14	15	16	17	18	19	20	2
1	1	1	1	1	0	1	1	0	1	1	0	0	1	1	0	1	1	0	1	1
2	1	1	1	1	1	1	1	0	1	0	1	1	1	1	0	1	0	1	1	1
3	0	0	0	0	1	1	0	1	1	1	1	0	0	0	1	1	1	1	1	0
4	1	1	1	1	1	1	1	1	1	1	1	1	0	1	1	1	1	1	1	1
5	1	1	1	1	0	1	1	1	1	0	0	0	1	0	1	1	1	0	1	1
6	1	1	0	0	1	1	1	1	1	1	0	1	1	1	1	1	1	1	1	1
7	1	1	1	1	0	1	0	1	0	1	1	0	1	1	0	0	1	1	1	1

Number of passing points	Number of freight trains	Number of train types	Number of route types
500	2000	5	100
Number of train drivers	Cost per hour	Taxi cost	Transfer time (min)
1240	40	120GBR/h	10

☑ Include passenger trains

Other transportations | Random generation | OK

Fig. 2 Freight trains and drivers

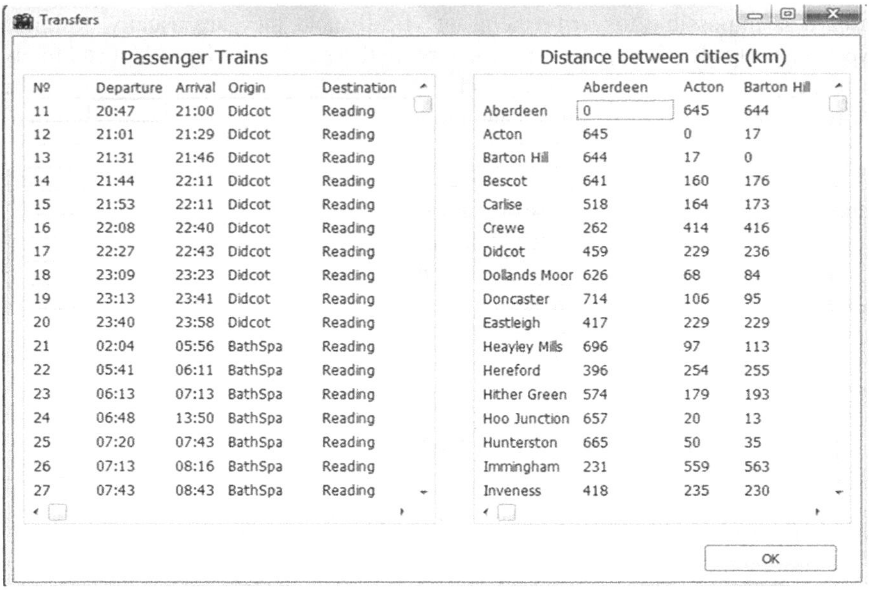
Transfers
Passenger Trains
Nº Departure Arrival Origin Destination
11 20:47 21:00 Didcot Reading
12 21:01 21:29 Didcot Reading
13 21:31 21:46 Didcot Reading
14 21:44 22:11 Didcot Reading
15 21:53 22:11 Didcot Reading
16 22:08 22:40 Didcot Reading
17 22:27 22:43 Didcot Reading
18 23:09 23:23 Didcot Reading
19 23:13 23:41 Didcot Reading
20 23:40 23:58 Didcot Reading
21 02:04 05:56 BathSpa Reading
22 05:41 06:11 BathSpa Reading
23 06:13 07:13 BathSpa Reading
24 06:48 13:50 BathSpa Reading
25 07:20 07:43 BathSpa Reading
26 07:13 08:16 BathSpa Reading
27 07:43 08:43 BathSpa Reading
Distance between cities (km)
Aberdeen Acton Barton Hill
Aberdeen 0 645 644
Acton 645 0 17
Barton Hill 644 17 0
Bescot 641 160 176
Carlise 518 164 173
Crewe 262 414 416
Didcot 459 229 236
Dollands Moor 626 68 84
Doncaster 714 106 95
Eastleigh 417 229 229
Heayley Mills 696 97 113
Hereford 396 254 255
Hither Green 574 179 193
Hoo Junction 657 20 13
Hunterston 665 50 35
Immingham 231 559 563
Inveness 418 235 230
OK

Fig. 3 Passenger trains and taxis

traction and route knowledge. In the boxes marked 'traction knowledge' and 'route knowledge', each row represents a driver and each column denotes either a route or traction code. The binary digits indicate whether a particular driver is capable of driving a certain train or knows a certain route. The program also captures all the passenger trains and distance between cities, which is needed to calculate any taxi costs (Fig. 3).

After all the necessary data have been uploaded, the GA is applied to construct an efficient schedule. The proposed algorithm overcomes the aforementioned challenges through a novel alternative chromosome representation and special decoding procedure. It allows the feasibility of chromosomes to be preserved at each iteration without the application of repair operators. As a result, the computational burden is considerably reduced.

3.2 Chromosome Representation

The chromosome is represented by a series of integers, where each number stands for the number of the trip (Fig. 4). The population of chromosomes is generated at random and then the trips are allocated in series to the diagrams using a specific decoding procedure, which is discussed below.

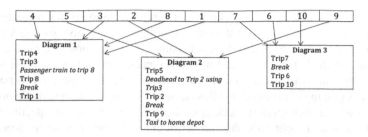

Fig. 4 Chromosome representation and decoding procedure

Starting from the leftmost gene, the procedure finds a driver with the necessary route and traction knowledge to operate that trip and creates a new diagram for him or her. Then the procedure checks if the same driver is able to drive on the next journey (i.e. the second gene). If it is possible, then that trip is added to his or her diagram. If the origin station for the current trip differs from the destination station of the previous trip, the algorithm first searches for passenger trains and the freight company's own trains that can deliver a driver within the available time slot to the next job location, e.g. Diagram 1, between trips 3 and 8 (Fig. 4). If no such trains have been found but there is a sufficient interval between the trips, then the algorithm inserts a taxi journey.

The information regarding driving times and the current duration of the diagrams is stored. Before adding a new trip, the algorithm inserts breaks if necessary. If the time expires and there are no trains to the home depot that a driver can drive, the deadheading activity completes the diagram, as in Diagram 2 (Fig. 4). If a trip cannot be placed in any of the existing diagrams, the procedure takes another driver from a database and creates a new diagram for him or her.

On rare occasions, a few diagrams might be left with only a few trips and a duration that is less than the minimum. This is due to the fact that other drivers are either busy at this time or located at different stations. In order to tackle this problem, a mechanism has been added for finding and assigning a driver from a remote depot with the lowest workload. This approach not only solved the problem of the short diagrams, but also helped in distributing the workload more equally across the depots. After implementation of this procedure, the algorithm has been tested on various data sets including real and randomly generated data. Neither of the chromosomes has been reported to violate the constraint.

The given representation has a visual resemblance to the flight-graph representation suggested by Ozdemir and Mohan [16], but the decoding procedures are different. The flight-graph representation generates trips based on a depth-first graph search, whereas in the proposed GA they are produced at random. Random generation is beneficial since it does not exclude situations where a driver can travel to another part of the country to start working in order to have even workload distribution across the depots, while depth-first search usually places only geographically adjusted trips together.

The advantage of the proposed chromosome representation is that it creates both the diagrams and schedule within the same algorithm, thereby giving the GA greater control over the solution. It also does not require the generation of a large amount of diagrams at the beginning. In addition, this representation does not leave under-covered trips and ensures that no unnecessary over-covering happens. It is possible that at the beginning of the algorithm this chromosome representation might produce schedules with a high number of deadheads. However, due to the specific fitness function and genetic operators, the number of chromosomes containing deadheads decreases rapidly with evolution.

3.3 Fitness Function

An adequate solution of the CSP requires the achievement of two conflicting objectives: high throttle time and low deviation from average diagram lengths. It is evident that with the increase in throttle time, the deviation from the average diagram length will be increased towards a minimum diagram length. This is due to the algorithm attempting to allocate a diagram for a single trip in order to achieve 100% throttle time.

Since GAs are a single-objective optimization method, a weighted sum approach has been applied in order to transform all the objectives into scalar fitness information [17]. The advantage of this technique is relative simplicity of implementation as well as high computational efficacy [18].

3.4 Selection

Preference was given to binary tournament selection as it is a comparatively simple and non-time consuming selection mechanism. It is also a popular selection strategy that is used in numerous GAs for CSP [9, 15, 16]. Binary tournament selection can be described as follows. Two individuals are selected at random form the population and the fittest among them constitutes the first parent. The same process repeats for the selection of the second parent.

3.5 Crossover and Mutation

Since one- or two-point crossover might produce invalid offspring by removing some trips or copying the same journey several times, a crossover mechanism utilizing domain-specific information has been designed. Firstly, the process detects genes responsible for diagrams with a high throttle time in the first parent. Then these genes are copied to the first child and the rest of the genes are added from the second

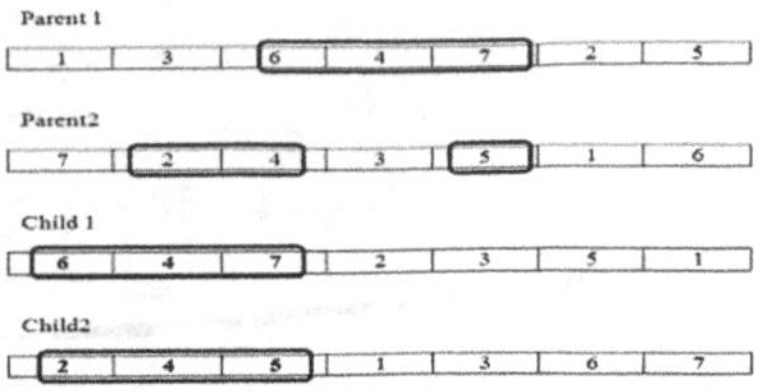

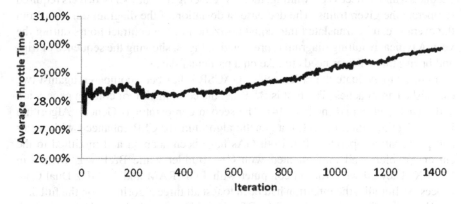

Fig. 5 Crossover

parent. The same procedure is then used to form the second child. The process is illustrated on the Fig. 5. By preserving the good parts of the chromosome accumulated through evolution, the implemented crossover was able to provide a schedule with a high throttle time much faster than traditional crossover that randomly the genes to propagate.

In order to maintain diversity in the population, randomly selected genes are mutated with 40 % probability. The mutation is performed by swapping two randomly identified genes. The mutation probability was determined through numerous tests and empirical observations.

4 Experimental Results

The proposed GA for CSP (referred to as GACSP) has been used to produce diagrams for the freight-train drivers. The GACSP has been tested on a full daily data set obtained from one of the largest rail-freight operators in the UK. The data instances comprise 2,000 freight-train legs, 500 cities, 39 depots, 1,240 drivers, 500,000 passenger-train links, and taxi trips connecting any of the stations at any time. Figures 6 and 7 illustrate a 3-h run of the algorithm and its achievement of

Fig. 6 Maximizing average throttle time

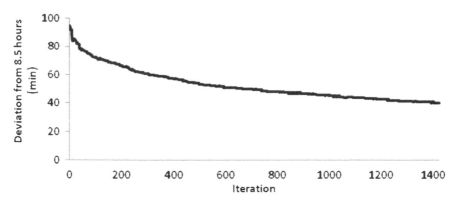

Fig. 7 Minimizing deviation from the average shift length of 8.5 h

Driver	Start time	End time	Activity	Origin	Destination
113	12:18	12:28	Book on	Westbury	Westbury
113	12:28	13:35	Driving	Westbury	Swindon
113	13:47	15:07	Driving	Swindon	Wootton Wawen
113	15:07	15:49	Break	Wootton Wawen	Wootton Wawen
113	15:59	17:29	Driving	Wootton Wawen	Swindon
113	17:37	21:00	Passenger Train	Swindon	Westbury
113	21:00	21:10	Book off	Westbury	Westbury
Diagram length: 8:52				**Throttle time: 46%**	

Fig. 8 A typical diagram, i.e. driver schedule

the main business objectives, i.e. maximized throttle time and minimized deviation from the average shift duration. Increasing the throttle time indicates a reduction in deadheads and unnecessary waiting, thereby reducing the number of drivers required to operate the given trains. The decrease in deviation of the diagram duration from the average can be translated into equal utilization of the contract hours during the year. A typical resulting diagram is presented in Fig. 8, showing the sequence of trips and breaks that a driver needs to take on a particular day.

In order to evaluate the efficiency of GACSP, it has been compared against two established approaches. The first is B&P, i.e. the combination of column generation and branch and bound methods [4]. The second comparator is Genetic Algorithm Process Optimization (GAPO), a genetic algorithm for CSP enhanced with repair and perturbation operators [9]. Both GAs have been adapted and modified to the current problem and implemented with C++ Builder while B&P was written in CPLEX. They all were run on computer with 4 GB RAM and 3.4 GHz Dual Core Processor. Initially, the intention had been to test all three algorithms on the full data set. However, after 12 h running of the B&P algorithm, no solution had been reached.

Table 1 Experimental results using the reduced data set

	B&P			GAPO			GACSP		
Computation time (min)	60	120	228	60	120	228	60	120	228
Number of diagrams	–	–	22	32	28	26	25	23	23
Throttle time (%)	–	–	63	50	56	59	60	62	62
Average number of deadheads per shift	–	–	1.36	2.21	1.85	1.60	1.66	1.47	1.47
Deviation from the average (min)	–	–	46	51	48	47	62	57	57

For the sake of comparison, the data size was reduced to six cities and 180 train legs, 500 passenger-train links. For the GA the population size was set as 20, crossover rate 90 % and mutation probability 40 %. As criteria for comparison, real business objectives such as throttle time, number of deadheads, average deviation from the desirable diagram length and computation time have been selected.

The computational results with the reduced data sets are displayed in Table 1. B&P obtained a solution in 228 min. Within 10 min, B&P had constructed 2,000 columns and solved LP relaxation without an integer solution. Further time was required for branching and generation of additional columns. In order to estimate efficiency of GACSP and GAPO, they have been run for the same period of time. GACSP obtained an entire feasible schedule within 10 s and after 1 h an acceptable schedule had been reached. Although the B&P algorithm ultimately achieved slightly better results, it has been tested on a problem of relatively small size. The computational time for linear programming algorithms usually grows exponentially with the increase in data size, so the B&P algorithm is likely to be impractical in environments where there is a crucial need to make fast decisions from large data sets.

As in other work [9], 3,308 columns have been generated for GAPO, which took 30 min of computational time. Unlike B&P and GACSP, this approach did not have an embedded ability to generate additional columns, limiting its capability to explore other possible diagrams. It was also observed that 70 % of the computational time was consumed by the heuristic and perturbation operators, whose aim was to restore the feasibility of the chromosomes. The repair operations were performed by scanning all available diagrams and selecting the best that could be inserted in the current schedule. GACSP overcomes this challenge by utilising the alternative chromosome representation that does not violate the validity of the chromosomes. Thus GACSP spends less time on each iteration and hence evolves more rapidly.

5 Potential Implementation and Integration Issues

The most common implementation problems with software for scheduling transit systems concern robustness [19], i.e. the ability of the schedule to adapt to different circumstances. An example of such circumstances might be the delay of the previous train, resulting in the driver being unable to catch the planned train. In our system, the transfer time regulates how much time is allocated for a driver to leave the previous train and start working on the next one. The larger the interval between trips, the lower the risk that the next freight train will be delayed by the late arrival of the previous one. On the other hand, a large transfer time decreases throttle time and requires more drivers to cover the trips. The best way to tackle this situation is to have an effective re-scheduling mechanism that makes changes in as few diagrams as possible.

In addition, the crew scheduling process is extremely complex. It is not always possible to model all the rules, nuances and exceptions of the schedule. For this reason, the system-generated diagrams have to be revised and amended by an experienced human planner until all the knowledge has been fully acquired.

Finally, although GAs are able to find an acceptable solution relatively quickly, they might also converge prematurely around a sub-optimal solution. Convergence can be controlled either by embedding variations in the selection procedure [17] or by changing the mutation rate [13].

6 Conclusions

In this paper, the complexities of CSP in the rail-freight industry in the UK have been described. Due to a high monetary cost of train crew, the profitability and success of the company might rely heavily on the quality of the constructed crew schedule. Given the wide geographical spread, numerous regulations, and severely constrained planning time, an IT system with an effective scheduling algorithm can equip a company with valuable decision-making support.

We have proposed a novel GA for crew scheduling (GACSP). Unlike other GAs for CSP, GACSP works with the entire schedule and does not restrict the algorithm in finding an optimal solution. The special chromosome representation and genetic operators are able to preserve the validity of the chromosomes without the need for additional repair operators or penalty functions. This capability enables the algorithm to consume fewer memory resources and to find a solution faster. In addition, the user can to retrieve a feasible schedule at any iteration.

It has been shown that although B&P was capable of finding an optimal solution from the mathematical perspective, time was its main weakness. In real-world operations, the cost for late optimal decision often can be much higher than that of a fast sub-optimal one. In this sense, the GA demonstrated excellent results as it provided a reasonable schedule nearly four times faster when using the reduced rail network. When faced with the scheduling task for the complete UK rail network, B&P had failed to find a solution at all after 12 h, whereas GACSP was able to find an adequate solution in 2 h. With further improvements of GACSP and possible hybridization with linear programming methods, its performance maybe further improved.

As future work, more domain specific rules will be incorporated into the chromosome generation process in order to achieve a better initial population. Moreover, it would be worthwhile to investigate a possible hybridization of a GA with the B&P

method. The hybridization might seize the advantages of both algorithms to reach a solution that is close to the mathematical optimum in a short computation time.

Acknowledgments This research has been supported by Sheffield Business School at Sheffield Hallam University. The authors would like to thank DB-Schenker Rail (UK) for the provision of data and information about real-world crew scheduling operations.

References

1. Barnhart, C., Hatay, L., Johnson, E.L.: Deadhead selection for the long-haul crew pairing problem. Oper. Res. **43**(3), 491–499 (1995)
2. Drexl, M., Prescott-Gagnon, E.: Labelling algorithms for the elementary shortest path problem with resource constraints considering EU drivers' rules. Log. Res. **2**(2), 79–96 (2010)
3. Duck, V., Wesselmann, F., Suhl, L.: Implementing a branch and price and cut method for the airline crew pairing optimization problem. Pub. Transp. **3**(1), 43–64 (2011)
4. Barnhart, C., Johnson, E.L., Nemhauser, G.L., Savelsbergh, M.W., Vance, P.H.: Branch-and-price: column generation for solving huge integer programs. Oper. Res. **46**(3), 316–329 (1998)
5. Derigs, U., Malcherek, D., Schafer, S.: Supporting strategic crew management at passenger railway model, method and system. Pub. Transp. **2**(4), 307–334 (2010)
6. Niederer, M.: Optimization of Swissair's crew scheduling by heuristic methods using integer linear programming models. In: AGIFORS Symposium (1966)
7. Levine, D.: Application of a hybrid genetic algorithm to airline crew scheduling. Comp. Oper. Res. **23**(6), 547–558 (1996)
8. Santos, A.G., Mateus, G.R.: General hybrid column generation algorithm for crew scheduling problems using genetic algorithm. In: Proceedings of the 2009 Congress on Evolutionary Computation, pp. 1799–1806 (2009)
9. Zeren, B., Özkol, I.: An improved genetic algorithm for crew pairing optimization. J. Intell. Learn. Sys. Appl. **4**(1), 70–80 (2012)
10. Souai, N., Teghem, J.: Genetic algorithm based approach for the integrated airline crew-pairing and rostering problem. Eur. J. Oper Res. **199**(3), 674–683 (2009)
11. Park, T., Ryu, K.: Crew pairing optimization by a genetic algorithm with unexpressed genes. J. Intell. Manuf. **17**(4), 375–383 (2006)
12. Kornilakis, H., Stamatopoulos, P.: Crew pairing optimization with genetic algorithms. In: Methods and Applications of Artificial Intelligence, pp. 109–120. Springer, Berlin (2002)
13. Kwan, R.K., Wren, A., Kwan, A.K.: Hybrid genetic algorithms for scheduling bus and train drivers. In: Proceedings of the 2000 Congress on Evolutionary Computation, vol. 1, pp. 282–292 (2000)
14. Costa, L., Santo, I.E., Oliveira, P.: An adaptive constraint handling technique for evolutionary algorithms. Optimization. **62**(2), 241–253 (2013)
15. Barnhart, C., Hatay, L., Johnson, E.L.: Deadhead selection for the long-haul crew pairing problem. Oper. Res. **43**(3), 491–499 (1995)
16. Ozdemir, H.T., Mohan, C.K.: Flight graph based genetic algorithm for crew scheduling in airlines. Inf. Sci. **133**(3–4), 165–173 (2001)
17. Hopgood, A.A.: Intelligent systems for engineers and scientists, 3rd edn. CRC Press, Boca Raton (2012)
18. Coello, A.C.: An updated survey of GA-based multiobjective optimization techniques. ACM (2000). doi:10.1145/358923.358929
19. Gopalakrishnan, B., Johnson, E.L.: Airline crew scheduling: state-of-the-art. Ann. Oper. Res. **140**, 305–337 (2005)

CR-Modified SOM to the Problem of Handwritten Digits Recognition

Ehsan Mohebi and Adil Bagirov

Abstract Recently, researchers show that the handwritten digit recognition is a challenging problem. In this paper first, we introduce a Modified Self Organizing Maps for vector quantization problem then we present a Convolutional Recursive Modified SOM to the problem of handwritten digit recognition. The Modified SOM is novel in the sense of initialization process and the topology preservation. The experimental result on the well known digit database of MNIST, denotes the superiority of the proposed algorithm over the existing SOM-based methods.

1 Introduction

The SOM is an unsupervised neural network [4] that usually contains a 2-Dim array of neurons weight, $\Psi = \{w_1, \ldots, w_q\}$. Assume that we are given the set of m input data vectors $A = \{x_1, \ldots, x_m\}$ where $x_i \in \mathbb{R}^n$, $i = 1, \ldots, m$. In the SOM network, Ψ, a weight $w_j \in \mathbb{R}^n$ is associated with the neuron j, $j = 1, \ldots, q$. For given $j \in \{1, \ldots, q\}$ define the following set (mapped data):

$$S_j = \{x_k : d(x_k, w_j) < d(x_k, w_l), \ l \neq j, \ l = 1, \ldots, q\} \tag{1}$$

where

$$d(x, y) = \|x - y\| = \left(\sum_{t=1}^{n} (x^t - y^t)^2 \right)^{1/2}, \ x, y \in \mathbb{R}^n$$

is the Euclidean distance. One data point x_i, $i \in \{1, \ldots, m\}$ at a time is presented to the network and is compared with all weight vectors. The nearest w_j, $j = 1, \ldots, q$

E. Mohebi (✉) · A. Bagirov
School of Science, Information Technology and Engineering,
Federation University Australia, Ballarat, Australia
e-mail: a.mohebi@federation.edu.au

A. Bagirov
e-mail: a.bagirov@federation.edu.au

© Springer International Publishing Switzerland 2014   225
M. Bramer and M. Petridis (eds.), *Research and Development
in Intelligent Systems XXXI*, DOI 10.1007/978-3-319-12069-0_17

is selected as the *best matching unit* (BMU) for the ith data point. This data point is mapped to the best matching neuron. Therefore,

$$S_j = S_j \cup x_i.$$

The set of neighborhood weights $N_c = \{w_l : p(c, l) \leq r, \ l \neq c\}$ around the BMU are updated as follows:

$$w_j := w_j + \alpha(\tau)h(\tau)(x_i - w_j). \tag{2}$$

where $p(c, l)$ is the distance between the BMU and the neighborhood neuron l in 2-Dim coordinates of the network topology and r is the predefined radius. Furthermore, $p(c, l) \in \mathbb{N}$ and $0 < p(c, l) \leq r$. The ordered neurons of the SOM preserve the topology of the mapped data and make the SOM to be very suitable for cluster analysis [10].

A Hierarchically Growing Hyperbolic Self-Organizing Map (H2SOM) [6] is proposed for incremental training with an automated adaptation of network size to improve the quantization error. The experimental results of H2SOM on MNIST data set denote the dominance of the proposed method over conventional SOM in the sense of accuracy. The paper [2] introduced a hybrid method based on SOM (CNN[1]-SOM) for classification of handwritten digit data set. Moreover, a rejection strategy is presented to change the topology of the map during training for improving reject quality. An interesting approach based on SOM (AOSSOM[2]) to the problem of MNIST hand written digits recognition is introduced in [11]. The AOSSOM learns organized linear manifolds based on gradient descent, which is reported as an improvement on Adaptive-Subspace Self Organizing Maps (ASSOM) [5] by updating the offset vector and the basis vectors of a linear manifold simultaneously at each learning step. The authors in [13] proposed a gradient-decent method (LMSOM[3]) to find multiple linear manifolds by using the self organizing maps to minimize the projection error function in the learning process. The LMSOM is tested on MNIST data set and the authors claimed that the proposed method overcome several shortcomings of ASSOM [5].

A neural model, referred to Locally Linear Online Mapping (LLOM), is introduced in [13] to the problem of handwritten digits recognition. The idea is to model nonlinear manifolds with mixtures of local linear manifolds via online learning. The mixture of local models is realized by using the Self Organizing framework and the online learning is desired for time reduction in large data sets. The classification accuracy of LLOM, which is presented by the authors, clearly outperforms ASSOM and AMSOM. A modification of the SOM, Binary Tree Time Adaptive Self-Organizing Map (BTASOM), is introduced and tested on MNIST data set [8]. The BTASOM is a dynamic binary tree in such a way that each of the tree's node contains a fixed

---

[1] Convolutional Neural Network.

[2] Adaptive-Subspace Self-Organizing Map.

[3] Linear Manifold Self Organizing Map.

predetermined number of neurons. The comparative results reported in [8] outline the efficiency of BTASOM in comparison with Growing Hierarchical SOM [1]. The authors in [3] proposed an Elastic Matching [9] concept for the problem of handwritten character recognition, where a character is transformed to match it with the template image with keeping its topology (GP + ASSOM + TM). To realize such transformation a Self Organizing Map is employed. Furthermore, the authors in [3] denote that using a hybrid technique based on Self Organizing Map and template matching improves the performance of template matching technique in handwritten character recognition.

As mentioned above, many SOM-based methods are proposed to the problem of handwritten digit recognition. None of the mentioned approaches pay attention to the initialization and topology of the SOM. In this paper, to improve the performance of handwritten digit recognition, first, we propose an initialization algorithm for the SOM based on split and merge algorithm. The high dense areas in input data space are detected by the proposed split and merge algorithm. Then neurons are generated in those detected areas. A new topology is presented for the SOM to restrict the adaptation of the neurons to those neurons which are located in the same high density areas. Such an approach leads to better local minimum of the quantization error than that of by the SOM. Moreover, a convolutional structure of Recursive Modified SOM is proposed to cope with recognizing diversity of styles and shapes of digits. The introduced recursive structure can learn various behaviors of incoming images. The proposed algorithm is tested on well known MNIST data set.

The rest of the paper is organized as follows. The split and merge algorithm and its definitions are presented in Sect. 2. The Modified SOM algorithm is presented in Sect. 3. In Sect. 4, the Recursive Modified SOM and its convolutional structure are introduced. Numerical results are presented in Sects. 5 and 6 concludes the paper.

2 Splitting and Merging Algorithms

In this section we discuss splitting and merging procedures in cluster analysis. Assume that we have a set of k cluster centers $\Lambda = \{c_1, \ldots, c_k\}$, $c_i \in \mathbb{R}^n$. These centers are solutions to the following problem:

$$\text{minimize} f = \sum_{i=1}^{k} \sum_{j=1}^{m} \|x_j - c_i\|^2 \text{ where } x_j \in C_i,$$ (3)

here c_i is the center point of the set C_i.

In some cases data points from the set C_i are not dense in some neighborhood of its center c_i. Given a radius $\varepsilon > 0$ we consider the following two sets for the cluster C_i:

$$\Phi_c^i(\varepsilon) = \{x_j \in C_i | \ d(x_j, c_i) \le \varepsilon\},$$ (4)

and

$$\Phi_s^i(\varepsilon) = \{x_j \in C_i | \ \varepsilon < d(x_j, c_i) \leq r_i\},$$

where

$$r_i = \max_j \{d(x_j, c_i) | \ x_j \in C_i\}, \ i = 1, \ldots, k.$$

Two clusters C_i and C_l are said to be well separated if $d(c_i, c_l) \geq (r_i + r_l)$. It is clear that for any cluster $C_i, i = 1, \ldots, k$ there exist $\varepsilon_i \in (0, r_i]$ such that $|\Phi_c^i(\varepsilon)| = \max(|\Phi_c^i(\varepsilon)|, |\Phi_s^i(\varepsilon)|)$ for all $\varepsilon \in (\varepsilon_i, r_i]$. Consider $\varepsilon_i = \beta r_i$ where $\beta \in (0, 1)$. If the ε_i is sufficiently small then data points from the cluster C_i are dense around its center c_i. The ε_i will be used to design a splitting algorithm for clusters whereas the definition of well separated clusters will be used to design a merging algorithm.

Splitting: The splitting procedure will be done using the parameter β and also special scheme to identify parts of a cluster where most of point reside. Assume that a set of k clusters, $\Omega = \{C_1, \ldots, C_k\}$ and a number $\beta \in (0, 1)$ are given. The number of points within the radius $\varepsilon_i = \beta r_i$ from the center of the cluster C_i is $q_c^i = |\Phi_c^i(\varepsilon_i)|$. We introduce the angle $\theta_{i,j}$ between the cluster center c_j and the data point $x_i \in C_j$ as follows assuming both $c_j \neq 0$ and $x_i \neq 0$:

$$\theta_{i,j} = \arccos \frac{\langle x_i, c_j \rangle}{\|x_i\| \|c_j\|}. \tag{5}$$

Now we introduce the following two sets:

$$\Phi_u^j(\varepsilon_j) = \{x_i \in C_j | \ \varepsilon_j < d(x_i, c_j) \leq r_j, \ 0 \leq \theta_{i,j} \leq \frac{\pi}{2}\}, \tag{6}$$

and

$$\Phi_d^j(\varepsilon_j) = \{x_i \in C_j | \ \varepsilon_j < d(x_i, c_j) \leq r_j, \ \frac{\pi}{2} \leq \theta_{i,j} \leq \pi\}. \tag{7}$$

The cardinalities of these sets are $q_u^j = |\Phi_u^j(\varepsilon_j)|$ and $q_d^j = |\Phi_d^j(\varepsilon_j)|$, respectively. Application of the splitting procedure to the cluster C_j depends on the values of q_u^j, q_d^j and q_c^j. If

$$q_c^j = \max\{q_d^j, q_c^j\} \tag{8}$$

then data points are dense around the cluster center and we do not split such a cluster, otherwise we split this cluster into two new ones.

Merging: Assume that the collection of k clusters, $\Omega = \{C_1, \ldots, C_k\}$, is given. It may happen that (also after applying the splitting algorithm) some clusters are not well separated. In this subsection we design an algorithm to merge clusters which are well separated from each other.

According to the definition of well separated clusters two clusters $C_j, C_p \in \Omega$ should be merged if

$$d(c_j, c_p) - (r_j + r_p) < 0. \tag{9}$$

The split and merge algorithm is presented as follows.

Algorithm 1 Split and Merge algorithm
Step 0. Input: A collection of k clusters $\Omega = \{C_1, \ldots, C_k\}$, the maximum number of iterations $\gamma_{max} > 0$ and the ratio $\beta \in (0, 1)$. Set $\gamma := 0$.
Step 1. Set $\gamma := \gamma + 1$.
Step 2. Select cluster $C_j \in \Omega$ and calculate its center c_j.
Step 3. Calculate $d(x_i, c_j)$ and also $\theta_{i,j}$ using (5) for all data point $x_i \in C_j$.
Step 4. For each cluster C_j calculate sets $\Phi_c^j(\varepsilon_i)$, $\Phi_u^j(\varepsilon_j)$, $\Phi_d^j(\varepsilon_j)$ using (4), (6) and (7), respectively.
Step 5. If (8) is satisfied then go to Step 7, otherwise go to Step 6.
Step 6. Split the cluster C_j into two new clusters. Update Ω and set $k := k + 1$.
Step 7. If all clusters C_j, $j = 1, \ldots, k$ are visited terminate, otherwise go to Step 2.
Step 8. Select cluster $C_j \in \Omega$ and calculate its center c_j.
Step 9. Select cluster $C_p \in \Omega$ and calculate its center c_p, where $j \neq p$.
Step 10. If the condition (9) is satisfied then go to Step 11, otherwise go to Step 12.
Step 11. Merge clusters C_j and C_p. Update the set Ω and set $k := k - 1$.
Step 12. If all cluster C_p, $p = 1, \ldots, k$ and $j \neq p$ are visited go to Step 13, otherwise go to Step 9.
Step 13. If all clusters $C_j \in \Omega$ are visited terminate, otherwise go to Step 8.
Step 14. If $\gamma > \gamma_{max}$, terminate, otherwise go to Step 1.

3 Modified SOM Algorithm

In this section we present a new algorithm based on Algorithm 1 to initialize the neurons of the SOM and moreover, define a modified topology of neurons at the initial points. The output of the Algorithm 1 are used as initial neurons of the Modified SOM algortihm. Algorithm 1 ensures that the initial neurons are located in distinct high density area of the input data space which is found through the steps 2–7. Steps 8–13 of Algorithm 1 guarantees that initial neurons are not close to each other.

SOM Modified Topology: The modified topology is based on initial neurons $\Psi = \{w_1, \ldots, w_q\}$ and generation of a set of e number of neurons $u_z \in \mathbb{R}^n$, $z = 1, \ldots, e$ using each individual neuron w_i, $i = 1, \ldots, q$ as follows:

$$g_i = \{u_z | w_i^t - \lambda \varepsilon_i \le u_z^t \le w_i^t + \lambda \varepsilon_i\}, \tag{10}$$

where $t = 1, \ldots, n, z = 1, \ldots, e$ and $\lambda \in \mathbb{R}$, $\lambda \ge 1$. One can see that all the neurons in the set g_i are close to w_i to cover up the dense area which is centered by neuron w_i. Then we define the following connectivity matrix for the new topology:

1 $con(i, j) \in \{0, 1\}$, $w_i, w_j \in \Psi$.
2 $con(i, j) \in \{0\}$, $u_i \in g_k$, $u_j \in g_p$, $k \neq p$.
3 $con(i, j) \in \{1\}$, $u_i \in g_k$, $u_j \in g_p$, $k = p$.
4 $con(i, j) \in \{1\}$, $u_i \in g_k$, $w_j \in \Psi$, $k = j$.
5 $con(i, j) \in \{0\}$, $u_i \in g_k$, $w_j \in \Psi$, $k \neq j$.

This new topology guarantees that neurons from one dense area are not connected with those from another dense area and therefore according to the Eq. (2) such neurons do not change each others weight. The Modified SOM algorithm can be summarized as follows.

Algorithm 2 Modified SOM algorithm

Step 0. (Initialization) Initialize the maximum number of iterations T of the network. Set the maximum number of iterations in the Splitting and Merging algorithm as γ_{max} and the value of the ratio $\beta_{max} > 0$. Set the initial value of iteration and ratio to γ_0 and β_0, respectively. Set the step length β_s and set the iteration counter $\tau := 0$. A set of m input data vectors $A = \{x_1, \ldots, x_m\}$.

Step 1. (Split and Merge). Apply Algorithm 1 to A for $\gamma_0 \to \gamma_{max}$ and $\beta_0 \to \beta_{max}$ to generate the set of $\Psi = \{w_1, \ldots, w_q\}$ which minimizes the function f in (3). This set is initial weights of neurons.

Step 2. Select data x_i, $i = 1, \ldots, m$ and find its closest neuron $w_c \in \{\Psi \bigcup_{i=1}^{q} g_i\}$, that is

$$c := \underset{j=1,\ldots,(q \cdot e)}{\text{argmin}} \|x_i - w_j\|. \tag{11}$$

Step 3. Update the set of neighborhood neurons $w_j \in N_c$ using the following equation:

$$w_j := w_j + \alpha(\tau)h(\tau)(x_i - w_j), \tag{12}$$

where

$$N_c = \begin{cases} g_i \cup w_i & \text{if } w_c = u_z \in g_i, \\ g_i \cup w_i \cup \Xi & \text{if } w_c = w_i, \ w_i \in \Psi, \end{cases}$$

subject to

$$\Xi = \{w_j | r_{ij} < r, \ i \neq j, \ w_j \in \Psi\}.$$

Step 4. If all input data are visited go to Step 5, otherwise go to Step 2.
Step 5. If $\tau > T$ terminate, otherwise set $\tau := \tau + 1$ and go to Step 2.

Note that the neighborhood function in Eq. (12) of Algorithm 2 is as follows.

$$h(\tau) = \exp\left(-\frac{r^2}{2\sigma(\tau)^2}\right), \tag{13}$$

subject to

$$\alpha(\tau) = \eta \frac{T - \tau}{\tau}, \ \eta \in \mathbb{R},$$

(14)

and usually $\eta \geq 1$.

One can see that the Step 2–5 in Algorithm 2 is similar to basic SOM. The exception is in Step 3 where the set N_c is defined in order to find near to global solution to the vector quantization problem.

4 Convolutional Structure of Recursive Modified SOM for Handwritten Digit Recognition

In this section, we present a semi-supervised tool for handwritten digit recognition using a Convolutional Structure of Recursive Modified SOM. The recursive neural network is a combination of neural model and feedback, in order to learn different behaviors of input images by one neural network. First, the idea is to have distinct sets of recursive neural networks, $\mathcal{M}_0, \ldots, \mathcal{M}_9$ for training a sequence of images with $0, \ldots, 9$ labels, respectively. Therefore,

$$\mathcal{M}_y = \{\Psi_h | h = 1, \ldots, \bar{n}^y\}, \ y = 0, \ldots, 9,$$

where Ψ_h is the result of training the set of B input images, P_b, $b = 1, \ldots, B$. P_b is a set of vectors indicating image pixels, $P_b = \{x_i | x_i \in \mathbb{R}^3\}$. We propose a recursive form of Algorithm 2 for training process, where the $P(0)$ is the first incoming image and the feedback of the network at time t is combined with the incoming picture at time $t + 1$ in order to be learned by the network. The Recursive Modified SOM Algorithm is as follows:

Algorithm 3 Recursive Modified SOM algorithm
Step 1. Let have a set of images $\aleph = \{P_1, \ldots, P_B\}$. Initialize parameters of the Algorithm 2. Set network $\Psi = \emptyset$.
Step 2. Select a $P_b \in \aleph$.
Step 3. (Training) Apply Algorithm 2 on the set $P_b \cup \Psi$ as input data vectors.
Step 4. (Update Ψ) Set the output neurons of Modified SOM into the parameter Ψ.
Step 5. If all $P_b \in \aleph$ are visited terminate, otherwise go to Step 2.

In Step 3 of Algorithm 3, the training is done on the union of new input image and the network obtained from the previous sequence. Usually all $P_b \in \aleph$ are from same label, therefore the final set Ψ is a trained network with a label as the input images. Assume that all $P_b \in \aleph$ are from the images of digit with label 0, therefore we add the network Ψ to the set $\mathcal{M}_0$,

$$\mathcal{M}_0 = \mathcal{M}_0 \cup \Psi,$$

and the network Ψ is a recognition tool for images with label 0 in the set $\mathcal{M}_0$.

Training: In the training phase, the cardinality of sets $\mathcal{M}_0, \ldots, \mathcal{M}_9$ are predefined. Assume that the cardinality of sets $\mathcal{M}_y$, $y = 0, \ldots, 9$ are set as $\bar{n}^y$, $y = 0, \ldots, 9$, therefore, we have $\bar{n}^y$ number of networks, $\Psi_1^y, \ldots, \Psi_{\bar{n}y}^y$, in the set $\mathcal{M}_y$. The set of training sample images $\aleph^y$ of label y are divided into $\bar{n}^y$ subsets $\aleph_1^y, \ldots, \aleph_{\bar{n}y}^y$ to be learned by the networks $\Psi_1^y, \ldots, \Psi_{\bar{n}y}^y$, in the set $\mathcal{M}_y$, respectively. It should be noted that

$$\aleph_i^y \cap \aleph_j^y = \emptyset, \quad i \neq j, \ i, j = 1, \ldots, \bar{n}^y.$$

Each subset $\aleph_i^y$ learned by Ψ_i^y, where $i = 1, \ldots, \bar{n}^y$, using Algorithm 3. The training Algorithm is defined as follows:

Algorithm 4 Training algorithm

Step 1. (Initilization) Let have a set of images $\aleph = \{P_1, \ldots, P_B\}$. Initialize the values $\bar{n}^y$ and set $\mathcal{M}_y = \emptyset$ for $y = 0, \ldots, 9$. Set label $y = 0$.

Step 2. (Selection and devision) Select all images with label y in $\aleph$ and put them in $\aleph^y$. Divide the set $\aleph^y$ into $\bar{n}^y$ number of distinct subsets $\aleph_j^y$, $j = 1, \ldots, \bar{n}^y$.

Step 3. (Training) Select a $\aleph_j^y$ and its corresponding network Ψ_j^y then send them as an input to Algorithm 3.

Step 4. The output of Algorithm 3 is a trained network Ψ_j^y, therefore,

$$\mathcal{M}_y = \mathcal{M}_y \cup \Psi_j^y.$$

Step 5. If all $\aleph_j^y$, $j = 1, \ldots, \bar{n}^y$ are visited go to Step 6, otherwise go to Step 3.
Step 6. If $y > 9$ terminate, otherwise $y = y + 1$ and go to Step 2.

After termination of Algorithm 4 the sets $\mathcal{M}_y$, $y = 0, \ldots, 9$ contain trained networks which are used for recognizing unknown images.

Testing: Assume a Modified SOM network Υ as a screen network is employed to train the input unknown images using Algorithm 2. Then, the network Υ is compared with all the networks in the sets $\mathcal{M}_y$, $y = 0, \ldots, 9$. The aim is to recognize the label, $\bar{y}$, of an input image by minimizing the following equation,

$$\bar{y} = \min_{\substack{y \\ j=1,\ldots,\bar{n}^y}} \mathcal{E}(\Psi_j^y, \Upsilon) \tag{15}$$

$$\mathcal{E}(\Psi_j^y, \Upsilon) = \sum_{i=1}^{|\Upsilon|} \min_k \|\hat{w}_i - w_k^y\|, \ w_k^y \in \Psi_j^y,$$

where $y = 0, \ldots, 9$. There are many neurons in the network Υ and also in all the networks in the sets $\mathcal{M}_y$, $y = 0, \ldots, 9$, which are dark/low-luminance and carry no information. Therefore, before using Eq. (15) for image recognition, we introduce a filtering parameter, δ, to remove those neurons in the sets of training networks

Ψ_j^y, $j = 1, \ldots, \bar{n}^y$, $y = 0, \ldots, 9$ and also in the screen network, Υ, which are low in luminance. The set of neurons that are extracted by filtering from the network Ψ_j^y are calculated as follows:

$$\bar{\Psi}_j^y = \left\{ w_k^y \mid \|w_k^y\| \geq \delta, \ w_k^y \in \Psi_j^y \right\}, \tag{16}$$

where $k = 1, \ldots, |\Psi_j^y|$, $j = 1, \ldots, \bar{n}^y$, $y = 0, \ldots, 9$.

Convolutional Structure: The convolutional structure learns the input image by different settings of parameters ε and γ in the network Υ, which is the output of Algorithm 1. The result is a set of derivative networks of Υ,

$$\mathscr{I} = \{\Upsilon_\varepsilon^\gamma \mid 0 < \varepsilon < 1, \ 1 < \gamma < \gamma_{\max}\}.$$

The parameter ε is defined in Sect. 2. This parameter is used to split the clusters with a low density center point. According to Eqs. (4)–(7), changing the ε to a high value reduce the number of initial neurons which, consequently, generates a new topology. If we run the Algorithm 1 with a low to a high value of ε, the number of initial neurons will change in a decreasing format in Algorithm 2. In the recognition procedure, the filtering process is applied by Eq. (16) on all networks $\Upsilon_\varepsilon^\gamma \in \mathscr{I}$. The result is a set of filtered networks, $\bar{\mathscr{I}} = \{\bar{\Upsilon}_\varepsilon^\gamma \mid 0 < \varepsilon < 1, \ 1 < \gamma < \gamma_{\max}\}$, with high luminance neurons.

In order to recognize an unknown image with its convolutional networks $\bar{\Upsilon}_\varepsilon^\gamma \in \bar{\mathscr{I}}$, we reformulate Eq. (15) to be applicable for these sets of networks along with the filtered training networks, $\bar{\Psi}_j^y$, $j = 1, \ldots, \bar{n}^y$, in the sets $\bar{\mathscr{M}}_y$, $y = 0, \ldots, 9$. The reformulation of Eq. (15), for comparing networks in the set $\bar{\mathscr{I}}$ with trained networks in the sets $\bar{\mathscr{M}}_y$, $y = 0, \ldots, 9$ to label an unknown input image is as follows:

$$\bar{y} = \min_y \min_{\substack{\varepsilon, \gamma \\ j=1,\ldots,\bar{n}^y}} \mathscr{E}(\bar{\Psi}_j^y, \bar{\Upsilon}_\varepsilon^\gamma) \tag{17}$$

$$\mathscr{E}(\bar{\Psi}_j^y, \bar{\Upsilon}_\varepsilon^\gamma) = \sum_{i=1}^{|\bar{\Upsilon}_\varepsilon^\gamma|} \min_k \|\hat{w}_i - w_k^y\|, \ w_k^y \in \bar{\Psi}_j^y, \tag{18}$$

where $y = 0, \ldots, 9$.

The handwritten digit recognition algorithm is presented as follows:

Algorithm 5 Handwritten digit recognition algorithm
Step 1. (Initilization) Set of unknown images $\varXi = \{P_1, \ldots, P_N\}$. Initialize the parameters ε, γ and get trained networks $\mathscr{M}_y$ for $y = 0, \ldots, 9$ by applying Algorithm 4. Set filtering parameter δ in Eq. (16). Set $\varepsilon_t = e_0$ and $\gamma_t = g_0$ as starting values with step lengths s_ε and s_γ, respectively. Set $minE = \infty$ and $\bar{y} = null$.
Step 2. Apply filtering Eq. (16) on the sets $\mathscr{M}_y$ to extract high luminance neurons. Set extracted networks to $\bar{\mathscr{M}}_y$ for $y = 0, \ldots, 9$, respectively.

Step 3. Select an images P_i in Ξ.
Step 4. (Training) Apply Algorithm 2 on the input image P_i. The result is the network $\Upsilon_{\varepsilon_t}^{\gamma_t}$.
Step 5. Calculate the network $\bar{\Upsilon}_{\varepsilon_t}^{\gamma_t}$ using Eq. (16).
Step 6. (Testing) Select a network $\bar{\Psi}_j^y$ in sets $\mathcal{M}_y$ for $y = 0, \ldots, 9$ and calculate Eq. (18) with $\bar{\Psi}_j^y$ and $\bar{\Upsilon}_{\varepsilon_t}^{\gamma_t}$ as input networks. If the result of $\mathcal{E}$ is less than $minE$, then $minE = \mathcal{E}$ and $\bar{y} = y$.
Step 7. If all networks in trained sets $\mathcal{M}_y$ for $y = 0, \ldots, 9$ are visited then go to Step 8, otherwise go to Step 6.
Step 8. Set all networks in trained sets $\mathcal{M}_y$ for $y = 0, \ldots, 9$ as un-visited. If $\varepsilon_t > \varepsilon$ set $\varepsilon_t = e_0$ and go to Step 9, otherwise $\varepsilon_t = \varepsilon_t + s_\varepsilon$ and go to Step 4.
Step 9. If $\gamma_t > \gamma$, set $\gamma_t = g_0$ and go to Step 10, otherwise $\gamma_t = \gamma_t + s_\gamma$ and go to Step 4.
Step 10. Label P_i as $\bar{y}$. If all $P_i \in \Xi$ are visited terminate, otherwise Set $\varepsilon_t = e_0$, $\gamma_t = g_0$ and go to Step 3.

In Step 1 of Algorithm 5, we initialize the parameters ε_t and γ_t, which are used to produce different networks, $\Upsilon_{\varepsilon_t}^{\gamma_t}$, from the input image. Though, we increment ε_t and γ_t using parameters s_ε and s_γ as step lengths, until reaching ε and γ, respectively. These conditions are holding in Step 8 and 9 of the Algorithm 5. In Step 2, the low luminance neurons are removed from the trained networks, which are the output of Algorithm 4, using Eq. (16). Steps 4 and 5 calculate the network from the input image P_i and make it ready for testing in Step 6. The Step 6 is executed for all trained networks in sets $\mathcal{M}_y$ for $y = 0, \ldots, 9$.

5 Numerical Results

To demonstrate the effectiveness of the proposed algorithm, the experiments is done on the well known MNIST data set. The Algorithms 1–5 have been coded in NetBeans IDE under Java platform and tested on a MAC OSX with 2.7 GHz core i7 CPU and 10 GB of RAM. The MNIST dataset consists of 60,000 training samples from approximately 250 writers and 10,000 test samples from a disjoint set of 250 other writers. We used the original 784-dimensional dataset which resembles 28×28 pixel grey level images of the handwritten digits.

The parameters $\bar{n}^y$, $y = 0, \ldots, 9$ are set according to the values, which are presented in the Table 1.

Table 1 Initialization of $\bar{n}^y$ in Algorithm 4

Image label	0	1	2	3	4	5	6	7	8	9
$\bar{n}^y$	5	7	3	21	1	2	19	18	14	22

Fig. 1 The networks
$\Psi_1^y \in \mathcal{M}_y$, $y = 0, \ldots, 9$ after
running Algorithm 4 on
training samples

The value $\bar{n}$ plays an important role in classification accuracy. Assume that, we have many misclassifications between digit 9 and 4. Then we set the values $\bar{n}^9$ and $\bar{n}^4$ in a such way to increase $|\bar{n}^9 - \bar{n}^4|$. Though, in the Table 1, the parameter $\bar{n}^y$ for image with label 9 is set to 22 while this parameter for the image 4 is set to 1. This significant difference in $\bar{n}$ leads to improvement in the misclassifications where image 9 misclassified as 4 and vice versa. The same criteria is hold for images with label 7 and 1, 3 and 8 and also for images with label 2 and 7. We have presented the networks Ψ_1^y, $y = 0, \ldots, 9$ in Fig. 1.

Recently, a number of handwritten digits recognition methods based on SOM have been proposed. Comparing the existing methods, the best accuracy of 98.73 % on MNIST dataset is reported in [2]. In our experiment on MNIST dataset, the accuracy of 99.03 % is obtained by using the Convolutional Rec-MSOM on the 10,000 test samples. Therefore, only 97 samples were misclassified. The proposed method in this paper boosted up the classification results up to 23.62 %, compared with the method presented in [2]. The comparative results of Convolutional Rec-MSOM and well known SOM-based handwritten digits recognition methods on MNIST dataset are presented in the Table 2.

The 97 misclassified samples by using the proposed method in this paper are presented in Fig. 2. In Fig. 2, the ID of each sample is printed at the top and the prediction is presented as label → prediction at the bottom. The results in Fig. 2 show that there are 6 digits with label 1 which have been predicted as a sample with label 9. These digits are a thick written number with label 1 which may caused the

Table 2 Comparison of accuracy obtained by SOM-based handwritten digits recognition methods on MNIST dataset

References	Method	Accuracy (%)
[8]	BTASOM	85.24
[3]	GP + ASSOM + TM	93.20
[6]	H²SOM	94.60
[7]	H²SOM	95.80
[11]	AOSSOM	97.30
[12]	LMSOM	97.30
[13]	LLOM	97.74
[2]	CNN-SOM	98.73
This paper	CR-MSOM	99.03

Fig. 2 The 97 misclassified samples from the 10,000 test samples

machine misclassifications. There are also 5 misclassification as $3 \to 8$. By analysis such errors, we found that, most of the errors are due to rotation, misalignment and broken structure in the handwritten numerals. For example, the samples with the ID 342, 4079, 4815 and 6167 are rotated which is caused by the people with bad habits in handwritings. The samples with the ID 620, 1045, 2381, 3845 and 6652 are related to the errors which are even hard for a human to predict without displaying their labels. Some of other errors are introduced by scanning procedure, like samples with ID 3061, 9630 and 9694. The confusion matrix of these misclassifications is presented in the Table 3.

Considering the misclassifications among 10 classes of images, which is presented in the Table 3, the max number of misclassification is 25 on the images with label 7. We present the set $\mathcal{M}_7$ in the Fig. 3.

The set of networks presented in the Fig. 3 are used to classify images with label 7. By comparing these networks with the digits with label 7 in the Fig. 2, one can see that the most errors are on the images which are written in a thick shape, like images

Table 3 Confusion matrix

Truth	Prediction									
	0	1	2	3	4	5	6	7	8	9
0			1				3	1	2	
1				4			1	1	1	6
2										
3		2	1					2	6	4
4										
5										
6	3	1	1			3			3	1
7		11	2	1	1				1	9
8	2		1	1			3			4
9	2	1		3			1	3	4	

Fig. 3 The set $\mathcal{M}_7$ used to classify images with label 7

with ID 1207, 6742, 7903, 7904 and 7928. The same criteria is true for the images with the label 1 in the Fig. 2. There are 1,028 images with label 7 included in the 10,000 test samples. It should be noted that by learning only 40 images with label 7, the set $\mathcal{M}_7$ misclassified 25 samples which is about 2.4 % of the total images with the label 7. This denotes the superiority of the proposed classifier where the number of training samples are limited.

6 Conclusion

In this paper, we developed a Convolutional Recursive Modified SOM for the problem of handwritten digits recognition. First, we presented a modified version of the Self Organizing Maps. The aim was to propose an initialization algorithm and a new topology which restrict the neighborhood adaptations to only those neurons that are not in different dense areas. Therefore, we introduced split and merge algorithm to initialize the neurons and based on those neurons we presented a topology for SOM to generate neurons in high dense areas of input data space and do not connect neurons which are in separate dense areas. The new topology also reduced the complexity of neighborhood adaptations from quadratic to a linear one, in comparison with the conventional SOM. We proposed a recursive form of Modified SOM for training process, in order to learn the diversity of shapes and styles of incoming images. Finally, a convolutional structure introduced to label the unknown handwritten digits images of MNIST data set. The experiments on the well known data set, MNIST, demonstrate the superiority of the proposed algorithm over the existing SOM-based methods in the sense of accuracy.

References

1. Batista, L.B., Gomes, H.M., Herbster, R.F.: Application of growing hierarchical self-organizing map in handwritten digit recognition. In: Proceedings of 16th Brazilian Symposium on Computer Graphics and Image Processing (SIBGRAPI), pp. 1539–1545 (2003)
2. Cecotti, H., Belaïd, A.: A new rejection strategy for convolutional neural network by adaptive topology. In: Kurzyński, M., Puchala, E., Woźniak, M., Zolnierek, A. (eds.) Computer Recognition Systems, Advances in Soft Computing, vol. 30, pp. 129–136. Springer, Berlin (2005)
3. Horio, K., Yamakawa, T.: Handwritten character recognition based on relative position of local features extracted by self-organizing maps. Int. J. Innovative Comput. Inf. Control 3(4), 789–798 (2007)
4. Kohonen, T.: Self-Organizing Maps. Springer Series in Information Sciences, Berlin (2001)
5. Kohonen, T., Kaski, S., Lappalainen, H.: Self-organized formation of various invariant-feature filters in the adaptive-subspace SOM. In: Neural Computation, pp. 1321–1344 (1997)
6. Ontrup, J., Ritter, H.: A hierarchically growing hyperbolic self-organizing map for rapid structuring of large data sets. In: Proceedings of 5th Workshop On Self-Organizing Maps, pp. 471–478 (2005)
7. Ontrup, J., Ritter, H.: Large-scale data exploration with the hierarchically growing hyperbolic SOM. Neural Netw.: Official J. Int. Neural Netw. Soc. 19(6–7), 751–761 (2006)
8. Shah-Hosseini, H.: Binary tree time adaptive self-organizing map. Neurocomputing 74(11), 1823–1839 (2011)
9. Uchida, S., Sakoe, H.: A survey of elastic matching techniques for handwritten character recognition. IEICE—Trans. Inf. Syst. E88-D(8), 1781–1790 (2005)
10. Yang, L., Ouyang, Z., Shi, Y.: A modified clustering method based on self-organizing maps and Its applications. Procedia Comput. Sci. 9, 1371–1379 (2012)
11. Zheng, H., Cunningham, P., Tsymbal, A.: Adaptive offset subspace self-organizing map: an application to handwritten digit recognition. In: Proceedings of Seventh International Workshop on Multimedia Data Mining, pp. 29–38 (2006)
12. Zheng, H., Cunningham, P., Tsymbal, A.: Learning multiple linear manifolds with self-organizing networks. Int. J. Parallel, Emergent Distrib. Syst. 22(6), 417–426 (2007)
13. Zheng, H., Shen, W., Dai, Q., Hu, S., Lu, Z.M.: Learning nonlinear manifolds based on mixtures of localized linear manifolds under a self-organizing framework. Neurocomputing 72(13–15), 3318–3330 (2009)

Dynamic Place Profiles from Geo-folksonomies on the GeoSocial Web

Soha Mohamed and Alia Abdelmoty

Abstract The growth of the Web and the increase in using GPS-enabled devices, coupled with the exponential growth of the social media sites, have led to a surge in research interest in Geo-folksonomy analysis. In Geo-Folksonomy, a user assigns an electronic tag to a geographical place resource identified by its longitude and latitude. The assigned tags are used to manage, categorize and describe place resources. Building data models of Geo-folksonomy data sets that represents and analyses, tags, location and time information can be helpful in studying and analysing place information. The aim of my research is to use the spatio temporal data available on the Social web, to extract dynamic place profile. Building a dynamic profile involves including the temporal dimension in the Geo-folksonomy. Indeed, adding the temporal dimension can provide an understanding of geographic places as perceived by users over time.

1 Introduction

Recently, media has become popular and attracts many researchers from a variety of fields. Social networks enable users to post their user generated content at any time and from any location. With the emergence of Web 2.0, tagging systems evolved. Users can assign keywords of their choice to web resources. These keywords are called tags, and the process of assigning keywords to resources is termed as tagging. A typical examples of tagging systems are Delicious, Flickr and Tagazania where uses enters tags of their choices to annotate resources (url, images, places, etc.). Consequently, with the emergence of new social tagging systems such as Twitter and Foursquare, the structure of tags changed. Traditional tags consist of one word,

S. Mohamed (✉) · A. Abdelmoty
Cardiff University, Cardiff, UK
e-mail: AlySA@cardiff.ac.uk

A. Abdelmoty
e-mail: A.I.Abdelmoty@cardiff.ac.uk

© Springer International Publishing Switzerland 2014
M. Bramer and M. Petridis (eds.), *Research and Development in Intelligent Systems XXXI*, DOI 10.1007/978-3-319-12069-0_18

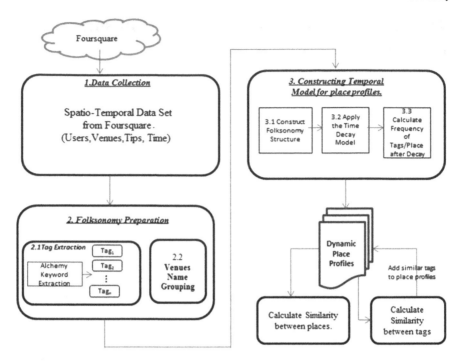

Fig. 1 Proposed model framework

while new tags may contain a sentence of more than one word. The result of tagging resources is called folksonomy. Folksonomies are created by tagging applications via users' interaction on web 2.0. It consists of three main entities: users, tags and resources. Geo-Folksonomies are a special kind of folksonomies where the users assigns tags to geographical place resources. Studying tags from users visiting places can enhance the understanding of the geographic places visited, and its characteristics. Understanding places and their use can help cities understand social dynamics of these places and improve service provision and planning.

This paper proposes a framework (Fig. 1) for constructing a dynamic place profile from the GeoFolksonomies. Mining the spatio-temporal data to extract the implicit semantics would give insight into the characteristics of users and place and interactions between them. In this work, Foursquare is used as a data source. Users can share their current location using venue check-in and/or leave a Tip about their experience in the visited venue. The framework involves different stages of tag extraction and folksonomy preparation in order to apply the temporal decay model for dynamic place profile construction. The derived place profile contains tags associated with the place and their weights. Furthermore, a temporal place similarity measure is used to measure the change of the similarity between places over time. Studying geofolksonmies was introduced before in [7, 20, 21] but they didnt consider how the places change over time. Furthermore, similar work was done in [12, 15] but they used the time factor to compare the folksonomy structure at different points of time and they didnt include the time in the equation of calculating the similarity between places and users.

The paper is structured as follows. A review of the related work on folksonomy analysis, geofolksonomy analysis, and the geo-temporal models are described in Sect. 2. The proposed model for dynamic place profile is explained in Sect. 3. Some results and analysis are discussed in Sect. 4.

2 Related Work

In this section a summary of the related work is introduced. The following subsections will explain the folksonomy analysis, geofolksonomy and the geotemporal models.

2.1 Folksonomy Analysis Methods

Formally, a folksonomy is a tuple $F = \langle U, T, R, A \rangle$ where U, T, R represent users, tags and resources respectively. The relationship 'A' relates U, T and R. Consequently, the folksonomy can be represented as a tripartite graph [9]. The vertices of the graph are the users, tags and resources. Alternatively, the folksonomy graph can be represented as a three dimensional adjacency matrix. However, to ease the manipulation, the tripartite graph can be decomposed to 3 bipartite graphs which are tag-user, tag-resource, and user-resource [1].

There are three measures of tag relatedness as stated in [4]: Co-occurrence, Cosine Similarity and Folk Rank.

In Co-occurrence measure, the tag-tag co-occurrence graph is defined as a weighted undirected graph whose set of vertices is the set T of tags, and two tags t1 and t2 are connected by an edge, iff there is a least one post. The weight of the edge is given by the number of posts that contain both t1 and t1 [5].

In Cosine similarity, the measure of tag relatedness is computed by using the cosine similarity of tag-tag co-occurrence distributions. Two tags are considered related when they occur in a similar context, and not when they occur together [18].

The FolkRank method is derived from the PageRank algorithm which reflects the idea that a web page is important if there are many pages linking to it, and if those pages are important themselves [2]. The same principle is employed for Folkrank, a resource which is tagged with important tags by important users becomes important itself. The same holds for tags and users [10].

2.2 Geo-folksonomy

Reference [7] introduced an approach for extracting place semantics embedded in geo-folksonomies. Social tags about places from Tagazania were used. In particular, perceptions of users about place and human activities related to places are captured to build place type and activity ontologies. The approach addresses the quality problems evident in the tags and place resources through a cleaning process; it also provides a place ontology model to capture the desired place semantics, and utilises external semantic resources and statistical co-occurrence methods to build the place ontology.

A suite of methods to extend folksonomies (SWE-FE) was presented by Rezel and Liang [20] SWE-FE utilizes the geospatial information associated with the three key components of the tagging system, tags resources and users. The authors extend the formal definition of folksonomy (user, tag, and resource) to contain the geographic location. They also extended the nave method to calculate the similarity between the users to include the distance as a factor in calculating the similarity. The geospatial folksonomy scheme described was implemented on GeoCENS. The authors argued that including the distance as a factor in the similarity can enrich the relationships among users and thus can provide better recommendation.

A model-based framework GeoFolk was introduced in [21] which combines both tags and spatial information to better content characterization. In Geofolk model, Bayesian statistics models were employed to represent Geodata, combined with tag co-occurrence patterns. The data set was taken from the publicly accessible CoPhIR2 dataset that contains metadata for over 54 million of Flickr resources. Each resource is annotated with one longitude and latitude and a set of tags. GeoFolk aims to explain the semantic relatedness of tags by means of latent topics in a probabilistic Bayesian framework. Latent Topics Models (LDA) are mechanisms for discovering the number of topics in a document in a probabilistic manner. In GeoFolk model, tag similarity/relatedness is estimated in a natural manner, by comparing tag distributions over latent topics. The author shows that GeoFolk works better than text-only analysis in tag recommendation, content classification and clustering. However, GeoFolk is not suitable for region clustering as it fails to find common topics in different geographical sites.

All the above work didn't consider the temporal dimension in the construction of folksonomy models. The following subsection introduces the temporal folksonomy that can be used to include the temporal dimension the in geo-folksonomy.

2.3 Geo-temporal Modeling

A recent work was done in [15] to extract topics from user generated tips assigned to venues in the Foursquare dataset. The latent Dirichlet Allocation (LDA) topic model discussed earlier was used. A "topic" consists of a cluster of words that frequently occur together. Using contextual clues, topic models can connect words with similar meanings and distinguish between uses of words with multiple meanings. Using this model, each venue is expressed as a mixture of a given number of topics. An individual's activity identity is a combination of the venues in which he checks in. User similarity measure is calculated using the distinct venue topic distribution of every user in the dataset. *Jensen-Shannon divergence* (JSD) is used to compute a dissimilarity metric among users' topic distribution. The resulting metric is bound between 0 and 1 where 0 indicates that the two users topic signatures are identical and 1 representing complete dissimilarity. The limitation of their work is that a 3 h time window was used as the temporal bound for users' activities collected, and the user similarity measure was used to compare users temporally based on this time window throughout the day. Further temporal factoring to reflect day of the week and month can enhance the ability of the model to discover similar users. Moreover, it is better to include time in computing the similarity between users and not only comparing similarities between user tags at different point of time.

Another recent work was claimed to be the first spatio temporal topic (STT) modelling for Location recommendation [12]. In their work, they addressed the problem of recommending right locations to users at the right time. Basically, they represented a check-in as a user, a location with a pair of coordinates, and a relative timestamp, which are all considered as observed random variables. Similar to LDA, a set of latent topics is defined. Each user is associated with a probability distribution over topics, which captures the user interests, and each topic has a probability distribution over locations, which captures the semantic relationship between locations. They performed an experimental evaluation on Twitter, Gowalla, and Brightkite data sets from New York. STT assumes that the topic distribution doesn't only depend on user topic distribution but also on the time's topic distribution. Moreover, the probability of recommending location depends on the given time's location distribution. The dynamic topic model they used captures the evolution of topics over time. The limitation of this model is that topics are modelled at each time slot, so the overall distribution of the topics in places can not be monitored.

3 The Proposed Model: Dynamic Place Profiles

The proposed model is composed of different components as shown in Fig. 1. Each component is explained in the following sections.

3.1 Data Collection

In this work, different location based social network applications like Foursquare, Twitter, Flickr were studied and compared to choose the appropriate application to collect data from. Foursquare was chosen as it contains venues in which users can

Table 1 Data sets

	Cardiff dataset	Bristol dataset
Venues count	1,627	2,082
Venus with tips	446	409
Tips count	1,084	1,411
Tags count	2,976	3,260
Users count	876	1,094

check-in and leave their tips. This data allows studying the three main aspects of human mobility which are geographic movement , temporal dynamics and interaction. The data collection system was implemented using Java and the Foursquare API. The two functions venuesSearch() and venueTips() were used to get all the venues in a specific longitude and latitude and the tips within each venue. Cardiff, Bristol were chosen for the data collection. The three data sets will be used in the evaluation of the proposed model. A summary of some statistics of the data collected in Cardiff and Bristol are in Table 1.

3.2 Tags Extraction

Tips are different from tags because they contain more than one word. So, there is a need to extract tags from tips as a pre-processing step. In this work, Alchemy API[1] is used to extract important keywords from the tips. The Alchemy API employ sophisticated statistical algorithms and natural language processing technology to analyse the data, extracting keywords that can be used to index content. The following are some examples of extracting keywords from tips.

Tip: Try the double chocolate brownie and enjoy
Keywords: double chocolate brownie, enjoy

Tip: Was in Millennium Stadium for Judgement Day on Easter Eve
Keywords: Millennium Stadium, Easter Eve, Judgement day.

3.3 Database Design

The database engine used in this research is MySQL V.5. The database is stored on the school server (emphesus.cs.cf.ac.uk). The Geofolksonomy database is designed

---

[1] http://www.alchemyapi.com/api/

Fig. 2 Database schema

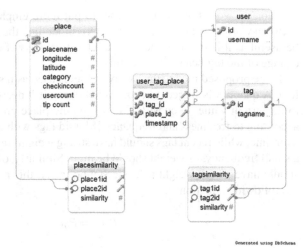

Generated using DbSchema

to support storing and searching of the collected Geo-Folksonomy datasets as well as the output of the Folksonomy analysis methods implemented. The data model of the database is shown in Fig. 2. The three distinct components of the geo-folksonomy are modelled using the Place table representing folksonomy venue resources, the Tag table representing folksonomy tags, and the User table representing folksonomy users. The user_tag_place table relates the three tables user, place, and tag. The timestamp is placed in the user_tag_place table in preparation to apply the time model to the data. The database also contains several tables for storing the output of the folksonomy analysis such as tags similarity and place similarity.

3.4 Dynamic Place Profile Construction

An important model used in temporal modelling is the decay model (also called The ageing theory) [6]. The Decay is an exponential function defined as the way in which a quantity naturally decreases over time at a rate proportional to its current value [8]. Tags and posts are considered to have a lifetime. When they are first defined, they are fresh and interesting to users then they decay over time and their interest value decreases [22]. The life cycle ageing theory was successfully used in a work done on Twitter in order to detect real time emerging topics by mining terms that frequently occur in specified time interval [3]. Yao et al. [22] used the decaying equation in order to track the time changing behaviour of a tag through the lifetime of all tagged posts. They used the temporally ordered sequential tag stream in order to mine burst or anomaly intervals of tags. Radinsky et al. [19] proposed a temporal semantic model to compute semantic relatedness of words. They applied the exponential weighting function on word time series in order to put more emphasis on recent correlations between words. Using the decay model for modelling the dynamic place profiles will be useful as it is important to maintain the freshness of the tags and to monitor the change of the tag behaviour over time.

In the proposed model, the Decay Model is used as it suits the nature of our problem. A quantity is subject to exponential decay if it decreases at a rate proportional to its current value. In constructing a dynamic place profile, tags should be ranked according to freshness and frequency [8]. Old tags with no repetitions should decay over time, while recent tags should have strong weight. In addition, if a recent tag has a small frequency, its weight should be high. Similarly, old tags with high frequency should have a high weight as well. Symbolically, this process can be expressed by the following equation:

$$P(t) = P_0 e^{-rt} \tag{1}$$

where:

P(t) = the amount of some quantity at time t.
P_0 = initial amount at time t = 0.
r = the decay rate.
t = time (number of periods).

In this work, we propose two ways of creating place profile. The first is the Dynamic Direct Place Profile, in which we add all the Tags associated with the place and its frequency, and the second way is the Dynamic In direct Place Profile in which the tag-tag similarity is computed using cosine similarity measure. The strength of association measure is then assigned to place based on tag similarity. The more similar a tag is to another, the stronger the relationship between that tag and the place.

3.4.1 Direct Dynamic Place Profile

Each place in the database is associated with a set of tags, and each tag has a frequency. As the time passes, the frequency of the tag should decay slightly to maintain the freshness of tags. Algorithm 1 is a summarized version of calculating the Tag Frequency after applying the decay equation. The algorithm has three inputs, TimeCluster in which the user can specify to cluster the data by hour, day, week, month or year. The second input is the Date, that is the point of Time the user want to calculate the Tag Frequencies. The third input is r which is the decay rate as mentioned previously. After applying this algorithm, the output is a list of Tag Frequencies of each place with the dynamic effect.

Algorithm 1: DYAMIC TAGFREQUENCY(*TimeClusters, Date, r*)

for each *place* ∈ *PlaceTable*

do
$\begin{cases} \textit{Cluster Tag\_User\_Place Table by TimeClusters} \\ \textit{Calculate TagFrequency within each TimeClusters} \\ \textbf{for each } T \in \textit{TimeStamps} \\ \quad \textbf{do } \begin{cases} \textit{Calculate No. of periods between T and Date} \\ \textit{ApplytheDecayEquation} \\ \textit{Calculate TagFrequencyafter Decay} \end{cases} \\ \textbf{for each } \textit{Tag} \\ \quad \textbf{do for all } \textit{TimeSlots} \\ \quad \textbf{do } \textit{SumUp TagFrequency} \end{cases}$

return *TagFrequency*

3.4.2 Indirect Dynamic Place Profile

In the Indirect place profile, the tag-tag Similarity Measure [7] is calculated using the cosine similarity measure [13]. The tag-tag Similarity is used to find the tags that are similar to the tags in the Dynamic Direct Place Profile. The similarity between two tags is defined by the following equation:

$$sim(T_1, T_2) = \frac{|P_1 \cap P_2|}{\sqrt{|P_1| \cdot |P_2|}} \tag{2}$$

where T_i represents a tag and P_i represents the set of instances of place resources associated with the tag T_i in the Geo-folksonomy after applying the dynamic tag frequency algorithm.

3.5 Place Similarity

The cosine similarity method is used to calculate the similarity between places in the database. The place similarity depends on the co-occurrence of tags in different place. It also depends on the weight of each tag. The geo-folksonomy dataset is used to assess the similarity of the place instances.

4 Analysis and Results

Figures 3 and 4 shows a comparison between tags similar to tag 'Bay' during years 2011 and 2012 in Cardiff dataset. The nodes represents the tags and the weights on the edges represent the similarity measure between two tags. The similarity measure is a number between 0 and 1. As shown in figures, more tags were added during 2012. The similarity measure between tags that were not mentioned in 2011 decreased and decayed in 2012, and the similarity of tags that are mentioned in both 2011 and 2012 increased. This shows the benefit of using the decay model over other temporal models as it doesn't forget the past tags unless they are never mentioned.

Figures 5 and 6 shows a map based comparison of the places similar to 'Millennium centre' in Cardiff during 2011 and 2012 respectively. The similarity between two places are calculated using the cosine similarity measure between the tags mentioned in the two places. The more similar the places are the bigger the radius of the circle.

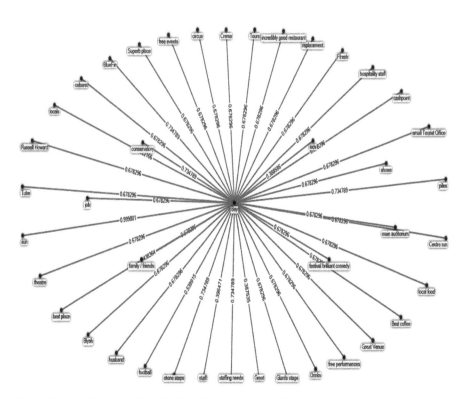

Fig. 3 Tags similar to tag 'Bay' in Cardiff dataset during 2011

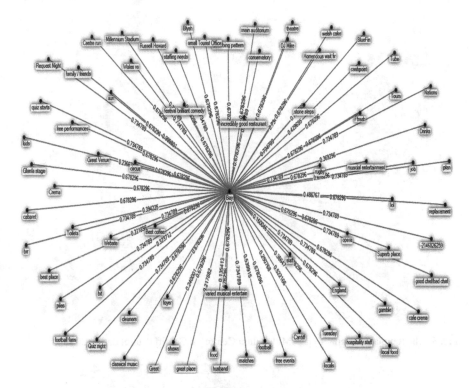

Fig. 4 Tags similar to tag 'Bay' in Cardiff dataset during 2012

5 Conclusion and Future Work

In this paper, a novel framework for dynamic place profile have been introduced. The aim of this research work is to study the spatiotemporal aspects of Social Web data and analyse their value in understanding user place characteristics visited by users. The framework developed have many potential applications and uses. In particular, it can be used to provide users with more personal search experience. It can be used by recommendation services and personalisation applications to provide users with relevant information. It can improve the recommendation services and lead to more targeted adverts and commercials. It can improve location-based service, by providing personalised access to local services. The next step is make an evaluation application in order to evaluate the place profile and to compare it against models developed for temporal geo-modeling mentioned in the related work.

Fig. 5 The top places similar to tag 'Millennium Centre' in Cardiff dataset during 2011

Fig. 6 The top places similar to tag 'Millennium Centre' in Cardiff dataset during 2012

References

1. Beldjoudi, S., Seridi, H., Faron-Zucker, C.: Ambiguity in tagging and the community effect in researching relevant resources in folksonomies. In: Proceedings of ESWC Workshop User Profile Data on the Social, Semantic Web (2011)
2. Brin, S., Page, L.: The anatomy of a large-scale hypertextual Web search engine. Comput. Netw. ISDN Syst. **30**(1), 107–117 (1998)
3. Cataldi, M., Di Caro, L., Schifanella, C.: Emerging topic detection on twitter based on temporal and social terms evaluation. In: Proceedings of the Tenth International Workshop on Multimedia Data Mining, p. 4. ACM, New York (2010)
4. Cattuto, C., Benz, D., Hotho, A., Stumme, G.: Semantic analysis of tag similarity measures in collaborative tagging systems. arXiv preprint arXiv:0805.2045 (2008)
5. Cattuto, C., Barrat, A., Baldassarri, A., Schehr, G., Loreto, V.: Collective dynamics of social annotation. Proc. Nat. Acad. Sci. **106**(26), 10511–10515 (2009)
6. Chen, C.C., Chen, Y.T., Sun, Y., Chen, M.C.: Life cycle modeling of news events using aging theory. In: Machine Learning: ECML 2003, pp. 47–59. Springer, Berlin Heidelberg (2003)
7. Elgindy, E.: Extracting place semantics from geo-folksonomies. Doctoral Dissertation, Cardiff University (2013)
8. Fonda, L., Ghirardi, G.C., Rimini, A.: Decay theory of unstable quantum systems. Rep. Progr. Phys. **41**(4), 587 (1978)
9. Halpin, H., Robu, V., Shepherd, H.: The complex dynamics of collaborative tagging. In: Proceedings of the 16th International Conference on World Wide Web. pp. 211–220, ACM, New York (2007)
10. Hotho, A., Jäschke, R., Schmitz, C., Stumme, G.: Folkrank: a ranking algorithm for folksonomies. In: Althoff, K.D. (ed.) LWA, vol. 1, pp. 111–114 (2006)
11. Hotho, A., Jäschke, R., Schmitz, C., Stumme, G.: Trend detection in folksonomies, pp. 56–70. Springer, Berlin (2006)
12. Hu, B., Jamali, M., Ester, M.: Spatio-temporal topic modeling in mobile social media for location recommendation. In: IEEE 13th International Conference on Data Mining (ICDM), 2013, pp. 1073–1078. IEEE (2013)
13. Markines, B., Cattuto, C., Menczer, F., Benz, D., Hotho, A., Stumme, G.: Evaluating similarity measures for emergent semantics of social tagging. In: Proceedings of the 18th International Conference on World Wide Web, pp. 641–650. ACM, New York (2009)
14. Mathioudakis, M., Koudas, N.: Twittermonitor: trend detection over the twitter stream. In: Proceedings of the 2010 ACM SIGMOD International Conference on Management of Data, pp. 1155–1158. ACM, New York (2010)
15. McKenzie, G., Adams, B., Janowicz, K.: A thematic approach to user similarity built on geosocial check-ins. In: Geographic Information Science at the Heart of Europe, pp. 39–53. Springer International Publishing, Berlin (2013)
16. Michlmayr, E., Cayzer, S., Shabajee, P.: Adaptive user profiles for enterprise information access. In: Proceedings of the 16th International World Wide Web Conference (2007)
17. Pham, X.H., Jung, J.J., Hwang, D.: Beating social pulse: understanding information propagation via online social tagging systems. J. UCS **18**(8), 1022–1031 (2012)
18. Quattrone, G., Ferrara, E., De Meo, P., Capra, L.: Measuring similarity in large-scale folksonomies. arXiv preprint arXiv:1207.6037 (2012)
19. Radinsky, K., Agichtein, E., Gabrilovich, E., Markovitch, S.: A word at a time: computing word relatedness using temporal semantic analysis. In: Proceedings of the 20th International Conference on World wide web, pp. 337–346. ACM, New York (2011)
20. Rezel, R., Liang, S.: SWE-FE: extending folksonomies to the sensor web. In: 2010 International Symposium on Collaborative Technologies and Systems (CTS), pp. 349–356. IEEE (2010)
21. Sizov, S.: Geofolk: latent spatial semantics in web 2.0 social media. In: Proceedings of the Third ACM International Conference on Web Search and Data Mining, pp. 281–290. ACM, New York (2010)
22. Yao, J., Cui, B., Huang, Y., Jin, X.: Temporal and social context based burst detection from folksonomies. In: AAAI (2010)

Planning and Optimisation

Hierarchical Type-2 Fuzzy Logic Based Real Time Dynamic Operational Planning System

Ahmed Mohamed, Hani Hagras, Sid Shakya, Anne Liret,
Raphael Dorne and Gilbert Owusu

Abstract Operational resource planning is critical for successful operations in service-based organizations as it underpins the process of utilizing resources to achieve a higher quality of service whilst lowering operational costs. The majority of service-based organizations use static operational planning. In recent years these, organizations have made attempts to switch to dynamic operational planners with the view of generating real-time operational plans. This paper proposes a hierarchical type-2 fuzzy logic based operational planner that can work in dynamic environments and can maintain operational plans in real-time. The proposed system outperformed ordinary heuristic-based systems and task dispatchers.

1 Introduction

Service providing companies use the process of efficient utilization of resources to realize a high quality of service whilst decreasing operational costs. This is often done with the assumption that the environment is static. Static resource planning

A. Mohamed (✉) · H. Hagras
School of Computer Science and Electronic Engineering, University of Essex, Wivenhoe Park,
Colchester CO4 3SQ, UK
e-mail: amhmoh@essex.ac.uk

H. Hagras
e-mail: hani@essex.ac.uk

S. Shakya · A. Liret · R. Dorne · G. Owusu
British Telecommunication Adastral Park, Ipswich, Martlesham, UK
e-mail: sid.shakya@bt.com

A. Liret
e-mail: gilbert.owusu@bt.com

R. Dorne
e-mail: anne.liret@bt.com

G. Owusu
e-mail: raphael.dorne@bt.com

© Springer International Publishing Switzerland 2014
M. Bramer and M. Petridis (eds.), *Research and Development
in Intelligent Systems XXXI*, DOI 10.1007/978-3-319-12069-0_19

generates schedules for short periods ranging from days to weeks by actively refining deployment plans. Service providing companies use dispatching systems to handle the dynamic aspects of the environment (e.g. a task cancellation). Real-time dynamic resource planning maintains the schedules generated by the static resource planning by adding new tasks to or modifying existing tasks in the schedule. In recent years, more efforts have been geared towards finding capabilities that can handle both the static and dynamic aspects of resource planning.

The process of static resource planning includes a full optimization run in the morning to produce a baseline plan for each resource for the day. This involves a large amount of computation as it deals with hundreds of resources and thousands of tasks whilst attempting different planning combinations to minimize the cost function. However, throughout the day the situation in the field changes, for example:

- New high priority tasks arrive
- Tasks get delayed or cancelled
- A resource goes off sick

Clearly, this requires constant updating of the baseline plan to take into account the dynamic changes in the environment. However, a full optimization run due to a single new event is not feasible as it is computationally very expensive. The goal then is to investigate a real-time optimization solution to repair and maintain the plans in a dynamic environment, within the limited computation time.

In this paper, we will present a hierarchical type-2 fuzzy logic based real-time dynamic operational resource planner for the workforce allocation. The proposed system avoids the curse of dimensionality that is usually faced in fuzzy logic systems by using a hierarchical fuzzy logic system. We also propose a matching algorithm that calculates the compatibility between resources and their compatible tasks. Next, the algorithm assigns the most compatible resources to the tasks. The proposed hierarchical fuzzy logic based system was able to achieve good results in comparison to the heuristic systems that are used in static and current real-time dynamic resource planning systems.

Section 2 presents an overview of the existing heuristic search-based operational resource planning tools in service-based organizations and its limitations. Section 3 gives a brief overview over type-2 fuzzy logic systems. Section 4 presents the proposed hierarchical fuzzy logic system. Section 5 presents the proposed hierarchical fuzzy logic type-2 operational resource planner. Section 6 presents the proposed hierarchical type-2 fuzzy logic based real-time dynamic operational resource planner. Section 7 presents the experiments and results. Finally, we present the conclusions and future work in Sect. 8.

2 Overview of the Existing Heuristic Search-Based Operational Resource Planning Tools in Service-Based Organizations

The vast majority of the existing resource planning tools available in service-based organizations employ heuristic searches based on crisp logic. These tools share the following process steps [1–4]:

- Constructing an initial solution for a period of time (it is called the baseline plan).
- A steepest gradient algorithm is employed to gradually modify and improve the solution until an optimum solution is reached.

The initial baseline plan is generated by assigning all available resources to their preferred area and skill choices. This guarantees minimal traveling time and maximal productivity. However, the present job volumes may not justify these assignments as poor resource utilization might occur [1–4].

The initial solution is therefore refined in a second planning step known as the optimization phase. In this phase, the current solution is varied in small steps. Three possible types of variation are considered: moves of a single resource to a new area and/or skill, swap moves which exchange the area and skill assignments between two resources, and replacement moves, i.e. moves in which one resource replaces a second resource so that this second resource is free to work somewhere else. The optimization mechanism consists of looping through all possible moves and executing the move that leads to the highest improvement, until no further improvements can be found. Central to this approach is an objective function, which determines the quality of each candidate solution.

2.1 Limitations of the Existing Heuristic Operational Resource Planner

Since the static resource planning is based on crisp logic, an in correct configuration of the weights of the objective function could lead to scenarios where the system might not be able to focus on more than one objective (for example the system might focus on allocating close tasks to the resource while neglecting allocating tasks with high importance). As some of the objectives might be conflicting, it might be difficult to find the right tradeoff.

For the real-time dynamic resource planning there are three possible solutions we can use but each solution has limitations:

- Running a full optimization run due to a single new event (e.g. new task coming), but this is computationally expensive.
- Periodic running of the optimization engine (i.e. full run) but again this is computationally expensive and impacts customer experience.

- Simple insertion of new tasks to the best available resource in the schedule though fast, will produce a poor schedule. This approach is commonly used by most service-based organizations in their real-time systems.

2.2 Improving the Existing Operational Resource Planner

The improved operational resource planning is divided into two parts: static and dynamic as shown in Fig. 1. The static part is responsible for scheduling pre-booked tasks where it employs a hierarchical fuzzy logic type-2 system with a matching algorithm that can handle the uncertainties in the inputs better than the heuristic crisp resource planning systems. It can also handle any conflicts between the objectives using the fuzzy rule base. The system then passes the SOD (Start of Day) schedule to the real-time dynamic scheduler where the following two processes run:

- The allocation process where the system waits for new tasks to arrive with the aim to allocate them to the most compatible resource and send a response to the customer that an appointment has been scheduled.
- The reallocation process which runs in the background for maintaining and improving the SOD schedule. This is done by having a fuzzy classifier that goes through

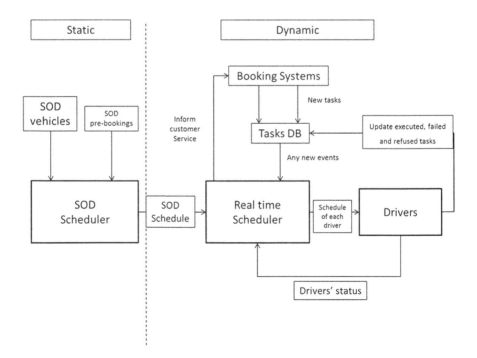

Fig. 1 The improved operational resource planning architecture

the tasks and selects which tasks should be considered for reallocation. The selected tasks are then passed to a hierarchical fuzzy logic type-2 system that is similar to the one used in the static part. The reallocation process keeps on running until the schedule's period ends and produces a schedule with a decreased operational cost and an improved quality of service.

3 Brief Overview Over Type-2 Fuzzy Logic Systems

The interval type-2 FLS depicted in Fig. 2a uses interval type-2 fuzzy sets (such as the type-2 fuzzy set shown in Fig. 2b) to represent the inputs and/or outputs of the FLS. In the interval type-2 fuzzy sets all the third dimension values are equal to one. The use of interval type-2 FLS helps to simplify the computation (as opposed to the general type-2 FLS) [5].

The interval type-2 FLS works as follows: the crisp inputs are first fuzzified into input type-2 fuzzy sets; singleton fuzzification is usually used in interval type-2 FLS applications due to its simplicity and suitability for embedded processors and real time applications. The input type-2 fuzzy sets then activate the inference engine and the rule base to produce output type-2 fuzzy sets. The type-2 FLS rule base remains the same as for the type-1 FLS but its Membership Functions (MFs) are represented by interval type-2 fuzzy sets instead of type-1 fuzzy sets. The inference engine combines the fired rules and gives a mapping from input type-2 fuzzy sets to output type-2 fuzzy sets. The type-2 fuzzy output sets of the inference engine are then processed by the type-reducer which combines the output sets and performs a

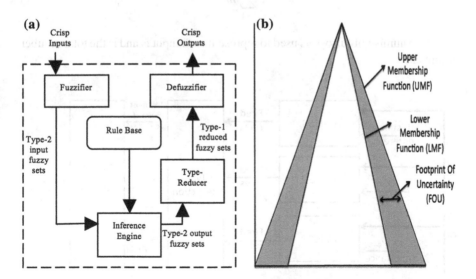

Fig. 2 **a** Structure of the type-2 FLS [5]. **b** An interval type-2 fuzzy set

centroid calculation which leads to type-1 fuzzy sets called the type-reduced sets. There are different types of type-reduction methods. In this paper we will be using the Centre of Sets type-reduction as it has reasonable computational complexity that lies between the computationally expensive centroid type-reduction and the simple height and modified height type-reductions which have problems when only one rule fires [5]. After the type-reduction process, the type-reduced sets are defuzzified (by taking the average of the type-reduced sets) to obtain crisp outputs. More information about the interval type-2 FLS and its benefits can be found in [5].

4 The Proposed Hierarchical Type-2 Fuzzy Logic System

The proposed type-2 Hierarchical Fuzzy Logic System (HFLS) used in our static and dynamic operational planning systems focuses on the input with the highest significance (in our case, it is the distance between a resource and a compatible task) by taking this input as a common factor with all the other inputs (as shown in Fig. 3). There is a separate fuzzy logic system where the inputs are the ones with the highest significance (distance in our case) and another input from the other available inputs. Each fuzzy logic system then outputs the fired rules by the incoming two inputs. The system then takes all the fired rules from all the fuzzy subsystems and passes them to the type-reducer and then the defuzzifier to calculate an overall compatibility score between a task and a resource. The same number of fuzzy sets will represent all the inputs in our fuzzy hierarchical structure. Hence, the number of rules increases linearly (will be a constant number) with the number of inputs as follows:

$$\sum_{i=1}^{N-1} m^2 \tag{1}$$

where the number of fuzzy sets used to represent each input is and is the total number of inputs.

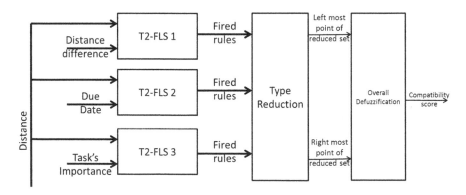

Fig. 3 1 level type-2 hierarchical fuzzy logic system

In our application we have 4 inputs, which result in 3 separate fuzzy logic systems. Each fuzzy logic system takes the distance and one of the other remaining 3 inputs as its inputs as shown in Fig. 3. Each of the inputs is represented by 3 fuzzy sets. Hence, each fuzzy logic system in our hierarchical structure will have 9 rules. Therefore, the total number of rules in our hierarchical structure will be 27 rules.

In Fig. 3, we can see that each fuzzy logic system outputs its fired rules that contribute to the overall type-reduction. This uses the center of sets type reduction. For each fired rule from the low level FLSs, we supply the upper and lower firing strengths and we also supply the centroids for the consequent of the fired rules and in order to calculate the type-reduced sets. The overall defuzzification stage then takes the average of the left most point and right most point of the type reduced set to produce the final crisp output [5].

The hierarchical fuzzy logic systems in [6] address the "curse of dimensionality" problem but they suffer from the intermediate inputs and rules problem (where it is difficult to describe the rules). In our case, the proposed hierarchical fuzzy logic system shown in Fig. 3 solves the "curse of dimensionality" problem where the rules grow linearly with the number of inputs and it has no intermediate inputs and rules. Another benefit of our proposed hierarchical type-2 fuzzy system is that it can be easily designed and it is modular (e.g. new fuzzy logic systems could be easily added or removed).

5 The Proposed Hierarchical Type-2 Fuzzy Logic Based Operational Planner

As shown in Fig. 4, the proposed system works as follows: an expert configures the system, then the data (resources and tasks) is uploaded by the system (from the database if static and from the schedule directly if dynamic), then the system sorts the tasks between the resources and vice versa. Afterwards, the compatibility scores between each resource and tasks that are compatible with this resource, and between each task and its resources, is calculated using the hierarchical type-2 fuzzy logic system mentioned in the previous section. The system then attempts to find the best matching couple to start the work simulation using the compatibility scores and a matching algorithm that takes the tasks and resources' points of view (i.e. when matching them together). If the system was unsuccessful in finding the best matching couple, then this indicates that the scheduling process is over. This system has been designed generically and allows tasks to have separate start and finish locations where necessary (for example the logistics task can be picked up and dropped off in different locations).

The proposed system uses the following 4 inputs to calculate the compatibility score between a task and a resource:

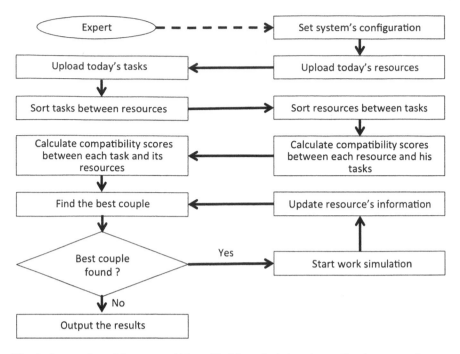

Fig. 4 An overview of the proposed hierarchical fuzzy logic based operational resource planner

- Distance between the resource and (start) location of the task
- Distance difference before adding the travel to the start and finish location of the task and after adding them to the resource's schedule
- Task's due date
- Task's importance

After calculating the compatibility scores using the previous inputs, a matching algorithm takes these compatibility scores and starts a matching procedure to match the best resources and tasks together. The behaviour of this matching algorithm can be configured by changing the rule base of the hierarchical type-2 fuzzy logic system.

6 The Proposed Hierarchical Type-2 Fuzzy Logic Based Real-Time Dynamic Operational Planner

Figure 5 shows a flowchart that describes the process of the proposed hierarchical type-2 fuzzy logic based real-time dynamic operational resource planner and it will be discussed in more detail in the following subsections.

In the first stage, an expert configures the system by setting two main parameters:

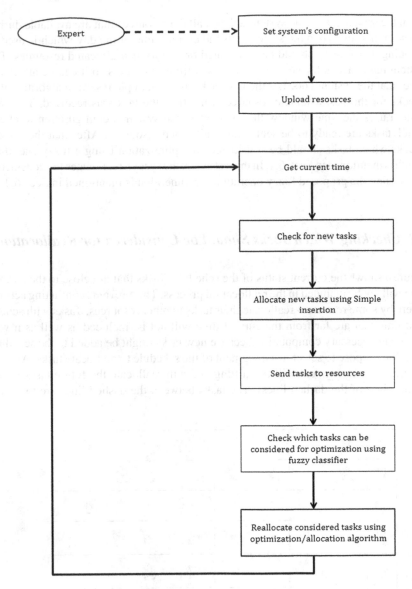

Fig. 5 An overview of the proposed hierarchical fuzzy logic based real-time dynamic operational resource planner

- The time window where the system checks if a task lies within this time window, this task will be sent to its allocated resource. In our experiments this time window was set to 60 min from the current time.
- The expert can tune the membership functions of the fuzzy classifier, which determines which tasks will be considered for reallocation.

In the second stage, the system uploads all the resources that are available during the system's run. Then the system gets the current time of the day which is used in deciding which tasks should be reallocated or sent to their allocated resources. The system then checks for new incoming tasks from customers so it can be allocated. After that the system allocates the new tasks using a simple insertion algorithm that checks for the best available resource at the time the task was received. Then the system uses the time window that was set in the system's configuration to check which tasks are ready to be sent to their allocated resources. After that the system checks which tasks should be considered for optimization using a fuzzy classifier, which is mentioned in Sect. 6.1. In the last stage, the system reallocates the considered tasks using the proposed fuzzy operational planner that is mentioned in Sect. 6.2.

6.1 Checking Which Tasks Should be Considered for Reallocation

Figure 6 shows the current status of the schedule. Tasks that are close to the current time will not be included in the reallocation process. This prevents conflicting actions where by some resources reallocate their tasks to other resources. Tasks with scheduled time that are far from the current time will not be included as well as it will require unnecessary computation because new tasks might be added to the schedule later. We propose to select just a segment of the schedule to reallocate tasks. As time progresses, this segment will be shifting and will reallocate the future tasks (tasks on the right of the dashed lines). The tasks between the dashed lines are the tasks

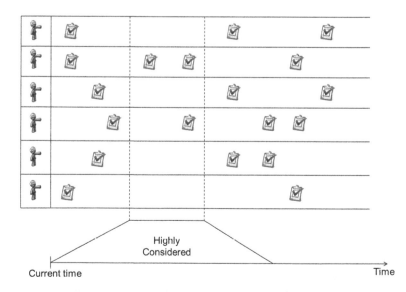

Fig. 6 Real-time dynamic scheduler

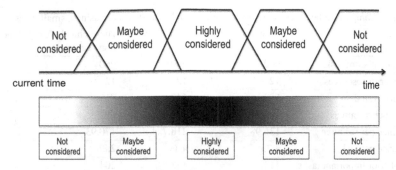

Fig. 7 Fuzzy tasks consideration classifier

that most likely will be considered for reallocation. These dashed lines are crisp and will not be easy to configure. Secondly, they do not give the ability to select which tasks should be considered for reallocation. To address the above challenges, we extend the solution to fuzzy-based logistics as shown in Fig. 7 where the tasks to be considered can be controlled using a fuzzy rule base.

Figure 7 shows the fuzzy aspect; if a task lies in the black area that means it is highly considered for reallocation, if it lies in the gray area that means it might be considered for reallocation and if it lies in the white area then it is not considered at all for reallocation. Figure 7 only uses time as an input but this can be easily extended to include other inputs such as importance or the tasks' density.

6.2 Reallocate Considered Tasks Using the Fuzzy Operational Planner

During this stage the system takes the tasks that were considered by the fuzzy classifier and uses the scheduler mentioned in Sect. 5 to reallocate these considered tasks to better resources. The aim is to decrease the operational costs and increase the quality of service.

7 Experiments and Results

We have run a number of experiments with the proposed system using simulated datasets from a city in England. Two datasets were used in these experiments (a smaller dataset representing part of the tested area and a larger dataset representing of the entire tested area). There were two constraints used in these experiments:

1. Tasks had to be done within a time window of 10 min from their target time.
2. Some tasks required resources with the ability of serving disabled people.

Table 1 Comparison between heuristics systems and the fuzzy system using a small dataset

Measure	Simple insertion	Hill climber	Hierarchical fuzzy type 2 logic scheduler
Total number of tasks	960	960	960
Number of resources	12	12	12
Tasks covered	508	640	712
Tasks per resources	42.33	53.33	59.33
Distance per resource	136.98	218.724	170.93
Distance per task	3.23	4.1	2.88
Number of important tasks	112	138	204
Average importance	552.5	553	613.7

Table 2 Comparison between heuristics systems and the fuzzy system using a big dataset

Measure	Simple insertion	Hill climber	Hierarchical fuzzy type 2 logic scheduler
Total number of tasks	1920	1920	1920
Number of resources	20	20	20
Tasks covered	876	876	1218
Tasks per resources	43.8	43.8	60.9
Distance per resource	12.462	181.163	157.604
Distance per task	2.75	4.13	2.58
Number of important tasks	182	220	388
Average importance	536.4	557.6	638

In our experiments, the proposed system was not compared against a type-1 fuzzy logic scheduler as we already proved that a type-2 fuzzy logic system deals with uncertainties better in [7].

As shown in Tables 1 and 2, we can see that the hierarchical type-2 fuzzy logic system was able to cover more tasks, travel less for tasks and do more important tasks compared to the other two heuristics systems.

8 Conclusions and Future work

In this paper, we presented a type-2 hierarchical fuzzy logic based approach for dynamic real-time operational resource planning.

We proposed an improved hierarchical fuzzy logic system, which is able to address the curse of the dimensionality problem found in ordinary fuzzy systems. The rules in our proposed system grow linearly rather than exponentially. In addition, there is no need to define the intermediate- or high-level rules in the hierarchical structure, as is the case with the existing hierarchical structures.

The hierarchical fuzzy based operational resource planner significantly outperformed the heuristics based systems and can result in significant uplift in the operations.

We have also presented a fuzzy classifier that selects which tasks should be considered for reallocation in real-time.

For the future work, we intend to integrate our fuzzy system with a fast optimization algorithm, such as the Big Bang Big Crunch [8, 9] for an improved performance of the system.

References

1. Owusu, G., Kern, M., Voudouris, C., Garyfalos, A., Anim-Ansah, G., Virginas, B.: On optimising resource planning in BT plc with FOS. In: International Conference on Service Systems and Service Management, Troyes, pp. 541–546 (2006)
2. Juedes, D., Drews, F., Welch, L., Fleeman, D.: Heuristic resource allocation algorithms for maximizing allowable workload in dynamic distributed real-time systems. In: Proceedings of the 2004 Symposium on Parallel and Distributed Processing (2004)
3. Aber, E., Drews, F., Gu, D., Juedes, D., Lenharth, A.: Experimental comparison of heuristic and optimal resource allocation algorithms for maximizing allowable workload in dynamic, distributed real-time systems. In: Proceedings of the 22nd Brazilian Symposium on Computer Networks-6th Brazilian Workshop on Real-Time Systems (WTR 2004) (2004)
4. Almeida, A., Aknine, S., Briot, J.: Dynamic resource allocation heuristics for providing fault tolerance in multi-agent systems. In: Proceedings of the 2008 ACM Symposium on Applied Computing, pp. 66–70 (2008)
5. Hagras, H.: A Hierarchical type-2 fuzzy logic control architecture for autonomous mobile robots. IEEE Trans. Fuzzy Syst. 12(4), 524–539 (2004)
6. Wang, D., Zeng, X., Keane, J.: A survey of hierarchical fuzzy systems. Int. J. Comput. Cogn. 4(1), 18–29 (2006)
7. Mohamed, A., Hagras, H., Liret, A., Shakya, S., Owusu, G.: A genetic interval type-2 fuzzy logic based approach for operational resource planning. In: IEEE International Conference on Fuzzy Systems, pp. 1–8 (2013)
8. Erol, O., Eksin, I.: A new optimization method: Big Bang-Big Crunch. Adv. Eng. Soft. 37(2), 106–111 (2006)
9. Jaradat, G.M., Ayob, M.: Big Bang-Big Crunch optimization algorithm to solve the course timetabling problem. In: Intelligent Systems Design and Applications (ISDA), pp. 1448–1452 (2010)

A Hybrid Algorithm for Solving Gate Assignment Problem with Robustness and Tow Considerations

C.H. Yu and Henry Y.K. Lau

Abstract In this paper, we propose a new method to schedule and evaluate the airport gate assignment. A mathematical model is built to show the relationship among three factors impacting on the gate assignment: robustness, tows, and passenger transfer distance. The stochastic uncertainty parameter is estimated by analyzing the historical data in Hong Kong International Airport (HKIA). Finally, an Artificial Intelligence-based hybrid meta-heuristic is designed to solve this problem.

1 Introduction

During the past 30 years, the air transport traffic has roughly doubled [1], which makes major international airports, especially those hub airports congested. Hong Kong International Airport (HKIA) is one of them. Among all the scarce and expensive resources, gate resource is the toughest one as redesign and construction of new terminal buildings requires government permission as well as huge investments and cannot be accomplished in a short timeframe. Therefore, gate arrangement optimization based on available resource is important. Improper gate assignment will affect the flight throughput, robust operation, passenger satisfaction, staff and facility cost.

Gate assignment problem (GAP) is to assign a gate to the flight activity (includes arrival, parking, and departure) for a set of flight activities. The normal hub airport has more than one hundred gates and more than one thousand flight activities to consider. The objectives of GAP include maximizing one-day throughput of the airport, maximizing airline preference, minimizing towing time, etc. Passenger service level is tightly related to airline revenue, thus should also be considered when making gate assignment. If passengers' satisfaction is taken into account, additional aims should be achieved such as minimizing passengers' waiting time, minimizing passengers' transfer distance.

C.H. Yu (✉) · H.Y.K. Lau
The University of Hong Kong, Pokfulam, Hong Kong
e-mail: ych1102@gmail.com

H.Y.K. Lau
e-mail: hyklau@hku.hk

© Springer International Publishing Switzerland 2014
M. Bramer and M. Petridis (eds.), *Research and Development
in Intelligent Systems XXXI*, DOI 10.1007/978-3-319-12069-0_20

Most previous researches focus on static gate assignment problems, which do not consider stochastic variations in real time operations. Some recent researches consider a gate reassignment problem (GRP), which re-schedules the disturbed schedule based on the original one. In this research, we take stochastic variations into account and try to maximize the robustness of schedule, which minimizes the possibility of later changes on the original schedules.

There are three cost factors in the objectives considered in this paper. Conflict cost measure the robustness of the original schedule as well as the effort saved to re-schedule the original assignment. Tow cost measure the staff and facility cost required when changing the position of aircraft. Passenger transfer cost measures the passengers' satisfaction level.

2 The Model

2.1 Conflict Cost

The flight conflict cost between two flights assigned to the same gate is defined as the overlap time between them. For example, if flight i arrives at gate k at 10:00 am, while flight j leaves gate k at 10:10 am, then the conflict time $cf_{ij} = 10$. The original schedule is usually planned without any gate conflict, however, late departure or early arrival will break this balance easily.

We analyze four consecutive days' data in HKIA from 5 May to 8 May with 2,191 arrivals and 2,253 departures. The delay distribution of arrival and departure is listed in Table 1.

The arrival delay (minute) distribution and departure delay (minute) distribution are displayed in Fig. 1.

Referring to the work of Kim [2], we model the delay using a Log-normal distribution. The negative delay is possible in arrival and departure, which denote the earlier occurrence. But Log-normal distribution is bounded to nonnegative values. A shift parameter is used to avoid negative value appearing in the Log-normal Equations. The probability density function of the distribution is given in Eq. (1).

Table 1 Delay distribution of HKIA flight data (5 May–8 May)

Index	Column 2	Column 3
Total flight	2,191	2,253
Mean/min	2.6	9.7
Median/min	−3	2
Max/min	1,153	208
Min/min	−78	−60
SD/min	55.3	28.3

Fig. 1 Arrival and departure
delay distribution

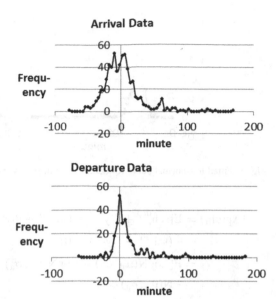

$$f_X(X; \mu, \sigma, c) = \frac{1}{(X-c)\sigma\sqrt{2\pi}} e^{-\frac{(\ln(X-c)-\mu)^2}{2\sigma^2}}, \quad \forall X > c \tag{1}$$

Using MatLab, we obtained the fitted parameters for both arrival delay data and departure delay data. The fitted curves are shown in Figs. 2 and 3.

The gate conflict between arrival flight i and departure flight j can be denoted as in Eq. (2)

$$cf_{ij} = \max(0, sch_j^{dep} + dly_j^{dep} - (sch_i^{arr} + dly_i^{arr})). \tag{2}$$

Assuming the arrival delay and departure delay are independent, we can calculate the expected conflict as in Eq. (3)

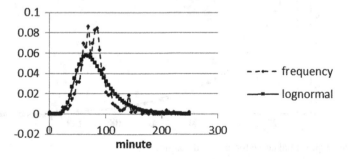

Fig. 2 Fitted log-normal curve for arrival with $\mu = 4.3187, \sigma = 0.399, c = -78$

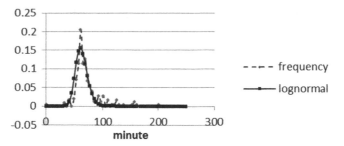

Fig. 3 Fitted log-normal curve for departure with $\mu = 4.0928, \sigma = 0.172, c = -60$

$$
\begin{aligned}
\text{Exp}(cf_{ij}) &= E[sch_i^{dep} + dly_i^{dep} - (sch_j^{arr} + dly_j^{arr})|sch_i^{dep} + dly_i^{dep} \\
&\quad - (sch_j^{arr} + dly_j^{arr}) > 0] \\
&= E[LogN(\mu_d, \sigma_d) - LogN(\mu_a, \sigma_a) - z|Log - N(\mu_d, \sigma_d) - Log \\
&\quad - N(\mu_a, \sigma_a) > z] \\
&= \int_0^\infty \int_{y+z}^\infty (x - y - z)f_d(x; \mu_d, \sigma_d, 0)f_a(y; \mu_a, \sigma_a, 0)dxdy
\end{aligned}
\tag{3}
$$

In which $z = sch_j^{arr} - sch_i^{dep} + c_a - c_d$

The calculation of this double integral is time-consuming, so we fit it to an exponential function. The fitted curve is shown in Fig. 4.

We obtained the approximate model as Eq. (4) using Mathematica:

$$
\text{Exp}(cf_{ij}) = f(Sep_{ij}) \sim 15.6 * 0.966^{Sep_{ij}}
\tag{4}
$$

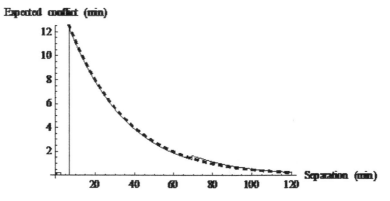

Fig. 4 Fitted exponential curve for expected conflict

We can approximate the conflict cost by relating it to the scheduled separation time between two flights that assigned to the same gate. Thus, the cost can be formulated as in Eq. (5)

$$z_1 = \sum exp\_cf = \sum_{i \in A \cup D} \sum_{j \in A \cup D, s_j > e_i} f(sep_{ij}) \sum_{k \in M} x_{ik} x_{jk} \qquad (5)$$

In which set A contains the arrival flights, set D contains the departure flights. S_j is the start time of flight j and e_i is the end time of flight i. M denotes gate set.

2.2 Tow Cost

In addition to arrival and departure, we also consider the third activity of an aircraft: parking activity. If the interval between successive arrival and departure served by the same aircraft is relatively long, it has to park at somewhere. Parking at its arrival gate until next departure is a choice, but will occupy that gate. If the successive activities are not assigned to the same gate, there will be a tow. When we tow the aircraft to some remote gate or parking apron area, it will not only lead to additional personnel and facility cost, but may also affect the taxiway operations. The tow cost is calculated by the frequency that successive activities are not assigned to the same gate. For example, in Fig. 5, two tows occur because we assign the parking activity to another gate.

According to this definition, we can formulate the tow cost according to Eq. (6)

$$z_2 = \sum_{i \in A \cup P} G(i, U(i)). \qquad (6)$$

In which $U(i) = j$, if is the successive activity of i, $U(i) = 0$ if there is no successive activity following i. For example, for one aircraft, the parking activity is the successive activity of arrival flight, while departure activity is the successive activity of parking. $G(i, j) = 0$ if i and j are assigned to the same gate, $G(i, j) = 1$ if they are assigned to different gates. In this paper, we assume there is a parking apron with unlimited capacity for parking activities only. Let P denotes the parking activities.

Assign the parking activity successive to the arrival will save a tow while occupy the gate and thus make the assignment on the gates more crowded, fewer separation time, therefore longer expected conflict time. Figure 6 illustrate this situation.

Fig. 5 Illustration of tows

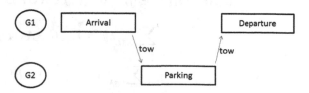

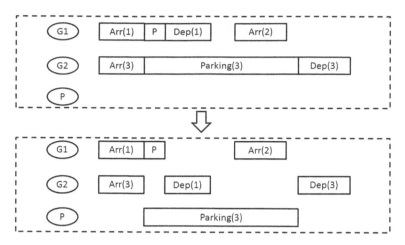

Fig. 6 Illustration of relationship between tows and separation time

2.3 Passenger Transfer Cost

The passenger satisfaction is measured by the total passenger transfer distance in this paper. The transfer distance is a traditional measurement for a gate assignment schedule which appears in early literature related to airport operation optimization problems. In this paper, passenger transfer cost is formulated as a quadratic objective.

$$z_3 = \sum_{i \in A} \sum_{j \in D} \sum_{k \in M} \sum_{l \in M} n_{ij} dist_{kl} x_{ik} x_{jl} \tag{7}$$

Therefore, our objective is to

$$\min z = \alpha z_1 + \beta z_2 + \gamma z_3. \tag{8}$$

2.4 Constraints

This gate assignment problem subjects to four constraints given in Eqs. (9–12).

$$x_{ik} x_{jk} (e_j - s_i)(e_i - s_j) \le 0, \quad (i, j \in A \cup P \cup D, i \ne j, k \in M) \tag{9}$$

$$\sum_{k \in M} x_{ik} = 1, \quad (i \in A \cup D) \tag{10}$$

$$\sum_{k \in M} x_{ik} \le 1, \quad (i \in P) \tag{11}$$

$$x_{ik} \in \{0, 1\}, \quad (i \in A \cup P \cup D, k \in M) \tag{12}$$

Constraint (9) ensures there is no time overlap within a gate. Equation (10) guarantee each arrival and each departure has exactly one gate assignment, while (11) indicates the parking activity could be assigned to a parking apron. Equation (12) indicates the decision variables are binary variables.

3 Solution Method—AI-Based Hybrid Meta-heuristic

Due to the NP-hard nature of the problem, although commercial optimization software CPLEX could give an optimal solution, its efficiency is quite low, especially when the problem size becomes large. The computation time is intractable for instance larger than 20 flights. A more efficient algorithm is therefore designed to solve larger size problems.

We proposed a hybrid algorithm which exploits the exploration capability of genetic algorithm (GA) as a population-based artificial intelligence method, and exploitation capability of large neighborhood search (LNS) methods to each individual in the population. GA is an evolutionary algorithm, which is inspired by the process of natural selection. It is originally developed by Holland [3]. The GA algorithms mainly consist of selection, crossover and mutation, and are well designed to solve scheduling and optimization problems. Since GA is inherently parallel; a number of processors can be deployed to improve the algorithm computational performance [4]. The other idea which benefits our algorithm is large neighborhood search, which has strong ability in deeply searching the individual solution space, trying to find near-optimal solution. Large neighborhood search (LNS) is proposed by Shaw [5] to solve the vehicle routing problems (VRP). Heuristics based on LNS have shown outstanding performance in solving various transportation and scheduling problems [6]. The main idea in LNS is iteratively destroying and repairing the solution, in order to search its neighborhood and find better candidate solutions. The LNS is embedded into GA to control the mutation operation.

3.1 Representation

We use an integer string to represent a solution. The value at th digit represent the gate assigned to th activity. The parking apron is numbered as $|M| + 1$, where M is the set of gates. An example with 9 activities in which 2nd, 5th, 8th are parking activities is shown in Fig. 7. There are 3 gates available, $g(i) = 4$ means the ith activity is assigned to parking apron. Note that only parking activities could be assigned to parking apron by our assumption.

i	1	2	3	4	5	6	7	8	9
$g(i)$	2	4	2	1	1	1	3	4	3

Fig. 7 Integer representation of an individual solution

3.2 Initial Population

Each initial individual solution is obtained by a simple greedy method. Firstly all the flight activities are sorted according to the start time and activity type. The parking activities are ranked at rear part, which means they have lower priority when assigning gates.

For each activity, check gates if this activity could be assigned to the gate without conflict with the other assigned activities. The remaining parking activities which cannot find a time slot are assigned to the parking apron area, which is assumed to have unlimited capacity.

To diversify our population pool and ensure feasibility of each individual solution (chromosome), each time we generate a new individual, the order of parking activities sequence is randomly disturbed, while the other two groups of activities, arrival and departure, are maintained high priority in assignment. The population is composed by $100^{|M|}$ individuals.

3.3 Fitness Function, Parent Selection and Crossover

The fitness of each individual is calculated according to Eq. (8). The solutions with lower fitness values are better.

The parent selection and crossover operations are similar to the work of Bolat [4]. Two parent pools are generated by randomly selecting individuals from the solution pool. The equal probability of all solution candidates being selected prevents premature that may occur if we always select the best individuals to generate offspring. In order to keep efficiency, each pool consists of two individuals. From the two pools, we select the best two individuals to crossover and generate one offspring.

In the crossover process, one random crossover length l is selected in $[1, |N| - 1]$. $N = A \cup P \cup D$ is the set of all activities. Then the child inherits the first l digits of the first parent and the $[l + 1, |N|]$ digits of the second parent. This crossover mechanism will possibly lead to infeasible solution, so we adopt a repair process. The first observed gene causing infeasibility will received a new digit randomly generated in $[1, |M|]$. If the feasible solution cannot be obtained after predefined limited trails, this child will be abandoned. The population size is maintained as $100|M|$.

3.4 Large Neighborhood Search (LNS) as the Mutation Operation

The mutation operation is used to explore the solution space of each individual. LNS is able to search the solution space of one individual solution and therefore is adopted as the mutation operation embedded in GA. The critical part of LNS is the design of its removal and insertion method. Two removal methods, Shaw remove and worst remove are adopted in this paper to destroy part of the individual solution.

Shaw remove is proposed by Shaw [5, 7]. The main idea is to randomly choose one activity as target activity, define distance between each pair of activities. The neighbors are the activities with small distance with the target activity. In this paper, the distance between two activities is defined as Eq. (13):

$$distance_{ij} = |s_i - s_j| + |t_i - t_j| \tag{13}$$

By using this distance definition, the nearer two activities are, the higher probability they could be occupy each other's previous position.

Worst remove is proposed by [8]. The main idea is to remove the activity by evaluating the difference of objective value between solutions before and after removal, the neighbors preferred are ones with large difference. In this paper, the objective value is based on Eq. (8).

Two insertion methods, greedy insertion and regret insertion methods are used to repair the solutions obtained after removal operations.

The greedy insertion is to calculate the lowest insertion c_i cost for all activities in unassigned activity bank, the one with $\min_i c_i$ is inserted first. Greedy methods usually have straightforward operations and thus relatively achieving higher efficiency, however may lead to myopia. The early insertion flight activities may prevent some later promising insertions.

The regret insertion incorporates look-ahead information by consider not only the lowest cost, but also the second lowest, to kth lowest. To illustrate this strategy, let k to be 3 and denote insertion cost of activity i in ascending order as c_{i1}, c_{i2}, c_{i3}. Then the regret cost $c_i = \sum_{j=1,2,3} c_{ij} - c_{i1}$. We choose and insert the activity with lowest regret cost repeatedly.

When we obtain the initial solution, due to the shortage of gates provided, some parking activities may be assigned to the parking apron area. Therefore, the removal and insertion methods also consider these activities, although with a lower probability.

The flowcharts of LNS part as well as the overall algorithm is illustrated in Fig. 8.

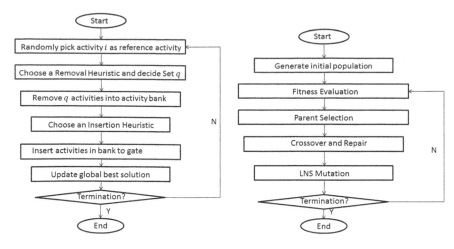

Fig. 8 The LNS algorithm and overall GALNS algorithm

4 Computational Results

All the experiments are carried out on a PC with an i7-2620M 2.70GHz. The mathematical model is solved using commercial software CPLEX, while genetic algorithm
with large neighborhood search (GALNS) is implemented in Java.

The data is generated similar to that in [9]. The difference lies in that we treat three
activities of an aircraft independently rather than treating an aircraft as one activity.
s_i^a (start time of arrival activity of aircraft i) is randomly generated in the interval
$[10i, 10i + 7]$, and the e_i^a (end time of arrival activity of aircraft i) is generated in the
interval $[s_i^a + 10, s_i^a + 30]$. s_i^p (start time of parking activity of aircraft i) $= e_i^a$, e_i^p (end
time of parking activity of aircraft i) is generated in the interval $[s_i^p + 10, s_i^p + 160]$.
s_i^d (start time of departure activity of aircraft i) $= e_i^p$, e_i^d (end time of departure activity
of aircraft i) is generated in the interval $[s_i^d + 10, s_i^d + 30]$.

The transfer passenger number between activity i and $j^{n_{ij}}$ is generated in the
interval $[1, 20]$ if $i \in A, j \in D$ and $s_j > e_i$. $n_{ij} = 0$ otherwise.

The gate distance matrix is symmetric and $dist_{ij}$ is generated in the interval
$[3, 30]$.

The results are listed in Table 2. Here the objective coefficients are set as $\alpha = \beta = \gamma = 1$.

Using CPLEX to solve the model is only practical when the problem size is less
than 21 activities. When the problem size grows, the computation time required by
CPLEX grows much faster than GALNS.

Table 2 Comparison between CPLEX and GALNS

Problem size	CPLEX		GALNS	
	Result	Time (s)	Result	Time (s)
9	3.98	0.2	3.98	0.05
15	473	0.5	473	0.1
21	1,017	64	1,017	0.7
24	–	–	646	1.3
30	–	–	1,268	3
60	–	–	4,215	12

5 Conclusion

In this paper, we have provided a new angle to measure the airport gate assignment, which is to comprehensively consider robustness, tows, and passenger transfer distance. The model is re-formulated as an equivalent network flow model. In order to deal with the realistic size problems, we proposed an AI-based hybrid algorithm to solve this problem, making up the inadequacy of CPLEX in solving large-size problems.

Future research will focus on improving the performance of GALNS for solving larger problems and to benchmark with other similar algorithms. Although the addition of LNS improves the search ability of GA, it slows down the computation speed. Decomposition-based method may apply to this algorithm for further improving its real-life application value.

References

1. Dorndorf, U., et al.: Flight gate scheduling: state-of-the-art and recent developments. Omega **35**(3), 326–334 (2007)
2. Kim, S.H.: Airport control through intelligent gate assignment (2013)
3. Holland, J.H.: Adaptation in natural and artificial systems: an introductory analysis with applications to biology, control, and artificial intelligence. U Michigan Press, Michigan (1975)
4. Bolat, A.: Models and a genetic algorithm for static aircraft-gate assignment problem. J. Oper. Res. Soc. **52**(10), 1107–1120 (2001)
5. Shaw, P.: Using constraint programming and local search methods to solve vehicle routing problems. In: Principles and Practice of Constraint Programming-CP98, pp. 417–431. Springer, Berlin (1998)
6. Pisinger, D., Ropke, S.: Large neighborhood search. In: Handbook of Metaheuristics, pp. 399–419. Springer, Berlin (2010)
7. Shaw, P.: A new local search algorithm providing high quality solutions to vehicle routing problems. Dept. of Computer Science, University of Strathclyde, Glasgow. APES Group (1997)
8. Ropke, S., Pisinger, D.: An adaptive large neighborhood search heuristic for the pickup and delivery problem with time windows. Transp. Sci. **40**(4), 455–472 (2006)
9. Ding, H., et al.: Aircraft and gate scheduling optimization at airports. In: Proceedings of the 37th Annual Hawaii International Conference on System Sciences. IEEE, New Jersey (2004)

An Iterative Heuristics Algorithm for Solving the Integrated Aircraft and Passenger Recovery Problem

Zhang Dong and H.Y.K. Henry Lau

Abstract Airline disruption incurred huge cost for airlines and serious inconvenience for travelers. In this paper, we study the integrated aircraft and passenger schedule recovery problem. To efficiently solve this problem, we proposed decomposition method to divide the whole problem into two smaller problems. An iterative heuristics strategy is proposed to improve solution quality by iteratively solving decomposed problems. Our algorithm is tested on the data set provided by ROADEF 2009. We simulate several airport closure scenarios and experimental results show that our algorithm can provide a high quality solution in the required time limit.

1 Introduction

Airline scheduling has always been the primary focus both in academic research and industrial application because it is essential for profit, service level, competitiveness of airline in the competing market. On-time performance of airlines schedule is key factor in maintaining satisfaction of current customers and attracting new ones. However, airline planned schedules are often subjected to numerous sources of irregularity such as adverse weather, air traffic congestion, aircraft mechanical problems, crew member's absence, propagated delay from upstream, longer passenger embarking and disembarking time and so on [1]. Delay and cancellation of flights are commonplace every day. Most disruptions in airline are attributed to two main causes [2]: (1) Airline resource shortages; (2) Airport and airspace capacity shortage. Huge loss caused by airline disruption attracted researchers from industries and academy to study airline schedule recovery problem which aims to re-allocate and re-schedule resources to minimize total delay and other objectives.

Once disruptions occurred, Airline Operations Control Center (AOCC) is responsible for centrally controlling airline operations of resources which include aircrafts,

Z. Dong (✉) · H.Y.K. Henry Lau
University of Hong Kong, Pokfulam, Hong Kong
e-mail: zd.pony@gmail.com

H.Y.K. Henry Lau
e-mail: hyklau@hku.hk

© Springer International Publishing Switzerland 2014
M. Bramer and M. Petridis (eds.), *Research and Development in Intelligent Systems XXXI*, DOI 10.1007/978-3-319-12069-0_21

crews and passengers. Operation controllers face a myriad of options to deal with various disruptions, including holding delayed flight legs until resources are ready; cancelling flight legs; swapping resources (aircraft, crew and passenger) between two flights; postponing the departure times of flight legs departing a hub airport to prevent connecting passenger missing their connections, etc. The primary objective of airline recovery is to minimize total passenger disruption cost and airline operating cost, and satisfying a set of airline and airport constraints.

Current practice of airline recovery is performed in a sequential manner, first recovering aircraft, second crew and finally passenger. Passenger recovery is assigned a low priority in current recovery practice as it is considered only after all the aircrafts and crews are recovered. However, passenger experience, which is primarily affected by flight on-time performance, promptly recovery scheme and disruption compensation, has strong impact on passenger satisfaction on airline's service. The objective of passenger recovery is to find a passenger re-assignment plan so that total passenger disruption cost is minimized. In this work, we propose an integrated airline recovery algorithm which is passenger-centric.

A large amount of literature can be found related to the airline recovery problems including Teodorović and Guberinic [3], Teodorović and Stojkovic [4, 5], Jarrah et al. [6], Cao and Kanafani [7, 8], Yan and Yang [9], Yan and Tu [10], Yan and Lin [11], Thengvall et al. [12], Jonathan et al. [13], Eggenberg et al. [14], Rosenberger et al. [15], Clarke [16] and Argüello et al. [17] and for crew recovery problem we refer to Wei et al. [18], Stojković et al. [19], Abdelghany et al. [20], Lettovský et al. [21], Nissen and Hasse [22].

Few researches can be found about passenger recovery. To our knowledge, Bratu and Barnhart [2] developed models to solve integrated aircraft and passenger recovery problem. Aircraft recovery problem is formulated as a network flow model and estimation of passenger delay and disruption costs is considered by using passenger itinerary short connections constraints. Crew recovery is partially considered by estimating cost of using reserve crews. Some practical constraints such as aircraft maintenance requirement are not taken into consideration in their work. Bisaillon et al. [23] proposed a large neighborhood search algorithm to solve integrated airline and passenger recovery. In their work, flights are allowed to be delayed and cancelled but no aircraft swap is considered. Disrupted passenger itinerary recovery is recovered by solving shortest path problem for them. Based on Bisaillon's work, an improved large neighborhood search algorithm is proposed by Sinclair et al. [24]. Jafari et al. [25] proposed an integrated model considering more practical constraints such as maximum flight delay, minimum aircraft connection time, minimum passenger connection time are considered. A methodology based on Abdelghany et al. [26] is adapted to solve the model.

In this research, we adopted idea of large neighborhood search and mathematical programming and proposed a heuristics algorithm solving the integrated aircraft and passenger recovery problem.

2 Problem Description

2.1 Aircraft Recovery Problem

We first introduce several parameters and decisions variables used in the following context. Notations and decision variables

A: set of aircrafts, which is indexed by a
F: set of flights which is indexed by f
FC: set of flight copies which is indexed by fc
I: set of itinerary which is indexed by i
P: set of candidate passenger itineraries, which is indexed by p
R_a: set of routes which is operated by aircraft a
P_i: set of candidate passenger which can serve itinerary i
Slot: set of airport slots which is indexed by s
P_{fc}: set of candidate passenger itineraries which contains flight copy fc
R_{fc}: set of routes which contains flight copy fc
Node: set of all the activity nodes, which is indexed by n
$n_{input}^{aircraft}$: aircraft input value at node n
n_{input}^{fc}: the set of input flight copies at node n
n_{input}^{ga}: the set of input ground arcs at node n
$n_{output}^{aircraft}$: aircraft output value at node n
n_{output}^{fc}: the set of output flight copies at node n
n_{output}^{ga}: the set of output ground arcs at node n
c_{fc}^{a}: the cost of flight copy fc flying by aircraft a
TPD: it represents total passenger delay. Calculation of this value
TNC: total number of cancelled passenger
Swap: total number of aircraft swap
α: unit delay cost for passenger
β: unit cancellation cost
γ_{af}: operating cost of aircraft a to flight f
Δ: unit swap cost

Decision variables

x_{fc}^{a}: binary decision variable which indicates whether the flight copy fc is flying by aircraft a
y_f: binary decision variable which indicates whether flight f is cancelled
w_p: integer decision variable which indicates the number of passenger flied in this candidate passenger itinerary p

When disruption occurs, there are three options available for the airline operation controller.

(1) Flight delay: A flight can be delayed for a certain period of time. Since passenger solution is not known in current stage, an estimated delay cost is generally

Fig. 1 Illustration of activity node

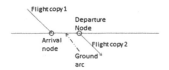

adopted. The estimated delay cost function is usually set as a non-decreasing function of length of delay.

(2) Flight cancellation: if no aircraft available, the airline operational controller can also cancel this flight. The cancellation cost is very huge and is serious loss of customer satisfaction. Therefore, it is the last choice for the airline.

(3) Aircraft swapping: if the flight can't depart since the assigned aircraft isn't available, assigning another available aircraft can mitigate delay. In the hub-and-spoke network, quite a lot of such swapping opportunities can be found. However, swapping aircraft must consider the mismatching impact in the following period.

In order to make an optimized solution, the aircraft recovery problem and its related recovery option can be modeled in a time-space network. Several constraints must be considered in making aircraft recovery solution.

Activity Nodes is defined at each network. The activity node contains two types of node, departure nodes and arrival nodes. Each flight copy corresponds to one arrival node and departure nodes at different airport. The ground arc corresponds to one arrival node and departure node at the same airport. The Fig. 1 illustrates the construction of nodes.

The aircraft Node is another kind of node related to the aircraft input and aircraft output. Each aircraft has its specific ready time and ready location. An aircraft input node is then created at the ready time and ready location. If this node is already created by flight copies, then the aircraft input value is increased by 1; otherwise, a new node is created. Similarly, by the end of recovery time window, the specific number of aircrafts must locate at the designated airports. The node at the designated airport and end of the recovery time widow is then constructed. If this node already exists, the aircraft output value is increased by the required aircraft number at that airport.

Each activity node contains the input flight copies, input ground arc, aircraft input value, output flight copies, output arc and aircraft output value. For a feasible flow network work, the flow network at each activity node must be balance which can be restricted by function (1).

$$n_{input}^{aircraft} + \sum_{fc \in n_{input}^{fc}} x_{fc} + \sum_{ga \in n_{input}^{ga}} z_{ga} = n_{output}^{aircraft} + \sum_{fc \subset n_{output}^{fc}} x_{fc} + \sum_{ga \in n_{output}^{ga}} z_{ga} \, \forall h \in \text{Node}$$

$$(1)$$

Three kinds of costs are considered in the aircraft recovery problem, which are aircraft operating cost, flight delay cost and flight cancellation cost. Aircraft operating cost is incurred when operating a certain flight copy by the specific aircraft fleet type. It is the product of aircraft hour operating cost multiplying the flight duration. When the flight is cancelled, this cost will not occur. The flight delay cost and flight cancellation cost are representing passenger disruption costs. In this stage, since we don't know the actual passenger itinerary, the flight delay cost and cancel cost can only be estimated. The notation c_{fc}^a is the summation of the operating cost of flight copy fc by aircraft fleet type a and the passenger delay cost $cancel_f$. represents the cancellation cost of flight f.

Based on the previous introduction, the mathematical model for the aircraft recovery problem can be modeled as model 1.

Model (1)

$$\min \sum_{fc \in FC} c_{fc} * x_{fc} + \sum_{f \in F} cancel_f * y_f \tag{2}$$

subject to

$$\sum_{fc \in FC_f} x_{fc} + y_f = 1 \quad \forall f \in F \tag{3}$$

$$n_{input}^{aircraft} + \sum_{fc \in n_{input}^{fc}} x_{fc} + \sum_{ga \in n_{input}^{ga}} z_{ga} = n_{output}^{aircraft} + \sum_{fc \in n_{output}^{fc}} x_{fc} + \sum_{ga \in n_{output}^{ga}} z_{ga} \forall h \in Node$$

$$\tag{4}$$

$$\sum_{fc \in FC_s} x_{fc} \leq Cap_s \quad s \in Slot \tag{5}$$

x_{fc}, y_f are binary decision variables. z_{ga} is positive integer decision variable.

The constraint (3) restricts that each flight is either cancelled or flewby one flight copy fc. The constraint (4) restricts the flow balance at each node. The constraint (5) is the air traffic control constraints. The number of flight copies assigned to each slot must be smaller than the maximum allowed capacity.

2.2 Passenger Recovery Problem

The aircraft recovery decision results in flight cancellation and flight scheduling modification. The passenger itineraries might be disrupted since scheduled flight is cancelled or not enough passenger connection time is left in the recovered flight schedule. The disrupted passengers are needed to be assigned to alternative itineraries. Several constraints must be satisfied for the candidate itineraries:

the candidate itineraries must start at the same airport as the original itinerary
the actual departure time of the candidate itinerary must be later than schedule departure time of the original itinerary
the candidate itinerary must terminate with same airport as the original itinerary
the delay time of the candidate is smaller than the maximum allowed delay time
the connection time between two flights in the candidate itinerary must be larger than the minimum passenger connection time.

Passenger disruption cost is categories into three types, which are cancellation cost, delay cost and discomfort cost. The cancellation cost is incurred when the passenger itinerary is disrupted and no candidate itinerary is assigned. The delay cost is incurred since the actual arrival time of the itinerary is later than scheduled arrival time. The delay cost is proportional to the delay time. The discomfort cost is incurred by the degrading itineraries from higher class to the lower classes. In general, airlines will provide three different classes, which are first class, business class and economic class. Besides to compensating the price difference, degrading the passenger buying the higher class tickets to a lower class may decrease the customer satisfactory which is essentially important to the long term development of the airlines. In the passenger recovery problem, this cost is also quantified and transferred to the standard cost.

The generating of the candidate itineraries for the disrupted one can be illustrated in the Fig. 2.

The original itinerary is constructed by the flight f1 and f3. Since the f1 is delayed and original passenger connection is disrupted. The candidate itinerary consisting flight f1 and f4 is then constructed. The constructing of candidate routes must satisfy the requirements we mentioned before.

The passenger recovery problem can also be modeled as a multi-commodity network model. Each passenger itinerary is modeled as a commodity in the network, the candidates itineraries are the possible paths for this itinerary. Each flight arcs in the network has the passenger capacity limit which restricts the flow decision on specific candidate itinerary.

Fig. 2 Illustration of passenger itinerary

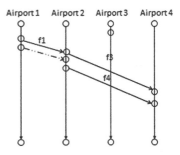

Model (2)

$$\text{Min} \sum_{i \in I} \sum_{p \in P_i} c_p * w_p + \sum_{i \in I} \text{cancel}_i * z_i \tag{6}$$

Subject to

$$\sum_{p \in P_i} w_p + z_i = N_i \quad \forall i \in I \tag{7}$$

$$\sum_{p \in P_f} w_p \leq \text{Cap}_f \quad \forall f \in F \tag{8}$$

w_p and z_i are positive integer decision variables.

The constraint (7) restricts that the passenger is either assigned to one candidate itinerary or cancelled. The constraint (8) restricts that passenger volume assigned to specific flight must be smaller than the capacity decided in the aircraft recovery solution.

3 Iterative Heuristics

3.1 Mathematical Model

In this section, we propose an integrated model in which aircraft recovery and passenger recovery problems are solved simultaneously. The model is formulated as following:

Model (2)

$$\min \sum_{fc \in FC} c_{fc} * x_{fc} + \sum_{f \in F} \text{cancel}_f * y_f + \sum_{i \in I} \sum_{p \in P_i} c_p * w_p + \sum_{i \in I} \text{cancel}_i * z_i \tag{9}$$

subject to

$$\sum_{fc \in FC_f} x_{fc} + y_f = 1 \quad \forall f \in F \tag{10}$$

$$n_{\text{input}}^{\text{aircraft}} + \sum_{fc \in n_{\text{input}}^{fc}} x_{fc} + \sum_{ga \in n_{\text{input}}^{ga}} z_{ga} = n_{\text{output}}^{\text{aircraft}} + \sum_{fc \in n_{\text{output}}^{fc}} x_{fc} + \sum_{ga \in n_{\text{output}}^{ga}} z_{ga} \forall h \in \text{Node}$$

$$\tag{11}$$

$$\sum_{fc \in FC_s} x_{fc} \leq \text{Cap}_s \quad s \in \text{Slot} \tag{12}$$

$$\sum_{p \in P_i} w_p + z_i = N_i \quad \forall i \in I \tag{13}$$

$$\sum_{p \in P_{fc}} w_p \le Cap_{fc}^a * x_{fc}^a \quad \forall fc \in FC \tag{14}$$

x_{fc}, y_f are binary decision variables. z_{ga} is positive integer decision variable. w_p and z_i are positive integer decision variables.

In this integrated model, the aircraft recovery problem and passenger recovery problem are both considered. The constraint relates the two problems together. It restricts that the volume of passenger assigned to the certain flight copy is smaller than the passenger capacity in the aircraft recovery problem. Different problem the single passenger recovery model we introduced before, here we adopt the flight copy since the aircraft assignment decisions are made for each flight copy.

Solving the model (3) is rather tricky considering two reasons, first for the single aircraft recovery, it is already difficult to get an integer solution, adding the passenger constraints cause it even harder to be solved; secondly, aircraft recovery generally needs a solution in minutes, however, for even medium scale data, it might not generate a feasible integer solution within 10 min computation using CPlex. In the following parts, we will describe an iterative based heuristics to solve this integrated model.

3.2 Iterative Scheme

As illustrated in Fig. 3, we propose a three stage algorithm to solve the integrated aircraft and passenger recovery problem. In the first stage, the aircraft recovery network flow model is solved. The flight cancellation decision and aircraft assignment decision is decided and transferred to the next stage. In the next stage, we will only consider the flight rescheduling and passenger reassignment problem. All the flights chosen in the first stage must be covered here. Since in the first, we already get a feasible solution, we can guarantee there is at least one feasible in the flight rescheduling problem based on input flight cancellation decision and aircraft assignment decision.

Fig. 3 Three stage algorithm

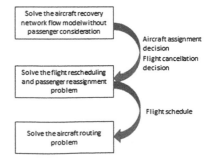

In the final stage, the aircraft routing problem is to be solved so that specific route for each aircraft is decided. In the following part, we will introduce the details about each stage.

3.2.1 Solving the Aircraft Recovery Network Flow Model

In the first stage, we direct build a model identical to the model (1). No passenger consideration is incorporated right now. We focus on the flight cancellation decision and aircraft assignment decision. According to our computational experiences, the flight cancellation cost dominates the whole recovery cost, therefore, the less cancelled flights generally mean the better solution quality. Solving the model (1) can provide a flight cancellation decision and aircraft assignment decision. By assigning a large cancellation cost, we can get a solution with least flights cancelled.

3.2.2 Solving the Flight Rescheduling and Passenger Reassignment Problem

In the second stage, we mainly want to decide the new flight schedule and passenger reassignment plan for all the uncancelled flights and disrupted passengers. The mathematical model is illustrated in model (3).

Compared to the model (2), there are three main differences:

(1) only non-cancelled flights are considered here
(2) no flight cancellation decision variable is considered and it forces that all the non-cancelled flights must be covered by at least one flight delay arc.
(3) for each aircraft fleet type, only flight legs assigned to this type is considered in its network.

Model (3)

$$\min \sum_{fc \in FC} c_{fc} * x_{fc} + \sum_{f \in F} cancel_f * y_f + \sum_{i \in I} \sum_{p \in P_i} c_p * w_p + \sum_{i \in I} cancel_i * z_i \quad (15)$$

subject to

$$\sum_{fc \in FC_f} x_{fc} = 1 \quad \forall f \in F \quad (16)$$

$$n_{input}^{aircraft} + \sum_{fc \in n_{input}^{fc}} x_{fc} + \sum_{ga \in n_{input}^{ga}} z_{ga} = n_{output}^{aircraft} + \sum_{fc \in n_{output}^{fc}} x_{fc} + \sum_{ga \in n_{output}^{ga}} z_{ga} \forall h \in Node$$

$$(17)$$

$$\sum_{fc \in FC_s} x_{fc} \le Cap_s \quad s \in Slot \quad (18)$$

Fig. 4 Illustration of limiting
flight delay options

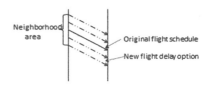

$$\sum_{p \in P_i} w_p + z_i = N_i \quad \forall i \in I \tag{19}$$

$$\sum_{p \in P_{fc}} w_p \leq Cap^a_{fc} * x^a_{fc} \quad \forall fc \in FC \tag{20}$$

x_{fc}, y_f are binary decision variables. z_{ga} is positive integer decision variable. w_p and z_i are positive integer decision variables.

Although the problem scale is decreased compared to the model (3), for several data set it is still hard to solve the model (3). For solving such large scale problem, one possible way is to decrease the solution space and then solve it. In this study, we propose two different solution space decreasing method.

Limiting the Flight Delay Options Only the limited number of flight copies are considered. In our first stage, in order to find a feasible flight cancellation decision, we generally assign many flight delay options to each flight leg. In order to achieve the better efficiency, the number of delay options is narrowed. We adopt the idea in the local search heuristics, each iteration the new flight delay options are generated around the flight schedule solution decided in the previous iteration.

As illustrated in Fig. 4, around the original flight schedule, we create four flight copies. The two flight copies depart in advance and two flight copies are delay options. In our algorithm, we define a neighborhood for each flight. Suppose the original departure time of f is d_f, the neighborhood is defined as $[d_f - \Delta, d_f + \Delta]$. The uniformly distributed flight copies are generated within the neighborhood area.

Limiting the number of candidate itineraries The second way to decrease the solution space is to decrease the number of itinerary variable in the model. In the original integrated model, each itinerary is considered as a commodity in the network. As the increasing of flight options, the number of possible itineraries increases exponentially. If all the possible itineraries are considered here, the problem scale is too large to be solved. In our algorithm, we separate all the itineraries into two groups, forced itinerary group and variable itinerary group.

The forced itinerary here is defined as the itinerary which must be served. The variable itinerary is defined as an itinerary candidate which might be selected in the solution. For the compulsory itinerary, we can transfer the original set partition model into a much concise model. For the variable itinerary, the original form in the integrated model is kept.

Lemma 1 *Assume that it is compulsory to take one itinerary, then the model can be transferred to equivalent one.*

(1) transferred the delay cost of itinerary into the last flight leg within the itinerary; suppose the itinerary contains the two flight leg, the delay option fc indicates the exact delay cost for itinerary. We incorporated this cost as a part of delay cost for this flight copy.
(2) add constraint into model

$$\sum_{fc \in FC_{f1}} \text{arr}_{fc} * x_{fc} + \text{connTime} \le \sum_{fc \in FC_{f2}} \text{dep}_{fc} * x_{fc}$$

(3) decrease the itinerary passenger number from the flight passenger capacity
(4) removing the decision variables and constraints related to this itinerary in the integrated model.

Proof First, the constraint forces that there is enough connection time between two flights f1 and f2, the procedure (3) reserves the flight capacity to this itinerary. If one itinerary is selected, the cost of this itinerary is only determined by the last flight and we already incorporate the itinerary cost in procedure (1).

According to the Lemma 1, we can remove all the itineraries in the compulsory set. Each time, constraint (1) will be generated and added into the new model. We can further compress model by eliminating redundant constraints. For two constraints
$\sum_{fc \in FC_{f1}} \text{arr}_{fc} * x_{fc} + \text{connTime}_1 \le \sum_{fc \in FC_{f2}} \text{dep}_{fc} * x_{fc}$ and
$\sum_{fc \in FC_{f3}} \text{arr}_{fc} * x_{fc} + \text{connTime}_2 \le \sum_{fc \in FC_{f4}} \text{dep}_{fc} * x_{fc}$
If f1 == f3 and f2 == f4, we can compress these two constraints into one
$\sum_{fc \in FC_{f1}} \text{arr}_{fc} * x_{fc} + \max(\text{connTime}_1, \text{connTime}_2) \le \sum_{fc \in FC_{f2}} \text{dep}_{fc} * x_{fc}$

The original integrated model is transferred into integrated model (4).

$$\min \sum_{fc \in FC} c_{fc} * x_{fc} + \sum_{i \in I_v} \sum_{p \in P_i} c_p * w_p + \sum_{i \in I_v} \text{cancel}_i * z_i \qquad (21)$$

subject to

$$\sum_{fc \in FC_f} x_{fc} + y_f = 1 \quad \forall f \in F \qquad (22)$$

$$n_{\text{input}}^{\text{aircraft}} + \sum_{fc \in n_{\text{input}}^{fc}} x_{fc} + \sum_{ga \in n_{\text{input}}^{ga}} z_{ga} = n_{\text{output}}^{\text{aircraft}} + \sum_{fc \in n_{\text{output}}^{fc}} x_{fc} + \sum_{ga \in n_{\text{output}}^{ga}} z_{ga} \forall h \in \text{Node}$$

$$\qquad (23)$$

$$\sum_{fc \in FC_s} x_{fc} \le \text{Cap}_s \quad s \in \text{Slot} \qquad (24)$$

$$\sum_{p \in P_i} w_p + z_i = N_i \quad \forall i \in I_v \tag{25}$$

$$\sum_{p \in P_{fc}} w_p \leq Cap_{fc}^a * x_{fc}^a \quad \forall fc \in FC \tag{26}$$

$$\sum_{fc \in FC_{f1}} arr_{fc} * x_{fc} + connTime_1 \leq \sum_{fc \in FC_{f2}} dep_{fc} * x_{fc} \quad \forall (f_1, f_2) \in compuConn$$

$$\tag{27}$$

x_{fc}, y_f are binary decision variables. z_{ga} is positive integer decision variable.

It is noted that as the compulsory itinerary set increasing, the problem scale of is decreased exponentially. However, the solution quality also deteriorates. It is needed to carefully trade-off between efficiency and solution quality.

3.3 Aircraft Routing Problem

After the flight schedule is fixed, the airline operational controllers need the specific aircraft routing solution. If no further information is provided, finding the aircraft information can be modeled as a vehicle scheduling problem which is proved to be NP hard problem. In this study, we propose a randomized greedy algorithm utilizing the information provided in the network flow solution.

Before introducing the algorithm, we first propose two criteria to judge the aircraft routing solution.

(1) the number of flights swapped in the new aircraft routing solution
(2) the number of aircraft not locating at the original location in the end of recovery

For a good aircraft recovery solution, these two criteria should be as low as possible. In the following part, a randomized greedy algorithm is proposed to search several good aircraft routing solutions.

3.3.1 Shortest Path Algorithm for Single Aircraft

Suppose we only consider one aircraft and want to search a best route for this aircraft. This problem can be modeled as a shortest path problem.

For each aircraft, a unique network is constructed according to the ground arc information and flight copy selection decision. We will illustrate the example of network in Fig. 5.

In the original time-space network for aircraft type e, the flights selected for the fleet type e and ground arcs with value larger than 0 are highlighted in broad arrow.

Fig. 5 Illustration of shortest
path for aircraft

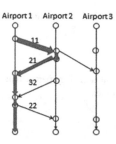

We find to find an aircraft path only using those selected arcs. We plan to propose a shortest path algorithm to find a route for the aircraft a. For each arc, we assign a value. If the arc is flight arc and this flight is originally flew by the aircraft a, we assign -1 to the flight arc; otherwise we assign the arc value as 0. The start node is the aircraft input node of a and it is unique node. The termination nodes are not unique. For any aircraft output nodes of fleet type e, it can be the termination node of a. Generally, airline prefers the termination node with location identical to the original termination location of the aircraft a. Therefore, we separate all the termination nodes into two subsets, set with location identical to the original termination location and set with location different from the original termination location.

The shortest path of the aircraft a is a path ending at one output node and consisting of most number of original assigned flights.

3.3.2 Randomized Greedy Algorithm

A randomized greedy algorithm is then proposed to find a routing solution for all the aircrafts.

Step 1: randomly sort all the aircrafts
Step 2: for each aircraft according to the sorted order, the algorithm (2) is executed to find the route
Step 3: update all the ground arc and flight arc information, then return to step (2).

4 Computational Results

Our algorithm is tested using data from ROADEF challenge in 2009. It contains 608 daily flights and 81 aircrafts. 11 different fleet types and 14 different aircraft configurations are included among all the aircrafts. The maximum disruption scenario spans two days and it involves nearly 1,000 flight.

We test our algorithm's ability to recovery schedule after airport closure. When airport capacity is affected by the storm or other extreme weather, the airport might

294 Z. Dong and H.Y.K. Henry Lau

Table 1 Parameter setting

Parameter	Value
NeighRange	5
DisIteSize	300

Table 2 Key performance index

Notation	Description	Unit
PDC	Passenger disrupted cost	US dollars
TPD	Total passenger delay	Minutes
CT	Computation time	Seconds

be closed for several hours. In the hub-and-spoke network, closure of hub airport can affect a lot of flights because of network propagation effect. In our study, we business airports which are ORY and CDG are selected to be closured for different periods. 24 different scenarios are designed to test the ability of algorithms dealing withdifferent disruption.

The MIP solver in our study is ilog Cplex and our cup is intel core i7-3770. The maximum computation time for all the computations is set to 600 s. The aircraft recovery network model is not solved optimally and the optimal computation time is 100 s. Table 1 shows the parameters adopted in the experiments.

To evaluate the solution quality, we propose 2 passenger centric key performance index in our experiments to test the performance of each algorithms. PDC is short for the total passenger disruption cost. It is the summation of passenger delay cost and passenger cancellation cost. TPD is short for the total passenger delay time. CT is short for the computation time of each algorithm (Table 2).

In the following context, the SA is short for the sequential algorithm. PIA is short for the Bratu's partial integrated algorithm. IA is short for our algorithm.

4.1 Airport Closure Scenarios

The detailed information about 24 different scenarios are illustrated in Tables 3 and 4.

As shown in the results, both the proposed integrated algorithm and partial integrated algorithm performs better than the sequential algorithm except that in several serial disruption scenarios, partial integrated algorithm performs worse than the sequential algorithm. Compared to the partial integrated algorithm, the proposed integrated algorithm performs stably. In most of scenarios, the proposed algorithm is able to obtained a solution with cost around 10 % less than the solution generated by the partial integrated algorithm.

Table 3 Disruption scenarios for the airport closure

	Description	Recovery time window
Scenario 1	ORY is closed from 8:00 to 10:00	8:00–28:00
Scenario 2	CDG is closed from 8:00 to 10:00	8:00–28:00
Scenario 3	ORY and CDG are closed from 8:00 to 10:00	8:00–28:00
Scenario 4	ORY is closed from 10:00 to 12:00	10:00–28:00
Scenario 5	CDG is closed from 10:00 to 12:00	10:00–28:00
Scenario 6	ORY and CDG are closed from 10:00 to 12:00	10:00–28:00
Scenario 7	ORY is closed from 12:00 to 14:00	12:00–28:00
Scenario 8	CDG is closed from 12:00 to 14:00	12:00–28:00
Scenario 9	ORY and CDG are closed from 12:00 to 14:00	12:00–28:00
Scenario 10	ORY is closed from 8:00 to 12:00	8:00–28:00
Scenario 11	CDG is closed from 8:00 to 12:00	8:00–28:00
Scenario 12	ORY and CDG are closed from 8:00 to 12:00	8:00–28:00

Table 4 Illustration of the detailed result for 24 scenarios

	SA			PIA			IA		
	PDC	TPD	CT	PDC	TPD	CT	PDC	TPD	CT
A1	121,720	779,235	0.7	55,829	957,220	17.4	59,890	813,735	18
A2	24,759	203,465	0.5	21,670	302,610	1.2	19,260	196,575	4.7
A3	154,796	1,012,785	0.7	84,304	1,233,615	28.9	82,977	1,049,315	39.6
A4	108,428	656,150	0.4	44,145	873,990	2	36,421	696,030	27
A5	17,583	88,065	0.4	7,429	132,390	0.7	6,510	126,610	1.2
A6	134,968	773,375	0.4	53,985	1,060,440	5.1	44,190	862,810	49.8
A7	145,440	709,415	0.6	46,247	910,305	7	46,199	749,490	14.7
A8	77,551	214,095	0.3	20,242	322,200	0.6	16,635	252,760	1.9
A9	203,803	967,015	0.3	67,585	1,265,070	17.1	60,662	983,165	24.9
A10	334,153	3,047,955	2.1	199,548	3,651,055	600.3	185,403	2,907,115	195.4
A11	216,434	957,225	0.6	109,514	1,514,810	8.3	75,857	996,110	52.3
A12	485,985	4,241,990	3.2	512,327	5,640,920	600.3	275,376	51,519,501	155.7

5 Conclusion

In this study, we propose an integrated algorithm for solving the aircraft and passenger recovery problems. A mixed integer programming model for the integrated problem is firstly formulated. Considering the NP-hard nature and large problem scale, an iterative heuristics is proposed. For each iteration, a diminishing solution space is constructed based on the incumbent solution. An equivalent model which is further simplified is then solved. The generated solution then serves as the incumbent solution and starts next iteration until the stopping criteria is satisfied. In the computational experiments, the data from ROADEF is adopted to test our algorithm. We simulate two different kinds of disruptions, which are airport closure and ground delay program. Each disruption kind includes 24 different scenarios. The sequential method and integrated algorithm proposed by Bratu is implemented to compare with our algorithm. The result is encouraging. In all the testing cases, our algorithm can provide better solution than the solution of the sequential algorithm. In the most of testing cases, our solution is better than Bratu's. Only in two cases, our solution is a slightly worse than Bratu's. The computational efficiency also satisfies our requirements and a solution is obtained within 5 min.

References

1. Ball, M., Barnhart, C., Nemhauser, G., Odoni, A., Laporte, G.: Air transportation: irregular operations and control. Handb. Oper. Res. Manage. Sci. **14**(1), 1–67 (2007)
2. Bratu, S., Barnhart, C.: Flight operations recovery: new approaches considering passenger recovery. J. Sched. **9**(3), 279–298 (2006)
3. Teodorović, D., Guberinić, S.: Optimal dispatching strategy on an airline network after a schedule perturbation. Eur. J. Oper. Res. **15**(2), 178–182 (1984)
4. Teodorović, D., Stojković, G.: Model for operational daily airline scheduling. Transp. Plann. Technol. **14**(4), 273–285 (1990)
5. Teodorovic, D., Stojkovic, G.: Model to reduce airline schedule disturbances. J. Transp. Eng. **121**(4), 324–331 (1995)
6. Jarrah, A.I.Z., Yu, G., Krishnamurthy, N., Rakshit, A.: A decision support framework for airline flight cancellations and delays. Transp. Sci. **27**(3), 266–280 (1993)
7. Cao, J.M., Kanafani, A.: Real-time decision support for integration of airline flight cancellations and delays part I: mathematical formulation. Transp. Plann. Technol. **20**(3), 183–199 (1997)
8. Cao, J.M., Kanafani, A.: Real-time decision support for integration of airline flight cancellations and delays part II: algorithm and computational experiments. Transp. Plann. Technol. **20**(3), 201–217 (1997)
9. Yan, S., Yang, D.H.: A decision support framework for handling schedule perturbation. Transp. Res. B: Methodol. **30**(6), 405–419 (1996)
10. Yan, S., Tu, Y.P.: Multifleet routing and multistop flight scheduling for schedule perturbation. Eur. J. Oper. Res. **103**(1), 155–169 (1997)
11. Yan, S., Lin, C.G.: Airline scheduling for the temporary closure of airports. Transp. Sci. **31**(1), 72–82 (1997)
12. Thengvall, B.G., Bard, J.F., Yu, G.: Balancing user preferences for aircraft schedule recovery during irregular operations. IIE Trans. **32**(3), 181–193 (2000)
13. Jonathan, F., Gang, Y., Michael, F.A.: Optimizing aircraft routings in response to groundings and delays. IIE Trans. **33**(10), 931–947 (2001)
14. Eggenberg, N., Bierlaire, M., Salani, M.: A column generation algorithm for disrupted airline schedules. In: A Column Generation Algorithm for Disrupted Airline Schedules. Technical report, Ecole Polytechnique Federale de Lausanne, edn (2007)
15. Rosenberger, J.M., Johnson, E.L., Nemhauser, G.L.: Rerouting aircraft for airline recovery. Transp. Sci. **37**(4), 408–421 (2003)
16. Clarke, M.D.D.: Development of heuristic procedures for flight rescheduling in the aftermath of irregular airline operations. In: Development of Heuristic Procedures for Flight Rescheduling in the Aftermath of Irregular Airline Operations. Massachusetts Institute of Technology, Flight Transportation Laboratory, Cambridge (1998)
17. Argüello, M.F., Bard, J.F., Yu, G.: A GRASP for aircraft routing in response to groundings and delays. J. Comb. Optim. **1**(3), 211–228 (1997)

18. Wei, G., Yu, G., Song, M.: Optimization model and algorithm for crew management during airline irregular operations. J. Comb. Optim. **1**(3), 305–321 (1997)
19. Stojković, M., Soumis, F., Desrosiers, J.: The operational airline crew scheduling problem. Transp. Sci. **32**(3), 232–245 (1998)
20. Abdelghany, A., Ekollu, G., Narasimhan, R., Abdelghany, K.: A proactive crew recovery decision support tool for commercial airlines during irregular operations. Ann. Oper. Res. **127**(1), 309–331 (2004)
21. Lettovský, L., Johnson, E.L., Nemhauser, G.L.: Airline crew recovery. Transp. Sci. **34**(4), 337–348 (2000)
22. Nissen, R., Haase, K.: Duty-period-based network model for crew rescheduling in European airlines. J. Sched. **9**(3), 255–278 (2006)
23. Bisaillon, S., Cordeau, J.-F., Laporte, G., Pasin, F.: A large neighbourhood search heuristic for the aircraft and passenger recovery problem. 4OR **9**(2), 139–157 (2011)
24. Sinclair, K., Cordeau, J., Laporte, G.: Improvements to a large neighbourhood search heuristic for an integrated aircraft and passenger recovery problem. Eur. J. Oper. Res. (2012)
25. Jafari, N., Hessameddin Zegordi, S.: Simultaneous recovery model for aircraft and passengers. J. Franklin Inst. **348**(7), 1638–1655 (2011)
26. Abdelghany, K.F., Abdelghany, A.F., Ekollu, G.: An integrated decision support tool for airlines schedule recovery during irregular operations. Eur. J. Oper. Res. **185**(2), 825–848 (2008)

Machine Learning
and Data Mining

A Framework for Brand Reputation Mining and Visualisation

Ayesh Alshukri, Frans Coenen, Yang Li, Andrew Redfern
and Prudence W.H. Wong

Abstract Online brand reputation is of increasing significance to many organisations and institutes around the globe. As the usage of the www continues to increase it has become the most commonly used platform for users and customers of services and products to discuss their views and experiences. The nature of this www discussion can significantly influence the perception and hence the success of a brand. Brand Reputation Mining (BRM) is a process to help brand owners to know what is being said about their brand online. This paper proposes a BRM framework to provide support for enterprises wishing to conduct brand reputation management. The proposed framework can be generically applied to collect, process and display the reputation of different brands. A key feature is the visualisation facilities included to allow the display of the results of reputation mining activities. The framework is fully described and illustrated using a case study. The concepts expressed in this paper have been incorporated into the "LittleBirdy" brand reputation management product commercially available from Hit Search Ltd.

A. Alshukri (✉) · A. Redfern
Hit Search, Liverpool Innovation Park, Liverpool L7 9NG, UK
e-mail: AAlshukri@hitsearchlimited.com; a.alshukri@liverpool.ac.uk

A. Redfern
e-mail: ARedfern@hitsearchlimited.com

A. Alshukri · F. Coenen · Y. Li · P.W.H. Wong
Department of Computer Science, University of Liverpool, Liverpool L69 3BX, UK

F. Coenen
e-mail: coenen@liverpool.ac.uk

Y. Li
e-mail: y.li48@liverpool.ac.uk

P.W.H. Wong
e-mail: pwong@liverpool.ac.uk

© Springer International Publishing Switzerland 2014
M. Bramer and M. Petridis (eds.), *Research and Development
in Intelligent Systems XXXI*, DOI 10.1007/978-3-319-12069-0_22

1 Introduction

Brand reputation has always been an important issue with respect to many organisations (both commercial and non-commercial), particularly in the context of consumer facing organisations. Recommendation and "word-of-mouth" are an important factor in how organisations are perceived and can have a substantial influence on the success, or otherwise, of an organisation; it is easy to tarnish a brand if negative social activities are associated with it [8]. With the advent of the www, and the prolific use of social media, it has become very important for organisations to manage their online reputation. A growing percentage of users maintain that blogs and online fora are credible ways of finding out about products and services [8]. Organisations that wish to preserve their reputation are therefore interested in knowing "what is being said about them" on social media. However, the increasing predominance of Consumer Generated Content (CGC) on the www (examples include blogs, news forms, message boards and web pages/sites), makes it virtually impossible for organisations to manually monitor the reputation of their brand based on human effort alone [10]. One solution is to automate the process.

This paper proposes a framework for conducting effective Brand Reputation Mining (BRM). In this context BRM is concerned with the identification of "mentions" expressed across electronic (social) media with respect to a particular company, institution or organisation, and determination of the sentiment of such mentions. The information gathered from such a BRM activity can be used by organisations to: (i) compare their performance against competitors, (ii) assess specific marketing strategies and (iii) gauge how a particular product or service is received in the market place. The successful conduct of BRM entails three broad challenges: (i) the identification and collection of mentions on www media (for example with respect to social networks, blogs and news sites); (ii) the application of data mining techniques to the gathered information in order to determine the sentiment associated with opinions expressed in the form of mentions or to group mentions according to the views expressed; and (iii) the presentation of the data mining results obtained in a manner that allows it to be acted upon. In the case of the third challenge visualisation techniques are seen to be the most appropriate solution. However, such visualisation also presents a significant challenge concerned with how best to present the data in a meaningful manner.

The proposed BRM framework can be generically applied to collect, process and display the reputation of brands belonging to organisations. The framework includes mechanisms for the collection of information from www sources such as social media, news and blogs. The framework also includes mechanisms for mining the collected data, this includes: (i) the application of sentiment and opinion mining techniques and (ii) the discovery of topical groupings (using hierarchical clustering). The most significant element of the framework is a collection of visualisation options whereby the BRM outcomes can be displayed so that users can effectively digest the collected information and obtain a view concerning their brand in the context of online media. The framework has been incorporated into a commercial brand reputation management system called LittleBirdy (www.littlebirdy.buzz).

The remainder of this paper is organised as follows. In Sect. 2 some related work is presented. Section 3 gives a formal description of the BRM problem, while Sect. 4 describes the proposed BRM framework. Section 5 then presents a case study and an evaluation of the framework. Some conclusions are presented in Sect. 6.

2 Related Work

Social media is providing a new form of communication platform which is both unregulated and unofficial. By definition social media content is outside of the direct control of organisations [3]. However, social media activity can also provide substantial opportunities for organisations. For example the ability for organisations to be speedily aware of, and able to identify, disgruntled customers according to activity on social media is seen to be particularly beneficial in that organisations can quickly implement some action in an attempt to safeguard their product [2]. Another example is the swift identification of the use of copyrighted material on www fora such as YouTube (for video sharing), Flikr (for photograph sharing) and SlideShare (for presentation sharing), amongst many others [2]. BRM is therefore seen as an important aspect of the modern commercial world. It is thus not surprising that there has been substantial recent work directed at BRM. Examples can be found in [4, 7–10].

The main focus of the work presented in Morinaga [4] is to determine the reputation of a company/brand by focusing on mining online opinions concerning their products. By using collected texts it is assumed that factual information about a product is not required, concentrating on opinion only so as to focus on the experience of a product that individuals are writing about. The main distinguishing factor between Morinaga's work and the BRM framework presented in this paper is the data mining approach. Morinaga does use sentiment analysis, but there is no topic discovery element as included in the BRM framework (which uses a hierarchical clustering approach for topic discovery).

In the work by Ziegler et al. [10] a system is described that uses RSS (Rich Site Summary) feeds to collect news related to large corporations. The feeds are categorised using taxonomies from the DMOZ open directory project www.dmoz.org. The BRM framework presented in this paper focuses on current trends in social media and blogs amongst other sources of information. This differs from Ziegler's work where the focus of the reputation mining effort is directed at large corporations and only uses news feeds as the information source (no social media). It is argued that restricting BRM activity to news feeds may not be the most appropriate method of collecting brand mentions, particularly if we wish to focus on user opinions.

In the work by Spangler [8] a system called COrporate Brand and Reputation Analysis (COBRA) is proposed in order to provide corporate brand monitoring and alerting. The main focus of the COBRA system is to identify product categories, topics, issues, and brands to be monitored. It does this by focusing on broad keyword based queries related to a brand in order to retrieve "sufficient" data. This approach can be very wasteful in terms of bandwidth usage and data storage. Distinction with the BRM framework proposed is that the framework's focus is on brand mentions, consequently the collected data is significantly less "noisy" (little irrelevant data is collected), hence the data mining applied at a later stage is more effective.

3 Formal Description

This section presents a formal description of the BRM problem. A *mention* of a particular brand comprises a body of text collected from an online social media site of some form. We use the notation m_i to indicate a particular mention. A data set comprising n mentions, collected over a period of time, is denoted using the notation $M = \{m_1, m_2, \ldots, m_n\}$. The content of each mention m_i is represented using a feature vector F_i (how this is generated will become clear later in this paper) founded on a feature-space model [6]. Each dimension in the feature space represents some attribute a_i whereby each attribute can take two or more values (if it could take only one value it would be a constant and therefore not of interest with respect to the BRM). We indicate the set of values for an attribute a_i using the notation $a_i.V = \{v_1, v_2, \ldots\}$, we indicate a particular value v_j associated with a particular attribute a_i using the notation $a_i.v_j$. Thus we have a global set A of attribute-value pairs. Thus each F_i is a subset of A ($F_i \subset A$). The complete data set M is therefore represented by a set F of feature vectors such that $F = \{F_1, F_2, \ldots, F_n\}$ (note that there is a one to one correspondence between M and F). The first challenge is thus to translate M into F.

Once F has been established the next stage is to apply some sentiment mining to the content of F. For each vector encoded mention in F we wish to attach a sentiment value. A coarse sentiment scoring was conducted using the label set $\{positive, negative, neutral\}$. Thus for each mention $m_i \in M$ there exists a corresponding sentiment label $s_i \in S$ (where $S = \{positive, negative, neutral\}$). The most appropriate sentiment label is derived using a classification algorithm which relies upon the features in F to assign a label to each mention. Details of the classification algorithm is presented in Sect. 4.2.1.

The next stage in the proposed BRM framework (see below) is to describe the identified mentions in terms of a hierarchical structure using a hierarchical clustering algorithm that uses the similarity of the features from F for each mention. Thus for each mention $m_i \in M$ there exists a "path" in the hierarchy $P = \{c_1, c_2, \ldots, c_x\}$, where x denotes the number of levels in the hierarchy and c_i denotes the cluster ID of a cluster belonging to level i. An example of a hierarchical clustering is shown in Fig. 1. The example shows a set of hierarchical clusters at various levels. Level 1 represents the complete collection of mentions (so the set M) and is indicated by cluster 0. Level 2 shows the complete collection of mentions segmented into two groups, cluster 1 and cluster 2. Level 3 shows that clusters 1 and 2 have been further segmented into two sub-groups each indicated by the notation 1.1, 1.2, 2.1 and 2.2 respectively. Thus a

Fig. 1 Example of a BRM
hierarchical clustering

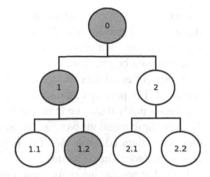

mention that is a member of cluster 1.1 is also a member of cluster 1 and of course
cluster 0 (as the latter contains all mentions). Figure 1 also shows a path (highlighted)
associated with a particular mention, $P = \{0, 1, 1.2\}$. Further details concerning the
hierarchical clustering algorithm is presented in Sect. 4.2.2. It should be noted that
this structure is independent of the sentiment score assigned to each mention, thus
each cluster in the hierarchy will contain mentions that feature different sentiment
labels.

4 Brand Reputation Mining Framework

This section presents details of the BRM framework which has been developed in
order to address the challenges of BRM (as presented in Sect. 1 above). The framework comprises three stages: (i) data collection, (ii) data mining and (iii) visualisation.
Each of these is discussed in detail in the following three sub-sections.

4.1 Data Collection

Two main approaches are used for data collection. The first is a server side RSS reader
and the second a social media related Application Programming Interface (API's).
Both approaches were utilised in order to collect and gather mentions at regular time
intervals. For the first relevant RSS feeds need to be identified with respect to the
brand in question. Examples include news website RSS feeds and industry specific
feeds which are updated on a regular basis. Social media platforms offer access, via
an API, so that relevant mentions can be obtained based on a submitted query. The
query in this case needs to include terms related to the brand in question, for example
brand name or related products and services.

The data is then processed and added to the collection of mentions M (at the start
$M = \emptyset$) in such a way that each $m_i \in M$ has associated with it details of: (i) the

source it was gathered from and (ii) the time it was gathered. Gathering data from the web inevitable leads to duplication and noise. For this reason a parsing function was applied to each chunk of collected data in order to act as a filter to "clean" the received data before storage. This function first confirmed that the mentions in the currently collected data "reference" the brand in question, if not these mentions were removed. The parsing function then searched for and removed duplicate mentions in the data. Finally it checked that the remaining mentions weren't already included in M, if so these mentions were also removed. The remaining mentions in the collected data were then added to M. The physical storage was provided using a scalable cloud database. The collection process continues in this manner with new data being collected at regular intervals, mentions that have expired (are older than some user specified time span) are removed from M so as to prevent M from containing out of date information and/or scaling to a size where it can no longer be processed in almost real time. Once a reasonably sized collection M has been obtained it can be processed, analysed and viewed at any time.

4.2 Data Mining

The objectives of the data mining stage were: (i) to conduct sentiment analysis and (ii) the grouping of mentions into related topics. To achieve these objectives two data mining techniques were used. The first technique comprised a sentiment classifier using the class label set {*positive, negative, neutral*}. The second technique was a hierarchical clustering algorithm which was used to identify groupings of related topics in a hierarchy structure. This idea was to segment the data into related topics on a level by level basis. Grouping the mentions into topics allows for top level analysis to be done. For example identifying what the currently most popular topics related to a particular brand are and what the comparative preponderance of these topics is. Further details of these two processes are given below in the following two sub-sections.

4.2.1 Sentiment Analysis

The sentiment analysis method used was based on an off the shelf, Naive Bayesian, sentiment classifier called uclassify (www.uclassify.com). The data set used to train the classifier was 40,000 Amazon product reviews from 25 different product genres [1]. This sentiment classifier was selected because: (i) it was directly related to the application of BRM, (ii) preliminary testing indicated that the recorded accuracy was acceptable compared with other sentiment classifiers considered, and (iii) the training data uses the same critical language style as that of someone commenting on a brand or service. The uclassify classifier requires input in the standard feature vector format, this a set F of the form described above. The classifier returns a value between 0 and 1, where 1 indicates a positive sentiment and 0 a negative sentiment.

Table 1 Pseudo code for the sentiment analysis algorithm

```
Algorithm Sentiment_Analysis (F)
 1: T = 0.2;
 2: s = null;
 3: p = Bayesian_Classifier(m);
 4: if p >= 1 − T then
 5:     s = positive; {positive sentiment}
 6: else
 7:     if p <= T then
 8:         s = negative; {negative sentiment}
 9:     else
10:         s = neutral; {neutral sentiment}
11:     end if
12: end if
13: return s;
```

Table 1 shows the pseudo code for allocating a sentiment label s, taken from the set $S = \{positive, negative, neutral\}$, to a mention m. The input is a feature vector F describing mention m (as defined above). The output is a sentiment label s to be associated with the mention. The sentiment mining algorithm first determines the polarity value associated with the mention m (line 3), the second part of the algorithm then determines the sentiment class label to be assigned to the mention using the polarity value p and a threshold T. If $p \geq 1 − T$ (line 4) then $s = positive$ (line 5). If $p \leq T$ (line 7) then $s = negative$ (line 8). Otherwise $s = neutral$ (line 10). It was found experimentally that $T = 0.2$ was the most appropriate value to be used in the context of the work presented in this paper. Although a sentiment class label set of size three was used it is clear from the above that a larger number of class labels (depending on the application domain and end user requirements) could equally well have been adopted.

4.2.2 Hierarchical Clustering

The purpose of the hierarchical clustering, as noted above, is to group mentions into a hierarchical format such that each mention belongs to one cluster in each level of the hierarchy. This idea is to reduce the potentially large number of mentions that may have been identified by "compartmentalising" them into smaller groups which can be readily understood and further analysed. In other words the purpose of the clustering is to identify topic hierarchies within the collection M. Once topic clusters have been identified statistical analysis can be conducted with respect to each cluster (topic or sub-topic). To obtain the desired hierarchical clustering a divisive (top-down) hierarchical clustering algorithm was used, the alternative would be a conglomerative (bottom-up) hierarchical clustering. Divisive hierarchical clustering operates in a breadth first manner by repeatedly dividing the candidates at each level

Table 2 Pseudo code for the hierarchical clustering procedure

```
Procedure divisive_cluster( M, k)
 1: if (|M| < k) then
 2:    exit
 3: end if
 4: C = set of k clusters {c₁,c₂,..., cₖ}
 5: if (Silhouette_Coefficient(C) > σ) then
 6:    exit
 7: end if
 8: for all cᵢ ∈ C do
 9:      M' = subset of M in cᵢ
10:      divisive_cluster( M', k)
11: end for
```

and each branch of the growing hierarchy into k clusters. In effect it can be thought of as an iterative k means process.

The pseudo code for this procedure is shown in Table 2. The input is the set of mentions M and the number of clusters k desired at each level. It is a recursive process in that the *divisive_cluster* procedure is called with respect to each discovered branch. The process continues until the number of mentions considered at some leaf node in the hierarchy is less than k (line 1) or until as a sufficiently cohesive clustering configuration was reached (line 5). This was measured in terms of the Silhouette Coefficient [5]. The Silhouette Coefficient of a cluster configuration is a measure of both the cohesiveness and separation of a configuration. It is a number between -1.0 and $+1.0$. The nearer the coefficient is to $+1.0$ the better the cluster configuration. Experiments were conducted using a variety of values for k and it was found that $k = 3$ produced the most effective results with respect to BRM.

Once the hierarchical clustering had been completed each cluster was allocated a label designed to describe its content. This was done by extracting the words that appeared most frequently in the feature vectors representing the mentions contained in a given cluster and identifying the words whose occurrence count was above some threshold or if there were more than three frequently occurring words selecting the top 3. The identified words thus constituted the cluster label.

4.3 Visualisation

The following section presents details on the visualisation stage of the BRM framework. The objective of the visualisation stage is to effectively communicate the brand reputation patterns and new knowledge that has been determined in the previous data mining stage (Sect. 4.2). The approach used to achieve this was to create a sequence of charts and diagrams, in a dashboard style interface, which made the information easily digestible by the end users of the system in a visually appealing manner. The dashboard interface comprised several different visualisations, these individual

Table 3 A list and description of the visualisation techniques provided with the BRM framework

Chart name	Representation	Function	Tool or API
Wheel charts			
Standard	Multi-level pie charts	Displays hierarchical cluster relationships and proportions	Fusion charts
Interactive	Multi-level pie charts	Created using D3 Library with some added interactive user exploration features	D3 Javascript Library
Advanced	Multi-level pie charts	Adds an extra layer of information related to the proportion of sentiments in each category	D3 Javascript Library
Treemap			
Treemap	Display hierarchical (tree-structured) data as a set of nested rectangles	Shows a visual representation of a data tree, clusters and sub-clusters. This chart can be used to explore various levels of the hierarchy	Google Chart API
Circle packing			
Standard	A hierarchical layout using recursive circle packing	Displays each category in a circle and shows the hierarchical relationships	D3 Javascript Library
Advanced	A hierarchical layout using recursive circle packing	Adds an extra layer of information in the form of the proportion of sentiments in each category with several pie charts	D3 Javascript Library
Timeline	A hierarchical layout using recursive circle packing and multiple data sets	Displays the changes in different data sets using animation; data sets may be from many different successive time periods	D3 Javascript Library

visualisations can be categorised as follows: (i) Wheel Charts, (ii) Treemaps and (iii) Circle Packings. Each category of visualisation was developed in order to highlight a certain aspect of the data mining effort. The complete list of data mining visualisations is given in Table 3. Each of the visualisations is described in more details below.

The **Standard Wheel Chart** shows all clusters (with labels) in one visualisation. The Inner core of the wheel shows all the data; while each ring, moving to the outer edge, shows the sub clusters (identified by the hierarchical clustering) of the previous ring. This chart can be used to instantly see a comparison of the number of mentions in each of clusters.

The **Interactive Wheel Chart** displays data in a similar way to the wheel chart explained above, but in this case adds a level of interactivity. When a user selects a cluster the chart is transformed so that this cluster becomes the "root node" in the centre of the chart. This allows for fast exploration of the hierarchical clustering outcomes. Large volumes of data can be explored by navigating clusters and sub-clusters in this manner.

The **Advanced Wheel Chart** displays the hierarchical clusters within each level as segments of a wheel. The cluster size of each sentiment category is displayed as a variation of the main colour for each cluster. Adding the sentiment information as an extra layer of data, means this chart can be used as an alternative interpretation to the wheel charts above while still maintaining the advantages of viewing large data sets effectively.

The **Treemap Chart** shows blocks of varying sizes and colours. Each block represents a cluster (with its label). The size of a block is a reflection of the number of mentions within the cluster. Each block has a colour, from a variable range, which is used to indicate the overall sentiment category of a cluster. The chart also features an interactive element whereby a desired hierarchical level can be selected so that the user can drill down further into the data. The treemap chart allows the user to quickly make comparisons between clusters and sentiments at a single glance.

The **Standard Circle Packing Chart** is constructed by representing the clusters within a level of the hierarchy as a pie chart. An algorithm is then used to recursively display all clusters and sub-clusters in one chart. The **Advanced Circle Packing Chart** is constructed based on the standard circle packing chart but in this case adding an extra layer of information which shows the sentiment categories and their relative cluster sizes. The **Timeline Circle Packing Chart** is a variation of the advanced circle packing chart described above, but with an animation element. This chart can display mentions at various points in time. This chart can also display changes to cluster sizes and sentiment categories as the chart cycles over various points.

5 Case Study

The most appropriate mechanism for gaining an appreciation of the operation of the BRM framework is by considering an example. This section therefore presents a case study which uses the presented BRM framework as applied to a specific "real world" brand reputation problem. Section 5.1 describes the scenario used and how the data set was derived. Section 5.2 presents the resulting visualisations produced by the BRM framework with respect to the brand used in this case study.

5.1 Data set

The particular data set used as a focus for the case study is the popular TV talent show "Britain's Got Talent" (BGT); other countries have a similar show. The show provides a forum for members of the public to audition, a panel of judges select the best acts which then go on to the next stage where each week a number of contestants get voted off; this continues until there is one winner. This particular brand was selected as it is an example of a popular brand that has a significant presence on social media where "debates" are conducted concerning the antics of the contestants, presenters and the panel of judges. The collated BGT data set comprised 14,280 Twitter mentions, collected from April to June 2013.

It should be noted that the BGT brand used in this work has a strong correlation with other household brands (in the context of the proposed BRM framework). Thus the brand name, and associated hash tags, are used on social media platforms to identify discussions about the BGT brand in the same manner that any other brand might be discussed. The opinions of the judges and contestants of the BGT brand discussed on social media can be considered to be analogous to products or services offered with respect to more traditional companies/brands. The case study is therefore designed to demonstrate that customer satisfaction can be derived from the analysis of such media using the proposed BRM framework.

5.2 Results

This sub-section describes the outputs from using the BRM framework with respect to the BGT brand. The first part of this sub-section presents the results from the hierarchical clustering algorithm. The second part concentrates on the visualisation of the combined hierarchical clustering and sentiment analysis.

The hierarchical clustering algorithm was applied to the BGT data set. The resulting output is shown in Fig. 2. From the figure it can be seen that the clustering algorithm identified some clear groupings regarding the BGT dataset. In particular it can be seen that the BGT brand has a clear set of top level sub topics related to the

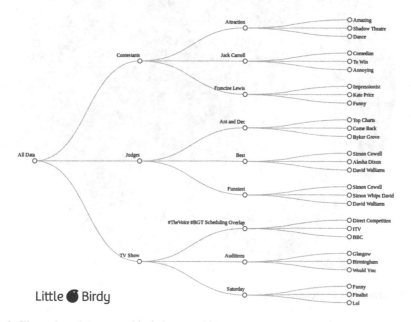

Fig. 2 Illustration of the hierarchical clusters within the BGT data set

judges, contestants, presenters and the TV show itself. From the sub-clusters it can be seen that the TV show topic contains sub-topics regarding dissatisfaction with the scheduling overlap of BGT and another popular TV talent show (The Voice). With respect to the generic application of the BRM framework, this could be a particular brand with clusters showing the most popular talking points reflected on social media. This could also highlight issues with competitor's products or services.

The **Advanced Wheel Chart** is shown in Fig. 3. The chart shows all the hierarchical clusters and the cluster sizes with respect to each sentiment. The cluster with the most fragmented sub-clusters can be seen very easily (clusters at the top of the chart in Fig. 3). The "TV show" and "Contestant" clusters have smaller sized sub-clusters representing more in-depth topics, for instance certain locations for auditions, and so on. The larger clusters represent topics of much broader interest. These cluster topics can be related to a more general aspect of the BGT brand.

The interactive element to this chart can also aid with the exploration of the information presented. This allows the user to drill down into certain topics of interest that may have had a previously unknown association with the brand. An example in the BGT case is that of the presenters topping the music charts with a re-released

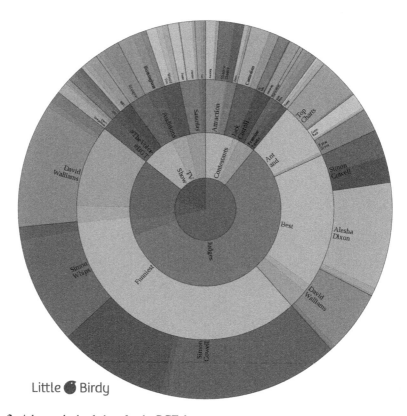

Fig. 3 Advanced wheel chart for the BGT data set

song. The sentiment categories are displayed in a pop-up box when the user hovers over a cluster.

The **Treemap Chart** is shown in Fig. 4. The chart shows blocks of varying sizes and colours. The size is a reflection of the number of mentions, colour is used to indicate the overall sentiment category of a cluster, in this case red is negative while blue is positive. This chart very clearly highlights the sentiment of the most popular clusters within the hierarchy. For example it can be seen that the "Simon Cowell" cluster (top right) has an overwhelming negative sentiment while the "Byker Grove" (a no longer running TV show in which the presenters were featured) cluster (bottom left) has mostly positive sentiment. The Treemap chart allows the user to quickly make a comparison of the clusters and their associated sentiment at a single glance. Thus it can be seen that the popularity of the judges is indicated by the size and colour of the blocks, this is shown in the case of judge "Simon Cowell" (negative sentiment) and "Byker Grove" (positive sentiment).

The **Circle Packing Chart** is shown in Fig. 5. The chart represents the clusters within a level of the hierarchy as a pie chart. The darkness of the colour of the pie chart reflects the sentiment categories, from darkest to lightest representing positive to negative. This chart includes an animation element. The animation can effectively display data at various points in time. This chart will display changes to cluster sizes and sentiment categories as the chart is "cycled" through time. It can thus be used to display changes in cluster size and sentiment of the various cluster topics over (say)

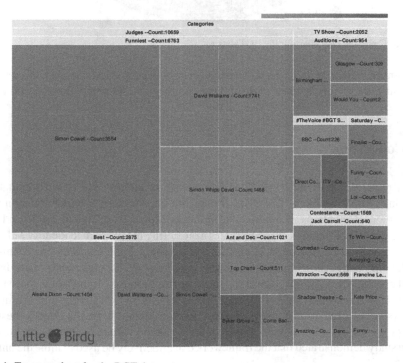

Fig. 4 Treemap chart for the BGT data set

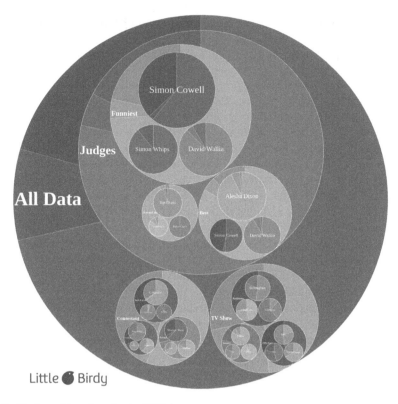

Fig. 5 Circle packing chart for the BGT data set

a number of weeks. In terms of more generic brand reputation mining this can be considered as a mechanism for displaying change in brand perception over time.

6 Conclusion

This paper has presented the BRM framework to address the Brand Reputation Mining (BRM) problem. The BRM framework encompasses the identification and collection of mentions of a particular brand as expressed on social media and the wider www. The framework incorporates a clustering mechanism to identify a cluster hierarchy within a collection of mentions concerning some brand and mechanisms to determine the sentiment associated with individual clusters (elements or attributes of the brand in question). The final stage of the proposed framework consists of a set of visualisation tools to display the results of the clustering and sentiment analysis. The evaluation comprised a case study to demonstrate the effectiveness of the proposed BRM framework in the context of brand reputation management. The case study demonstrated that the framework can effectively be used to mine the online reputation of brands. Analysis of the visualisation demonstrated previously unknown patterns and provided for an insight into the brand that would have previously been lost or very difficult to discover via manual analysis.

Acknowledgments This work was supported by the UK Technology Strategy Board Knowledge Transfer Partnerships programme (Grant No. KTP008896).

References

1. J. Blitzer, M. Dredze, and F. Pereira. Biographies, bollywood, boom-boxes and blenders: Domain adaptation for sentiment classification. ACL, 2007.
2. Kaplan, A.M., Haenlein, M.: Users of the world, unite! the challenges and opportunities of social media. Bus. Horiz. **53**(1), 59–68 (2010)
3. Mangold, W.G., Faulds, D.J.: Social media: the new hybrid element of the promotion mix. Bus. Horiz. **52**(4), 357–365 (2009)
4. Morinaga, S., Yamanishi, K., Tateishi, K., Fukushima, T.: Mining product reputations on the web. In: The eighth ACM SIGKDD international conference on Knowledge discovery and data mining, pp. 341–349. ACM, New York, NY, USA (2002)
5. Rousseeuw, P.J.: Silhouettes: a graphical aid to the interpretation and validation of cluster analysis. J. Comput. Appl. Math. **20**, 53–65 (1987)
6. Salton, G., Wong, A., Yang, C.S.: A vector space model for automatic indexing. Commun. ACM **18**(11), 613–620 (1975)
7. Skopik, F., Truong, H., Dustdar, S.: Trust and Reputation Mining in Professional Virtual Communities, pp. 1–15. Springer, Berlin (2009)
8. Spangler, S., Chen, Y., Proctor, L.: COBRA Mining web for corporate brand and reputation analysis. Web Intell. Agent Syst. **7**, 246–254 (2009)
9. Yi, J., Niblack, W.: Sentiment mining in WebFountain. In: Proceedings of 21st International Conference on Data Engineering, ICDE 2005, pp. 1073–1083 (2005)
10. Ziegler, C.N., Skubacz, M.: Towards automated reputation and brand monitoring on the web. In: Web Intelligence, 2006. WI 2006, IEEE (2006)

A Review of Voice Activity Detection Techniques for On-Device Isolated Digit Recognition on Mobile Devices

M.K. Mustafa, Tony Allen and Lindsay Evett

Abstract This paper presents a review of different Voice Activity Detection (VAD) techniques that can be easily applied to On-device Isolated digit recognition on a mobile device. Techniques investigated include; Short Time Energy, Linear predictive coding residual (prediction error), Discrete Fourier Transform (DFT) based linear cross correlation and K-means clustering based VAD. The optimum VAD technique was found to be K-means clustering of Prediction error which gives a recognition rate of 86.6 %. This technique will be further used with an LPC based speech recognition algorithm for digit recognition on the mobile device.

1 Introduction

Speech Recognition systems on Mobile Phones can be classified into two basic approaches—the client-server and on-device implementations. Currently, the dominant implementation on a mobile platform is in a client-server mode of operation; as can be found in modern Android and Apple iPhone smartphones. The principle behind how speech recognition is performed on these phones is hidden due to the market competition involved. However, it is known that the speech recognition of these phones is achieved by sending the speech signal over the internet to a speech recognition server. This server recognizes the spoken phrase, converts it into text and then either sends it directly back to the phone (Android platform) or, in the case of the iPhone SIRI application, sends it to the SIRI server where the semantic information contained within the text phrase is extracted and processed in order to give relevant feedback to the user.

M.K. Mustafa (✉) · T. Allen · L. Evett
Nottingham Trent University, Nottingham NG1 4BU, UK
e-mail: mohammed.mustafa2009@my.ntu.ac.uk

T. Allen
e-mail: tony.allen@ntu.ac.uk

L. Evett
e-mail: lindsay.evett@ntu.ac.uk

© Springer International Publishing Switzerland 2014
M. Bramer and M. Petridis (eds.), *Research and Development in Intelligent Systems XXXI*, DOI 10.1007/978-3-319-12069-0_23

317

Although client-server is the most common speech recognition implementation there are obvious, and not so obvious, disadvantages associated with this method of operation.

1.1 The Client-Server Approach Disadvantages

- There is an obvious need for 24/7 connection to the internet. In addition the availability of different connection modes (3G, 4G, GPRS) provided by the mobile network operator also determines the functionality of the speech recognition system.
- Less obvious is the need for efficient assignment of Automatic Speech Recognition (ASR) decoders to computing resources in network based server deployment (the efficiency at peak loads, when a large number of ASR decoders are used in multi-user scenarios) [1].
- The need to include implementation of acoustic adaptation and normalization algorithms within a client-server framework is also a performance requirement [1].

One way to address the shortcomings of the client-server implementation would be to perform speech recognition on the mobile device itself.

1.2 On-device Implementation Challenges

- Memory on the mobile platform is limited as compared to a computer platform. Though there have been recent improvements in the memory capacity of mobile devices, this can still not be compared to the computer platform.
- Computational Speed on the mobile platform is limited as compared to a computer based device. This hampers the processing time required to complete a speech recognition process as the computational requirement of a speech recognition engine is quite high.
- Power is limited on mobile devices. Speech recognition process requires a good deal of computation which rapidly drains the power of the mobile platform.

1.3 Aim

The overall aim of this research is to develop a standalone on-device implementation of isolated digit speech recognition. The work presented in this paper addresses voice activity detection within speech samples.

2 Voice Activity Detection

Voice Activity Detection (VAD) in audio streams is a classic technique for reducing the computation cost and response time of speech recognition systems. By only passing detected speech frames into the recognition algorithm, the computational overhead associated with processing the silence section of speech are minimized [2].

Observing the speech signal image shown in Fig. 1, it is easy to spot the regions with speech and the regions without speech in the speech signal. Segmenting this speech signal into frames of speech and silence will allow the system to only process the speech frames whilst ignoring the silence frames. Having a robust algorithm for VAD is also crucial for a good recognition performance within noisy environments [3].

VAD techniques have been used in a range of speech processing activities such as speech coding, speech recognition, hands free telephony, echo cancellation etc. In GSM-based systems it has been used to conserve battery power by truncating transmissions when speech pause is detected [4]. A large number of different approaches for voice activity detection have been proposed, each depending on different features of speech. These include; Short Time Energy (STE) based approach [5, 6], Zero crossing count [6], Discrete Fourier Transform [7], Linear Predictive coding, Prediction error, MFCC.

The algorithms reviewed for Voice activity detection, in this paper, are based on:

- Short Time Energy (STE)
- Linear Predictive Coding Residual [Prediction Error (PE)]
- Discrete Fourier Transform (DFT)

VAD algorithms can be divided into two parts;

- The Acoustic feature extraction part
- Decision Module part [8, 9]

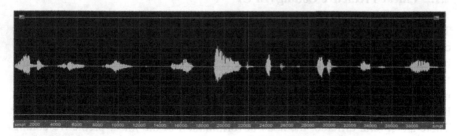

Fig. 1 Time domain speech signal

2.1 Feature Extraction Algorithms

To extract the speech features; the speech signal is first framed using a window function. The Hamming window is the predominant window function used in speech recognition [10–12]. Windowing is performed such that each frame contains 10–30 ms of speech. This approximates the time period during which the vocal tract maintains a fixed set of characteristics.

The following feature extraction algorithms used in this work are performed on the windowed sample.

2.1.1 Short Time Energy Analysis (STE)

This is computed by splitting the signal into frames of M samples and then computing the total squared values of the samples within each frame. The splitting of the signal is achieved by using a suitable window function to split the signal into desired frames [10, 11, 13]. The STE of a signal is computed using Eq. 1;

$$E_n = \sum_{m=-\infty}^{\infty} (x[m]w[n-m])^2 \qquad (1)$$

where E_n is the Energy of the nth frame of the speech signal and x[m] is the amplitude of the speech signal in the time domain. w[n − m] is the window within which speech samples exist.

The STE algorithm is said to be very good for detecting word boundaries [6]. Many experiments have shown that the addition of energy information as another model feature improves system information and as such helps in voice activity detection [5, 6, 14]. STE is considered as a very important feature in automatic speech recognition [14, 15].

2.1.2 Linear Predictive Coding (LPC)

The main idea behind this method is that a speech sample is approximated as a linear combination of past speech samples [16, 17]. Linear prediction is mostly used in low bit rate transmission of speech or storage [11, 13]. This is achieved by minimizing the sum of squared differences between the actual speech samples and the linearly predicted ones. LPC has also become the predominant technique for estimating the basic speech parameters such as formants, spectra, vocal tract area functions etc. [18]. There are several different methods of formulation of the linear predictive coding analysis. The prediction of a Speech sample x[n] is given in Eq. 2;

$$\tilde{x}(n) = \sum_{i=1}^{p} \alpha_i x(n - i) \tag{2}$$

where $\tilde{x}(n)$ is the sample being predicted, $x(n - i)$ is the previous sample, P is the order and α_i is the prediction coefficient.

2.1.3 LPC Residual (Prediction Error)

The LPC prediction error is a by-product signal of the LPC process. The prediction error decreases with the increase in P (order). The Prediction error is a very good approximation of the excitation source [11, 13]. Based on the ascertained assumptions above, the prediction error is large for voiced speech at the beginning of each pitch period. This can then be used to estimate or rather calculate the pitch period by subtracting between the large samples of the error signal. This can be used to identify pitch periods within samples [10, 11]. The equation for the Prediction error is given as;

$$e(n) = x(n) - \tilde{x}(n) = x(n) - \sum_{i=1}^{p} \alpha_i x(n - i) \tag{3}$$

where $e(n)$ is the error of the sample n and $x(n)$ is the time domain sample from the signal. $\tilde{x}(n)$ and $x(n - 1)$ and α_i are as defined in Eq. 2.

2.1.4 Discrete Fourier Transform (DFT)

The DFT basically decomposes the speech signal into a set of Cosine and Sine waves with basis function of each being;

$$\text{Cosine} = c_k[i] = \cos(2\pi ki/N) \tag{4}$$
$$\text{Sine} = s_k[i] = \sin(2\pi ki/N) \tag{5}$$

where $c_k[i]$ is the Cosine wave representation of the same $k[i]$ and $s_k[i]$ is the Sine wave representation of $k[i]$. This then synthesizes the sine and cosine waves to give a DFT Spectrum. The DFT spectrum is said to contain the raw information about a speech signal [19].

Due to the computational load of computing DFTs a faster technique was implemented by Cooley–Tukey [20] known as the Fast Fourier Transform (FFT). This technique is used to compute the DFT in a fast manner. The DFT spectrum for each frame contains two parts, the other part is a symmetrical version of the first part.

3 Experiments

3.1 Database

There are a total of 175 speech samples used in this work. These are divided into two sets, with each sample within the sets containing the digits 1–9 spoken isolated. All the speech samples have a sampling rate of 8 kHz which is the requirement for them to be used over a standard telecommunication channel [19]. The samples are also in a 16 bit encoding format.

- The first set has a total of 15 samples from 5 Speakers, with each speaker having 3 samples, was recorded using the mobile phone.
- The second set is from the Centre for Spoken Language Understanding (CSLU2002) standard database, with a total of 160 samples from 10 speakers [21].

3.2 Device Used

The HTC Desire smartphone, used in this work, had the following specifications:

- Processor: Single Core, 1,000 MHz
- System Memory: 576 MB RAM/ 512 MB ROM

This 2011 mobile device is perhaps a little below average in specifications when compared to more recent smartphones, but does represent the "average" specification phone that is likely to be in use with the general population.

3.2.1 STE and LPC Residual Threshold VAD Algorithm Comparison

The STE Algorithm sets a threshold based on the maximum amplitude of a particular speech signal. The threshold is taken as a percentage of the maximum amplitude, with the STE being in raw power numbers. The algorithm also uses the frame distance to separate the words. The Frame distance is the distance between the frames containing the respective digits. The entire speech signal containing the digits 1–9 was cut into frames of 128 samples. The Frame distance variable is used to accommodate for the space between the utterances of the digits i.e. the break between the digits 1 and 2, the break between 2 and 3 and all the way to 8–9. The algorithm identifies the word boundaries and the identified word is then played to verify the identified boundaries were correct. The algorithm was used to see if it could effectively work for all the samples with a specified threshold and inter-word (frame) distance length.

The LPC residual raw power was used as a feature to produce a VAD algorithm. This was based on the similarity between the raw powered short time energy signal and the LPC residual signal. The energy profile of both signals is very similar.

Table 1 LPC residual versus STE of 10 samples (90 digits) CSLU database

Feature	Threshold	Frame distance	Recognized (%)
LPC residual	0.81	>7	80
STE	0.81	>7	73.30

Table 2 STE algorithm for 10 samples (90 digits) device recorded

Feature	Threshold	Frame distance	Recognized (%)
STE	0.81	>7	63.30

The LPC residual algorithm results were then compared to the equivalent STE algorithm results to see if there is any improvement in the results as compared to the traditional short time based energy algorithms. A threshold was set based on the maximum amplitude within the sound sample. The threshold applied was the same as that used with the STE algorithm. The same 10 samples from the CSLU database were used for this algorithm.

From Tables 1 and 2 it can be seen that the LPC residual based VAD outperforms the traditional Short time energy (STE) based approach. However, these raw powered signal algorithm are not easy to threshold due to the amplitude of these energy profiles. Thresholds can be very tricky because the amplitude of the speech signal is different for different users and even worst the same user can speak at different amplitudes. Hence, there was a need to develop more sophisticated VAD algorithms.

3.2.2 DFT Cross Correlation VAD Algorithm

This technique correlates the DFT frames against each other, thereby resulting in a higher correlation value for frames containing speech and low value for frames without [7]. The Linear cross correlation between two variables X and Y is given in Eq. 5;

$$LCC(x,y) = \frac{N\left(\sum_{i=1}^{N} x_i y_i\right) - \left(\sum_{i=1}^{N} x_i\right)\left(\sum_{i=1}^{N} y_i\right)}{\sqrt{N\left(\sum_{i=1}^{N} x^{2i}\right) - \left(\sum_{i=1}^{N} x_i\right)^2}\sqrt{N\left(\sum_{i=1}^{N} y^{2i}\right) - \left(\sum_{i=1}^{N} y_i\right)^2}} \quad (6)$$

where x_1 and y_1 are two vectors of N samples.

A DFT cross correlation based VAD algorithm was developed to build on the work done by Tashan et al. [7]. The original cross correlation concept applied to DFT frames. Further work was undertaken to apply the concept to other features; the LPC and PE to be precise. The linear predictive coding and prediction error order

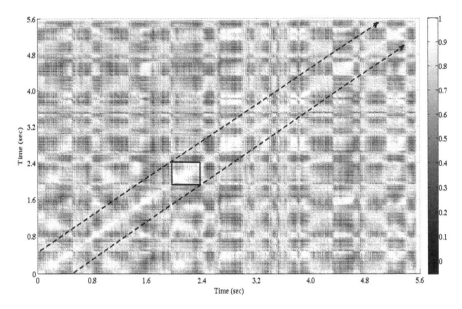

Fig. 2 DFT cross correlation

used is 64 and 21. The 64 order was used based on the size of the DFT frames. The identified digits are found in the light region within the diagonal centre starting from 0.

From Fig. 2 the voiced regions are found in the lighter regions of the diagonal center of the matrix; representing a higher correlation between the frames i.e. the square box in Fig. 2. In this paper, the DFT results can be seen to be poor. The results of the LPC and PE cross correlation results were also inconclusive. investigation of these results indicated that the reason for the differences between our results and those presented in [7] is because the cross correlation algorithm presented in [7] was used for speaker verification and as such the data applied to it was limited to three digits only. That paper uses the digits "5", "8" and "2" for verification on the system and these digits have a distinctive set of phonemes which aids the performance of the DFT Cross Correlation algorithm. Our results were based on the digits 1–9, which are less distinctive.

3.2.3 Log PE Based Fuzzy K-means Clustering VAD Algorithm

A previous work based on isolated word recognition used a modified K means algorithm to cluster test patterns into words [22]. The K means algorithm was applied to the log PE signal which has a relatively minimally flat envelope. The K means algorithm was used to cluster the speech data into two clusters: speech and silence. K means assigns feature vectors to clusters by the minimum distance assignment principle [23]. K means is based on the minimization of the mean squared error.

$$J = \sum_{j-1}^{k} \sum_{i=1}^{n} ||x_i{}^{(j)} - c_j||^2 \qquad (7)$$

where $||x_i{}^{(j)} - c_j||^2$ is a chosen distance measure between a data point $x_i{}^{(j)}$ and the cluster centre c_j [24]. The steps involved in the standard K means clustering include:

- Choose initial cluster centers (randomly).
- Assign each sample vector to the cluster with minimum distance assignment principle.
- Compute new average as new center for each cluster.
- If any center has changed, then go to step 2, else terminate [23].

To minimize the computational time of the K means algorithm, two modifications were applied.

3.2.4 Modified K-means Prediction Error VAD Algorithm

The first modification applied to the standard K means clustering technique was in the choice of random centroids. A good choice of initial cluster centre affects the performance of the algorithm [23] because the first step in the algorithm is a choice of random centroid points. Due to the computational processing speed limitation of mobile platforms, the choice of random centroids was replaced with data specific choices. The minimum and maximum mean within the means of the feature vectors were chosen as the initial centroids. This choice helped in significantly reducing computation time because the number of iterations involved in the whole process was reduced.

The second modification was based on the assignment principle. The K means algorithm iterates through each feature vector, assigning it into the respective clusters. For VAD purposes, only two clusters are required. The first cluster (C_A) represents frames with speech and the second cluster (C_B) representing silence or unvoiced regions of the speech signal. The assignment of a frame is done based on the shortest Euclidean distance between the mean of the frame and the mean of the respective clusters which is referred to as the centre of the cluster. The modification applied was to introduce a bias to this assignment module.

$$C_A = \left(\sum_{i=1}^{N} X_i \right) \div N \qquad (8)$$

$$C_B = \left(\sum_{i=1}^{N} X_i \right) \div N \qquad (9)$$

$$\text{bias} = (C_A - C_B)^* 0.33 \qquad (10)$$

$$C_A = C_B + \text{bias} \qquad (11)$$

Table 3 PE K-means
clustering results

Feature	Frame distance	Recognized (%)
Prediction error (PE) K means	≥ 4	86.60

where X_i is the mean of the ith frame within the cluster and N the total number of
frames within the cluster.

Table 3 it is worthy of note that the frame distance has been reduced in this
experiment in order to increase the sensitivity of the algorithm in recognizing the
silence between the respective frames containing speech. However, for the sake of
simplicity the decision module was not made to be very rigorous. Further work will be
done to ensure a more rigorous decision module because there were frames that were
discarded despite being recognised because the algorithm uses only two components
(assignment principle bias and frame distance) for making its decision.

4 Analysis

It is clear from Table 4 that the conventional methods of voice activity detection
such as the short time energy are not as effective at determining voice activity within
speech signals as the PE K-means algorithm. However, computation time is another
key factor in deciding which VAD algorithm is best for on-device speech recognition.
Table 5 contains the computation time for the different feature extraction algorithms
as well as the VAD algorithms.

It is clear from Table 5 that the computational time involved in extracting the DFT
spectrum and then computing the cross correlation from it is considerably more than
the computational time involved in computing the PE based K-means.

Despite the apparent minimum computation time involved in computing the STE
based algorithm, it should be noted that to perform any aspect of speech recognition
either the LPC features or the DFT features would have to be extracted in addition
to the STE (2.6 + 3.4332 in the case of the LPC or 2.6 + 14.9648 for the DFT). Thus
the STE + LPC extraction time (6.0332) makes the use of the STE VAD almost the
same time with that of the PE K means VAD (6.684); which automatically includes
the LPC feature extraction process. However, the PE K-means VAD outperforms the
STE VAD algorithm in respect of recognition accuracy (86.6 % compared to 73.3 %).

The results presented in this paper clearly show the PE K means clustering VAD
algorithm to be a faster and more effective method of identifying usable speech

Table 4 PE K-means versus
threshold PE versus threshold
STE

Algorithms		
PE K-means (%)	Raw power PE (LPC/PE) (%)	STE (%)
86.60	80	73.30

Table 5 Computation time of signal with 44,330 samples

Feature extraction computation times (s)		
DFT	LPC/PE	ZCC/STE
14.9648	3.4332	1.3128
Algorithm computation time (s)		
DFT cross correlation	K-means	STE
49.7546	6.684	2.6

Note algorithm computation time includes the respective feature extraction

frames for recognition. As the LPC/LPC Residual (PE) based VAD is effectively a by-product of the LPC process there is no additional impact on the computational time in extracting the feature vectors for subsequent speech recognition. However, how the choice of LPC features (as opposed to DFT or MFCC) affects the speech recognition process has not been ascertained at this point. The recognized frames in the K means algorithm will be referred to as usable speech at the moment pending a proper evaluation of the frames in the recognition stage. An additional advantage of the PE K-means VAD process is that it requires no training and as such every particular clustering or extraction done is always going to be speech signal specific. The dynamism of another signal or a different previously used signal will not affect the process as it treats each data as an individual data set and calculates the centroids and does the entire clustering and processing on that particular data set. The method can be applied to different isolated speech data sets for voice activity detection.

5 Conclusion

The work presented in this paper is based on isolated digit recognition on the mobile platform. However, we believe that the PE K means method or technique can be applied to continuous speech recognition with a few adjustments to the algorithms sensitivity. To apply the PE K-means to a continuous SR system, there are two possible implementations that can be applied. Simply eliminating the frame distance decision module can aid in recognising frames with speech and then further processing can be done on it. However, for a much more sophisticated process, a third decision variable dealing with recognized frames that were not classified with the frame distance variable will aid in continuous speech recognition.

6 Future work

Current work is being undertaken to use the recognized usable LPC frames from the PE based K-means to perform digit recognition using an Echo State Network (ESN).

Acknowledgments I will like to thank the Petroleum Technology Development Fund (PTDF) for their continued support and sponsorship of this research. My parents, supervisor, as well as other colleagues who helped in experiments and algorithms.

References

1. Huggins-Daines, D., Kumar, M., Chan, A., Black, A.W., Ravishankar, M., Rudnicky, A.I.: Pocketsphinx: A free, real-time continuous speech recognition system for hand-held devices. In: Proceedings of IEEE International Conference on Acoustics, Speech and Signal Processing 2006, ICASSP 2006, vol. 1, pp. I–I. IEEE (2006)
2. Ali, S.A., Haider, N.G., Pathan, M.K.: A LPC-PEV based VAD for word boundary detection. Int. J. Electr. Comput. Sci. **12**(02) (2012)
3. Cournapeau, D., Kawahara, T., Mase, K., Toriyama, T.: Voice activity detector based on enhanced cumulant of IPC residual and on-line EM algorithm. In: Proceedings of INTER-SPEECH06 (2006)
4. Wu, B., Wang, K.: Voice activity detection based on auto-correlation function using wavelet transform and teager energy operator. Comput. Linguist. Chin. Lang. Process. **11**, 87–100 (2006)
5. Enqing, D., Guizhong, L., Yatong, Z., Yu, C.: Voice activity detection based on short-time energy and noise spectrum adaptation. In: 6th International Conference on Signal Processing, vol. 1, pp. 464–467. IEEE (2002)
6. Rabiner, L.R., Sambur, M.R.: An algorithm for determining the endpoints of isolated utterances. Bell Syst. Techn. J. **54**, 297–315 (1975)
7. Tashan, T., Allen, T., Nolle, L.: Speaker verification using heterogeneous neural network architecture with linear correlation speech activity detection. Expert Syst. (2013). doi: 10.1111/exsy.12030
8. Huang, H., Lin, F.: A speech feature extraction method using complexity measure for voice activity detection in WGN. Speech Commun. **51**, 714–723 (2009)
9. Ghaemmaghami, H., Baker, B.J., Vogt, R.J., Sridharan, S.: Noise robust voice activity detection using features extracted from the time-domain autocorrelation function. In: Proceedings of Interspeech 2010
10. Plannerer, B.: An introduction to speech recognition. Munich (2005)
11. Rabiner, L.R., Schafer, R.W.: Digital processing of speech signals. In: IET (1979)
12. Kesarkar, M.: Feature extraction for speech recognition. Electronic Systems, Department of Electrical Engineering, IIT Bombay (2003)
13. Rabiner, L., Juang, B.: Fundamentals of speech recognition (1993)
14. Zheng, F., Zhang, G., Song, Z.: Comparison of different implementations of MFCC. J. Comput. Sci. Technol. **16**, 582–589 (2001)
15. Waheed, K., Weaver, K., Salam, F.M.: A robust algorithm for detecting speech segments using an entropic contrast. In: 45th Midwest Symposium on Circuits and Systems, 2002, MWSCAS-2002, vol. 3, pp. III-328–III-331. IEEE (2002)
16. Vaidyanathan, P.: The theory of linear prediction. Synth. Lect. Signal Process. **2**, 1–184 (2007)
17. Jones, D.L., Appadwedula, S., Berry, M., Haun, M., Janovetz, J., Kramer, M., Moussa, D., Sachs, D., Wade, B.: Speech processing: theory of LPC analysis and synthesis (2009)
18. Hachkar, Z., Mounir, B., Farchi, A., El Abbadi, J.: Comparison of MFCC and PLP parameterization in pattern recognition of Arabic alphabet speech. Can. J. Artif. Intell. Mach. Learn. Pattern Recognit. **2**, 56–60 (2011)
19. Tashan, T.: Biologically inspired speaker verification. Submitted to Nottingham Trent University (2012)
20. Cooley, J.W., Tukey, J.W.: An algorithm for the machine calculation of complex Fourier series. Math. Comput. **19**, 297–301 (1965)

21. CSLU Database: Available at http://www.cslu.ogi.edu/corpora/isolet/
22. Wilpon, J., Rabiner, L.: A modified K-means clustering algorithm for use in isolated work recognition. Acoust. Speech Signal Process. IEEE Trans. **33**, 587–594 (1985)
23. Looney, C.G.: A fuzzy clustering and fuzzy merging algorithm. CS791q Class notes (1999)
24. A Tutorial on Clustering Algorithms: Available Online at http://home.deib.polimi.it/matteucc/Clustering/tutorial_html/kmeans.html

Short Papers

Ontology-Based Information Extraction and Reservoir Computing for Topic Detection from Blogosphere's Content: A Case Study About BBC Backstage

M. Arguello-Casteleiro and M.J. Fernandez-Prieto

Abstract This research study aims at detecting topics and extracting themes (subtopics) from the blogosphere's content while bridging the gap between the Social Web and the Semantic Web. The goal is to detect certain types of information from collections of blogs' and microblogs' narratives that lack explicit semantics. The approach presented introduces a novel approach that blends together two young paradigms: Ontology-Based Information Extraction (OBIE) and Reservoir Computing (RC). The novelty of the work lies in integrating ontologies and RC as well as the pioneering use of RC with social media data. Experiments with retrospect data from blogs and Twitter microblogs provide valuable insights into the BBC Backstage initiative and prove the viability of the approach presented in terms of scalability, computational complexity, and performance.

1 Introduction

The blogosphere has emerged as a vast pool of people's thoughts, opinions, and views. Some organisations encourage discussion in the blogosphere and use it as a gauge of public opinion on various issues. An example of such organisations is the British Broadcasting Corporation (BBC). The research study presented here uses retrospective blogosphere's content (blogs and Twitter microblogs) collected a few months before the closure of BBC Backstage [1]—a five-year initiative intended to radically open up the BBC that was also part of the BBC's commitment to use open standards.

Topic modelling has been applied to blogosphere's content, e.g. [2]. Some authors have investigated the use of external sources in topic models, such as [3, 4]. There is a number of works that have studied the addition of formal semantics in blogs [5] and Twitter microblogs [6–8].

M. Arguello-Casteleiro (✉)
University of Manchester, Manchester M13 9PL, UK
e-mail: m.arguello@computer.org; arguellm@cs.man.ac.uk

M.J. Fernandez-Prieto
University of Salford, Salford M5 4WT, UK

© Springer International Publishing Switzerland 2014
M. Bramer and M. Petridis (eds.), *Research and Development
in Intelligent Systems XXXI*, DOI 10.1007/978-3-319-12069-0_24

333

This paper investigates autocategorisation, i.e. automatic indexing and tagging, of blogosphere's content (blogs and microblogs) using specialised semantic lexicons that can support semantically complex queries. Our research work investigates the viability of detecting topics and extracting specific distinct subtopics (also known as themes), which are significant for a particular organisation or company (e.g. BBC), from blogs and microblogs. More concretely, the study presented here looks for answers to the following questions: (1) how can topics and subtopics relevant to BBC Backstage be represented? And (2) given the differences between traditional blogs and Twitter microblogs (a.k.a. tweets), can the same topic and subtopic detection approach be successfully applied to both blogs and microblogs? To answer those questions, this research proposes to blend together two young paradigms: Ontology-Based Information Extraction (OBIE) [9] and Reservoir Computing (RC) [10].

2 Approach Overview

Blog data is formatted in HTML and contains HTML tags like hyperlinks, text, and multimedia (e.g. photos). The main difference between microblogs and traditional blogs (a.k.a. web logs) is the amount of textual content. While a Twitter microblog (a.k.a. tweet) is limited to 140 characters, a traditional blog may include long substantive comments. Another difference is that microblogging is stream-based, i.e. each message stands on its own and forms part of a stream.

The current study is aligned with [11] and intends to perform a semantically meaningful analysis of social media data. However, the lack of standard vocabularies or terminologies directly related to BBC Backstage forces a grosser (i.e. less fine grain) content analysis than the one described in [11] for medicine. The current study obtains relevant terms (single word or multi-word) and hyperlinks (URLs) that are embedded in the BBC Backstage Web pages. While [11] talks about *semantic concepts*, the current study talks about *semantic groups*, where several *semantic concepts* belong to a single *semantic group*. For BBC Backstage five *semantic groups* have been identified: *general*; *prototype*; *data*; *ideas*; and *event*. The interconnection among the *semantic groups* cannot be easily captured by means of heuristic rules. Hence, the use of reservoir computing, which is an emerging trend in artificial recurrent neural networks, can be beneficial when dealing with this time-independent static pattern classification problem. Thus, this research proposes to combine two relatively young paradigms:

1. *Ontology-Based Information Extraction* (*OBIE*)—OBIE has recently emerged as a subfield of information extraction, where ontologies play a pivotal role. Ontologies are widely used to represent knowledge and can be seen as consensual models of meaning, and thus, the backbone for the Semantic Web. Due to the crucial role played by ontologies in the information extraction process, OBIE is a field related to knowledge representation and has the potential to assist the development of the Semantic Web [9]. This research study proposes an OBIE system architecture (see Fig. 1) that processes hypertext (blogs and microblogs) through a mechanism

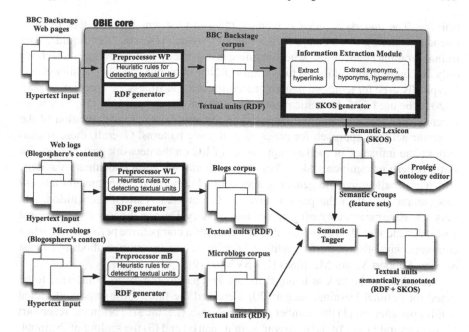

Fig. 1 Ontology-based information extraction (OBIE) system architecture overview

guided by semantic lexicons (lightweight ontologies) to semantically annotate certain types of information and presents the output using current Semantic Web standards such as Resource Description Framewok (RDF) [12] and Simple Knowledge Organisation System (SKOS) [13]. At the core of the OBIE system is the information extraction module (see Fig. 1) that is paramount for obtaining the semantic lexicons (i.e. *semantic groups*). One difference with the general architecture of an OBIE system presented in [9] is to split the 'ontology generator' component into two (see Fig. 1): the RDF generator and the SKOS generator.

Following the Turtle syntax for RDF, the following expression in SKOS represents 'barcamp', which is a subtopic or theme (i.e. *semantic concept*) of the BBC Backstage main topic (i.e. *semantic group*) '*Event*'.

```
bbcBackstage:barcamp rdf:type skos:Concept;

    skos:prefLabel "barcamp"@en;
    skos:altLabel "barcampmanchester2"@en;
    skos:altLabel "barcampnortheast2"@en;
    skos:altLabel "bcman2"@en;
    skos:broader bbcBackstage:event;
    skos:inScheme bbcBackstage:bbcBackstageEvents.
```

2. *Reservoir Computing* (*RC*)—In 2007, Reservoir Computing (RC) [10] was the coined term for a new paradigm in artificial recurrent neural networks (RNNs)

training. The underlying idea is using a random recurrent network that remains unchanged during the training, the so-called "dynamic reservoir" or "liquid", with trainable output connections, where learning is restricted to the outgoing connections only. RC outperforms classical fully trained RNNs in many tasks [14]. Although RC is typically used for temporal pattern processing, such as time-series prediction [15], in 2009 the use of RC for static pattern recognition was investigated [16, 17]. More recently, Emmerich et al. [18] presented an attractor-based implementation of the reservoir network approach for processing of static patterns. Overall, these studies explore the influence of the key ingredients of RC on the network performance for static pattern recognition tasks. They deal with classification benchmark datasets, which are challenging and generally accessible. Furthermore, they unveil that very few parameters affect the performance of RC for static pattern recognition, and thus extensive parameter tuning proves unnecessary. Despite the low computational cost, Alexandre et al. [16] proved that RC exhibits a competitive performance when compared with other powerful traditional nonlinear classification methods like Least Squares Support Vector Machines (LS-SVMs) [19].

According to the task at hand, there are key parameters of RC that need to be tuned for optimal learning. Jaeger [20] identified the three most important global control parameters: (1) the number of internal units (i.e. the size N) in the reservoir; (2) the spectral radius of the reservoir weight matrix; and (3) the scaling of the input. The experiments carried out by Alexandre and Embrechts [17] suggest that: (a) the values of the reservoir's spectral radius and the connectivity of the neurons in the reservoir do not seem to influence the computational performance of the ESNs for static classification tasks; and (b) the size N of the reservoir should not exceed the size of the training set, which is well aligned with the heuristic tuning strategy proposed by Jaeger [20] for N.

We propose using RC for classifying blogosphere's content (see Fig. 2) into topics and themes (subtopics). This study exploits the lessons learnt from the above-mentioned studies in parameter tuning for RC with good computational performance; however, this is the first study that uses RC for static pattern recognition with blogosphere's data.

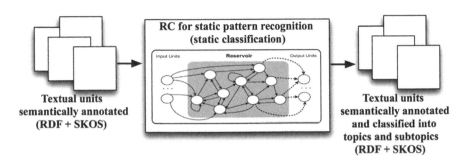

Fig. 2 Overview of the static classification task using reservoir computing (RC)

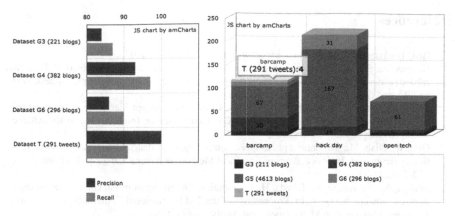

Fig. 3 On the *left* precision and recall obtained for the main topic *Event*. On the *right* number of blog posts and tweets for the three most popular themes under the main topic *Event*

3 Experimental Results

Seven datasets with retrospective blogosphere's data related to BBC Backstage were made available. Dataset G1 (221 blogs) and dataset G2 (14 blogs) were used for training. Dataset G3 (211 blogs), G4 (382 blogs), G6 (296 blogs) and T (291 tweets) were used for testing. Dataset G5 (4,613 blogs) is too large for making feasible a human/manual peer-review. In information extraction (as well as in information retrieval), precision and recall are the most used metrics for performance measure [9]. Due to space limitations, out of the five BBC Backstage *semantic groups* identified, we only illustrate the results obtained for the *semantic group* 'Event'. Figure 3 shows: (i) on the left hand side the precision and recall obtained for the main topic *Event*; (ii) on the right hand side the number of blog posts and tweets for the three most popular themes under the topic *Event*.

4 Conclusion

Discovering new information extraction techniques or combining existing ones in novel ways can have an important role in analysing blogosphere's content. This paper investigates the novel combination of OBIE and RC to extract information and classify blogosphere's content (blogs and microblogs) according to certain topics (i.e. *semantic groups*) and themes (i.e. *semantic concepts*). The precision and recall obtained from the experiments conducted show that the combination of OBIE and RC is a fruitful one. The BBC Backstage domain is a challenging one, for there is no taxonomy or controlled vocabulary available beforehand, and thus, BBC Backstage Web pages need to be processed to obtain *semantic lexicons*. This is a major disadvantage in comparison with other domains (e.g. healthcare), where lexicons and terminologies are already available.

References

1. BBC Backstage. http://backstage.bbc.co.uk. Accessed 9 May 2010
2. Kireyev, A., Palen, L., Anderson, K.M.: Applications of topic models to analysis of disaster-related twitter data. In: Proceedings of NIPS Workshop on Applications for Topic Models: Text and Beyond (2009)
3. Chemudugunta, C., Smyth, P., Steyvers, M.: Combining concept hierarchies and statistical topic models. In: Proceedings of the 17th ACM Conference on Information and Knowledge Management, pp. 1469–1470 (2008)
4. Guo, W., Diab, M.: Semantic topic models: combining word distributional statistics and dictionary definitions. In: Proceedings of Empirical Methods in Natural Language Processing, pp. 552–561 (2011)
5. Shakya, A., Wuwongse, V., Takeda, H., Ohmukai, I.: OntoBlog: informal knowledge management by semantic blogging. In: Proceedings of the 2nd International Conference on Software, Knowledge Information Management and Applications (2008)
6. Weichselbraun, A., Wohlgenannt, G., Scharl, A.: Augmenting lightweight domain ontologies with social evidence sources. In: Proceedings of the Workshop on Database and Expert Systems Applications (2010)
7. Abel, F., Gao, Q., Houben, G.J., Tao, K.: Semantic enrichment of Twitter posts for user profile construction on the social web. In: Proceedings of the 8th Extended Semantic Web Conference (2011)
8. Nebhi, K.: Ontology-based information extraction from Twitter. In: Proceedings of the Workshop on Information Extraction and Entity Analytics on Social Media Data, pp. 17–22 (2012)
9. Wimalasuriya, D.C., Dou, D.: Ontology-based information extraction: an introduction and a survey of current approaches. J. Inf. Sci. **36**, 306–323 (2010)
10. Verstraeten, D., Schrauwen, B., D'Haene, M., Stroobandt, D.: An experimental unification of reservoir computing methods. Neural Netw. **20**, 391–403 (2007)
11. Denecke, K., Nejdl, W.: How valuable is medical social data? Content analysis of the medical web. Inf. Sci. **179**, 1870–1880 (2009)
12. RDF. http://www.w3.org/TR/rdf11-primer/. Accessed June 2014
13. SKOS. http://www.w3.org/TR/skos-reference/. Accessed 28 June 2010
14. Lukosevicius, M., Jaeger, H.: Reservoir computing approaches to recurrent neural network training. Comput. Sci. Rev. **3**, 127–149 (2009)
15. Wyffels, F., Schrauwen, B.: A comparative study of reservoir computing strategies for monthly time series prediction. Neurocomputing **73**, 1958–1964 (2010)
16. Alexandre, L.A., Embrechts, M.J., Linton, J.: Benchmarking reservoir computing on time-independent classification tasks. In: Proceedings of International Joint Conference on Neural Networks, pp. 89–93 (2009)
17. Alexandre, L.A., Embrechts, M.J.: Reservoir size, spectral radius and connectivity in static classification problems. Proc. Int. Conf. Artif. Neural Netw. **5768**, 1015–1024 (2009)
18. Emmerich, C., Reinhart, R.F., Steil, J.J.: Recurrence enhances the spatial encoding of static inputs in reservoir networks. Proc. Int. Conf. Artif. Neural Netw. **6353**, 148–153 (2010)
19. Suykens, J.A.K., Vandewalle, J.: Least squares support vector machine classifiers. Neural Process. Lett. **9**, 293–300 (1999)
20. Jaeger, H.: A tutorial on training recurrent neural networks, covering BPPT, RTRL, EKF and the "echo state network" approach. Fifth revision: Dec 2013, http://minds.jacobs-university.de/sites/default/files/uploads/papers/ESNTutorialRev.pdf. Accessed May 2014

A Study on Road Junction Control Method Selection Using an Artificial Intelligent Multi-criteria Decision Making Framework

P.K. Kwok, D.W.H. Chau and H.Y.K. Lau

Abstract With the increasing number of vehicles on roads, choosing a proper Road Junction Control (RJC) Method has become an important decision for reducing traffic congestion and cost. However, the public awareness of environmental sustainability and diverse voices from different stakeholders make such decision a knotty one. In this paper, an artificial intelligent decision-making framework using Hierarchical Half Fuzzy TOPSIS (HHF-TOPSIS) is proposed for RJC method selection. Compared with the existing qualitative comparison method suggested in the Design Manual for Roads and Bridges, this method can provide a more efficient and objective approach to reach the best compromise against all relevant objectives.

1 Introduction

There exists many RJC methods, varying from a simple roundabout to a 4-level graded interchange [7], but all of these methods have their own pros and cons that best suit particular environment. Since building a road junction is costly, difficult to reverse and involves many conflicting criteria, like cost versus efficiency, selecting a suitable RJC method is crucial yet difficult. Currently, in United Kingdom, the RJC method selection process as suggested by the Department for Transport is to first list out the evaluation criteria which is followed by filling in the characteristic of the alternatives. After that the problem size is reduced by comparing the options, two at a time, eliminating the least favorite in turn [6]. The major drawback of this approach is that it is done in a qualitative manner and involves many human judgments. Hence, the result tends to be subjective. This limits the opportunity in obtaining an optimal solution and cripples transparency. Also, concerning the increasing number of stake-

P.K. Kwok (✉) · D.W.H. Chau · H.Y.K. Lau
The University of Hong Kong, Pokfulam Road, Pokfulam, Hong Kong
e-mail: kwokalex@hku.hk

D.W.H. Chau
e-mail: derek025@hku.hk

H.Y.K. Lau
e-mail: hyklau@hku.hk

© Springer International Publishing Switzerland 2014
M. Bramer and M. Petridis (eds.), *Research and Development in Intelligent Systems XXXI*, DOI 10.1007/978-3-319-12069-0_25

holders involved, an increasing number of factors are going to be included in RJC method selection. Decision making solely based on qualitative comparison becomes complicated and is difficult to reach a general consensus [5].

Therefore, in this paper, an artificial intelligent decision-making framework for RJC method selection using Hierarchical Half Fuzzy TOPSIS (HHF-TOPSIS) is proposed to provide a more objective and public traceable selection method in a limited data environment especially for new-town urban planning by direct inputing objective KPIs for the criteria. Compared with other domains such as supplier evaluation, the implementation of Fuzzy TOPSIS in RJC method selection is rather new. This paper is developed into 3 sections: Sect. 1 the introduction, the proposed selection framework is illustrated after a review on TOPSIS in Sect. 2; and conclusion is drawn in Sect. 3.

2 Hierarchical Half Fuzzy TOPSIS (HHF-TOPSIS)

TOPSIS (Technique for Order Preference by Similarity to Ideal Solution) is one of the well-known Multiple Criteria Decision Making (MCDM) methods proposed by Hwang and Yoon in 1981 [3]. It ranks alternatives basing on their similarities to the positive ideal solution (PIS) and remoteness to the negative ideal solution (NIS) [2]. However, traditional TOPSIS has a drawback that the performance ratings and importance weights of the criteria in its evaluation progress are both in crisp values. This makes traditional TOPSIS difficult to reflect most real-world situations, since human judgments with preferences are usually vague. Hence, it is essential to incorporate a fuzzy set theory in the traditional TOPSIS method to facilitate decision making [2]. Researches in applying fuzzy TOPSIS include Yong [8] applied fuzzy TOPSIS in plant location selection. In most of the literatures both criteria rating and the importance weights are represented in terms of linguistic variables, like "good" or "not good".

Even with the incorporation of fuzzy set to traditional TOPSIS, or the fuzzy TOPSIS approach, it is still not suited for choosing the RJC method which the criteria are represented in terms of exact numerical marks. Forcing the score to fuzzy values will affect trustworthiness and objectiveness of the decision making result. Furthermore, unlike the Analytic Hierarchy Process (AHP) method, the traditional TOPSIS method does not consider the hierarchy of attributes and alternatives [4]. In contrast, regarding the complexity of decision making activities today, Bao et al. [1] claims that there is increasing needs in breaking the problem into a hierarchical structure to facilitate the decision making process. He applied his improved hierarchical fuzzy TOPSIS for road safety performance evaluation. In this paper, the following 10-step HHF-TOPSIS is proposed for choosing the RJC method.

Step 1: The are many RJC methods in the Design Manual for Roads and Bridges, ranging from mini-roundabout to a 4-level interchange [7]. The decision committee should first identify all possible methods for evaluation. Possible methods mean methods not against the requirement of government policies or law.

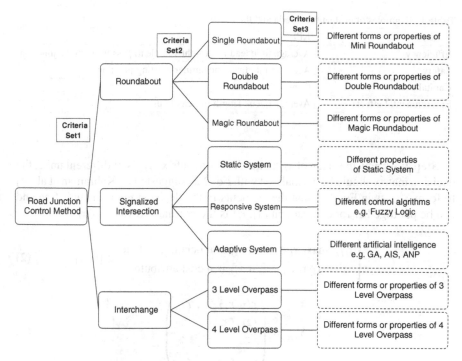

Fig. 1 The decision hierarchy of road junction control method selection

Step 2: To cut the problem into a smaller size so as to make the decision making more efficient, the possible methods selected in Step 1 are grouped according to their similarities in nature. An example is given in Fig. 1.

Step 3: There are many factors, like cost and environment that need to be considered in order to select a suitable road junction design. Decision committee should gather all these requirements and set up objective and measurable evaluation criteria with the corresponding Key Performance Indicators (KPIs) for decision making. Table 1 shows some examples of evaluation criteria and KPIs. After getting KPIs of the alternatives, matrix S_{mn} as in Eq. 1 is constructed, for m possible RJC methods and n criteria. x_{ij} is the KPI for RJC method i under criterion j. (Unlike other fuzzy TOPSIS, the method here only requires the direct input of KPIs into the decision matrix.)

$$S_{mn} = \begin{pmatrix} x_{11} & x_{12} & \cdots & x_{1n} \\ x_{21} & x_{22} & \cdots & x_{2n} \\ \vdots & \vdots & \ddots & \vdots \\ x_{m1} & x_{m2} & \cdots & x_{mn} \end{pmatrix} \qquad (1)$$

Table 1 Examples of evaluation criteria and their KPIs

Criterion	KPI
Efficiency	Average time lead for a light vehicle to pass through the junction
Safety	Accident rate from the similar existing junction
Capital cost	Construction cost
Environmental effects	Average queue length in peak hour

Step 4: As now the value of x_{ij} in the decision matrix S_{mn} is of different units, the decision matrix should be normalized with Eq. 2 as suggested by Kahraman et al. [4] to form S'_{mn} as in Eq. 3 so that the elements in the decision matrix is unit-free and can be put together for comparison [4]. r_{ij} is the normalized value of x_{ij}.

$$r_{ij} = \begin{cases} x_{ij}/(\max_i x_{ij}) & \text{if } x_{ij} \text{ is a benefit attribute} \\ (\min_i x_{ij})/x_{ij} & \text{if } x_{ij} \text{ is a cost attribute} \end{cases} \in [0, 1] \qquad (2)$$

$$S'_{mn} = \begin{pmatrix} r_{11} & r_{12} & \cdots & r_{1n} \\ r_{21} & r_{22} & \cdots & r_{2n} \\ \vdots & \vdots & \ddots & \vdots \\ r_{m1} & r_{m2} & \cdots & r_{mn} \end{pmatrix} \qquad (3)$$

Step 5: After identifying the evaluation criteria, surveys can be carried out to all stakeholders to obtain the objective importance weights for the criteria (Delphi method can be used.). As this step requires human judgments (usually vague), linguistic variables, which can be expressed in positive triangular fuzzy numbers as in Table 2, are preferred to crisp values when designing the questionnaires. The important weights determined should be normalized by Eq. 4 so that the weights fall between 0 and 1 and matrix $\tilde{W}'$ is formed as show in Eq. 5. $\tilde{y}_j$ and $\tilde{w}_j$ are the importance weight and the normalized importance weight for criterion j respectively. (Although normalization may cause rank reversal in fuzzy numbers, the result interpretation will not be affected.)

Table 2 The importance weights

Linguistic variables	Triangular fuzzy numbers
Completely not important	$(0.0, 0.1, 0.2)$
Less important	$(0.2, 0.3, 0.4)$
Medium	$(0.4, 0.5, 0.6)$
Somehow important	$(0.6, 0.7, 0.8)$
Very important	$(0.8, 0.9, 1.0)$

$$\tilde{w}_j = \tilde{y}_j / \sum_{j=1}^{n} \tilde{y}_j \in [0, 1] \tag{4}$$

$$\tilde{W}' = [\tilde{w}_1 \tilde{w}_2 \dots \tilde{w}_n] \tag{5}$$

Step 6: Weighted normalized fuzzy decision matrix is constructed by multiplication using Eq. 6 and form matrix $\tilde{A}$.

$$\tilde{A} = \begin{pmatrix} r_{11} \times \tilde{w}_1 & r_{12} \times \tilde{w}_2 & \dots & r_{1n} \times \tilde{w}_n \\ r_{21} \times \tilde{w}_1 & r_{22} \times \tilde{w}_2 & \dots & r_{2n} \times \tilde{w}_n \\ \vdots & \vdots & \ddots & \vdots \\ r_{m1} \times \tilde{w}_1 & r_{m2} \times \tilde{w}_2 & \dots & r_{mn} \times \tilde{w}_n \end{pmatrix} = \begin{pmatrix} \tilde{v}_{11} & \tilde{v}_{12} & \dots & \tilde{v}_{1n} \\ \tilde{v}_{21} & \tilde{v}_{22} & \dots & \tilde{v}_{2n} \\ \vdots & \vdots & \ddots & \vdots \\ \tilde{v}_{m1} & \tilde{v}_{m2} & \dots & \tilde{v}_{mn} \end{pmatrix} \tag{6}$$

Step 7: FPIS ($A^+ = (\tilde{v}_1^+ \tilde{v}_2^+ \dots \tilde{v}_n^+)$), the point consisting all the best scores for the criteria in the solution set and FNIS ($A^- = (\tilde{v}_1^- \tilde{v}_2^- \dots \tilde{v}_n^-)$), the point consisting of all the worst scores, are determined by Formula 7.

$$\tilde{v}_j^+ = \max_i v_{ij} \quad \text{and} \quad \tilde{v}_j^- = \min_i v_{ij} \tag{7}$$

Step 8: The distance between each alternative and FPIS (d_i^+) and between FNIS (d_i^-) is calculated by Eqs. 8 and 9 as suggested by Chen [2] and their closeness coefficient (CC_i), an index showing the similarity of the alternatives to the ideal solution, is calculated by Eq. 10. (d_i^+ and d_i^- are squared to widen the spread of the result.)

$$d_i^+ = \sum_{j=1}^{n} \sqrt{[(v_{j1}^+ - v_{ij1})^2 + (v_{j2}^+ - v_{ij2})^2 + (v_{j3}^+ - v_{ij3})^2]/3} \tag{8}$$

$$d_i^- = \sum_{j=1}^{n} \sqrt{[(v_{j1}^- - v_{ij1})^2 + (v_{j2}^- - v_{ij2})^2 + (v_{j3}^- - v_{ij3})^2]/3} \tag{9}$$

$$CC_i = (d_i^-)^2 / [(d_i^+)^2 + (d_i^-)^2] \in [0, 1] \tag{10}$$

Step 9: After obtaining the CC_i, the alternatives can be ranked according to their CC_i values. The best solution should be the one which tends to be FPIS and far from FNIS. Hence, the higher the CC_i value is, the higher rank (better) the solution is.

Step 10: The best alternative chosen in Step 9 is kept for the next level of decision while Step 3–9 is repeated until the final decision is made.

3 Conclusion

With increasing number of factors needed to be considered in a construction plan for road junctions, choosing a proper RJC method becomes more difficult, especially that a wrong decision may lead to serious transportation problems, such as heavy traffic congestion, accidents, queues etc., and waste of resources. In this paper, a RJC method selection framework using HHF-TOPSIS artificial intelligent MCDM method is proposed to facilitate the choice of a suitable RJC method in a multiple objectives environment by directly inputting objective KPIs for the criteria. Future research needs to be done on the feasibility of extending this framework to choose the form of an adjacent road junction based on the result of the first junction so as to form a road network. This is especially useful in urban planning in which the starting point is usually the major junction of the city.

References

1. Bao, Q., Ruan, D., Shen, Y., Hermans, E., Janssens, D.: Improved hierarchical fuzzy TOPSIS for road safety performance evaluation. Knowl. Based Syst. **32**, 84–90 (2012)
2. Chen, C.T.: Extensions of the TOPSIS for group decision-making under fuzzy environment. Fuzzy Sets Syst. **114**(1), 1–9 (2000)
3. Hwang, C.L., Yoon, K.: Multiple Attribute Decision Making: Methods and Applications. Springer, New York (1981)
4. Kahraman, C., Ates, N.Y., Çevik, S., Gülbay, M., Erdogan, S.A.: Hierarchical fuzzy TOPSIS model for selection among logistics information technologies. J. Enterp. Inf. Manage. **20**(2), 143–168 (2007)
5. Mendoza, G.A., Macoun, P., Prabhu, R., Sukadri, D., Purnomo, H., Hartanto, H.: Guidelines for Applying Multi-Criteria Analysis to the Assessment of Criteria and Indicators. CIFOR, Indonesia (1999)
6. The Highway Agency, The Scottish Office Development Department, The Welsh Office Y Swyddfa Gymreig, The Department of the Environment for Northern Ireland: Choice between option for truck road schemes. In: Design Manual for Roads and Bridges, vol. 5. Department of Transport (1982). Available via Department of Transport http://www.dft.gov.uk/ha/standards/dmrb/vol5/section1/ta3082.pdf. Cited 22 May 2013
7. The Highway Agency, Scottish Executive, Welsh Assembly Government Llywodraeth Cynulliad Cymru, The Department for Regional Development Northern Ireland: Layout of grade separated junctions. In: Design Manual for Roads and Bridges, vol. 6. Department of Transport (2006) Available via Department of Transport http://www.dft.gov.uk/ha/standards/dmrb/vol6/section2/td2206.pdf. Cited 22 May 2013
8. Yong, D.: Plant location selection based on fuzzy TOPSIS. Int. J. Adv. Manufact. Technol. **28**(7–8), 839–844 (2006)